OPERATIONS RESEARCH

(For Enginering and Managemer

By:

Prof. VIVEK KUMAR
M.Tech., LMISTE, LMISNT, MIE, Ph. D, (Pursuing) – IIT, Delhi
Professor & Head
Department of Mechanical & Automation Engineering
Amity School of Engineering & Technology
Amity University UP, Noida

NUPUR KASHYAP
M. Tech., Ph. D – MNNIT

Software programs contributed by :
ELA KASHYAP SHARMA
B.Tech. (IT), MS (IT) from University of Melbourne, Australia
Assistant Professor, Department of Information & Technology,
Amity School of Engineering and Technology,
Amity University UP, Noida.

Edited by :
PRERNA VIVEK

S.K. KATARIA & SONS®
Publisher of Engineering Books
4885/109, Prakash Mahal, Dr. Subhash Bhargav Lane,
Opposite Delhi Medical Association, Daryaganj, New Delhi–110002
Phone: +91-11-43551243, Mobile: +91-9871775858
e-mail: katariabook@gmail.com, katariabooks@yahoo.com
Website: www.skkatariaandsons.com
(e-books, subscriptions and bulk buying also available on our website)

Published by:

S.K. Kataria & Sons®

4885/109, Prakash Mahal, Dr. Subhash Bhargav Lane,
Opposite Delhi Medical Association, Daryaganj, New Delhi–110002
Phone: +91-11-43551243 Mobile: +91-9871775858
e-mail: katariabook@gmail.com, katariabooks@yahoo.com

First Edition: February 2006, Fifth Edition: July, 2012; Reprint: 2024

Price:₹565/-

Laser Typesetting :
PLUS COMPUTERS, Shahdara, Delhi

Printed at :
New A.S. Printing Press, Delhi

Preface to the Fifth Edition

After publication of previous four editions of the book, I have revised the book keeping in mind the requirements of the undergraduate and postgraduate students of engineering and management and the valuable feedback received from the various segments of academic circle. The content of this book covers syllabus of most of the technical institutes / universities.

Along with more number of model papers, previous year's exam papers and questions from various competitive and entrance exams, I have included software programs to solve the OR problems. These software programs will help students to understand and formulate the problems / case studies in OR lab.

I am thankful to my colleagues Dr. Nitin Upadhye, Mr. Gaurav Gupta and Mr. Rahul Sindhwani for their input to improve the content of the book.

There is still scope for further improvement. The author would welcome feedback and suggestions, if any, in this regards.

Prof. Vivek Kumar

Preface to the First Edition

This course – Operations Research is designed for 8th Semester (Final Year) Students of Mechanical Engineering by Uttar Pradesh Technical University, Lucknow. Main objective of this book is to educate and familiarize students with different techniques of optimization.

Main features of the book are:

1. Theory is written in concise form with systematic approach to solve the problems.
2. Exhaustive Tutorial sheets/Exercises are given for the practice of students.
3. Several solved examples are given after each topic to clear the concepts.
4. Solutions of questions asked in previous exams are given in respective chapters.
5. Objective type questions are given at the end of each chapter.

I am grateful to many of my colleagues for their comments and encouragement, namely, Prof. Parmanand (HOD – CS/IT), Prof. J.K. Rai, Prof. K.M.Soni, Dr. Babita, Prof. N. Swaminathan. I extend my deep gratitude to Dr. G.P. Govil, HOD-Mechanical, for his extraordinary support.

My colleagues Prof. C. Pandey, Prof. Sanjay Kumar, Prof. S.N. Ali, have helped me in various ways at Galgotia's College of Engineering & Technology.

I am specifically thankful to my friends Prof. Anoop Dixit and Prof. Sudhir Kumar Singh for providing me help in a chapter – Stochastic Inventory Models.

I express my sincere thanks to our Chairman Mr. Suneel Galgotia and Director Dr. Krishna Kumar for providing me conducive atmosphere in Galgotia's College.

Last but not the least, I would like to thank my wife Smt. Prerna who not only took care of my needs during this period but helped me in writing the manuscript and my kids Rishabh, Chaitanya and Richa for their invaluable co-operation.

I sincerely wish to express my appreciation to publisher Sh. Sanjeev Kataria for taking initiative and encouraging me to write the book. I am grateful to him for bringing out this book in a very short time. I am thankful to Laser typesetter Mr. Manoj Tyagi and his Staff for nice presentation of the book. Any constructive criticism and suggestion for the improvement of the book will be welcome with gratitude.

Prof. Vivek Kumar

In the beloved memory

of

my mother

Late Smt. Phool Vati Devi

Contents

UNIT-II

UNIT-III

Unit 1

Chapter 1

Introduction to Operations Research

1.1. INTRODUCTION

In order to take certain decisions, one has to develope his skills in a way that he is able to take a correct decision at a proper time. Decision maker is faced with several interacting variables. Over some of the complex factors, he has no control. In such conditions, decisions taken on the basis of past experience and intuitions are misleading. Therefore, understanding the use of scientific approach to decision making is of great importance. Operations Research provides a scientific approach to problem solving for executive management.

During World War II, UK was facing problem of allocating scarce resources in an effective manner to various military operations. Therefore, UK and USA military management invited several interdisciplinary scientists to apply a scientific approach to many strategic and tactical problems. Their efforts provided intellectual support to the military commands and hence they were instrumental in winning the war. The word "Operations Research" came from the context in which it was used and developed—Research on Military Operations. Operations Research is also known as operational analysis, operations evaluation, systems analysis, systems research, quantitative methods, management science, decision science.

1.2. OPERATIONS RESEARCH DEFINED

With the development of the subject, definitions of operations research have been changed. A few definitions are given below:

(*i*) Operation research is defined as an art of winning the war without actually fighting it.

(*ii*) Operation research is a scientific method of providing executive departments with a quantitative basis for decisions regarding operations under their control. **—Morse and Kimbal**

(*iii*) Operation research is the art of giving bad answers to problems to which otherwise worse answers are given. **—T.L. Satty**

(*iv*) Operation research is the application of scientific methods, techniques and tools to problems involving the operations of systems so as to provide these in control of the operations with optimum solutions to the problem. **—Churchman, Arnoff**

(*v*) Operation research is a scientific approach to problem solving for executive management. —**H.M. Wagner**

In all definitions operation research is concerned with optimization problem. Optimisation means best possible decision in any set of circumstances.

The first step in mathematical optimisation problems is to express the given quantity as a function of several variables, to be maximised or minimised, subject to possibly one or more constraints (like equalities and inequalities). Next step is to choose a mathematical method (model) to solve the optimisation problem. Such methods are called *Optimisation techniques* or *Algorithms.*

For example

$$Z = f(x) \qquad \text{Objective function (maximise or minimise)}$$

subject to

$$g(x) \begin{bmatrix} \leq \\ = \\ \geq \end{bmatrix} b_{ij} \quad \text{constraints}$$

1.3. CHARACTERISTICS OF OPERATIONS RESEARCH

1. **Inter disciplinary Approach:** The optimum solution of a problem is found by an inter disciplinary team comprising experts from different disciplines such as Mathematics, Statistics, Economics, Management, Physics, Computer Science, Engineering, etc.
2. **Systems Approach:** In operation research each problem is examined in its entirety. One part of organisation (activity) has some effect on every other part of the organisation. Optimum operation of one part of the system may not be the optimum operation of all the activities. Therefore, manager must identify all possible interactions and determine their impact as a whole. The decision which is best for the organisation as a whole is called an optimal decision.
3. Operation research utilises scientific techniques to solve a problem.
4. Operation research tries to optimise total return by reducing the cost or loss and maximising the profit.
5. In operation research, we cannot obtain perfect solution but the quality of decision is improved.
6. To solve complex problems computers are used.

1.4. MODELS IN OPERATIONS RESEARCH

A model is defined as an idealized representation of some real life system (problem, process, operation, object or event). The objective of a model is to facilitate the analysis of the behaviour of the system for the purpose of improving its performance, *e.g.* map, organisation charts, balance sheet, PERT network, photographs, road maps etc.

Classification of Operations Research Models

(*i*) According to the Structure

(*a*) **Iconic model:** It is physical representation by scaling the system up or down such that properties of the system remain same, *e.g.* toy aeroplane, photograph of a person, a globe of earth, blue print of a building.

(*b*) **Analogue model:** A set of different properties is used to represent properties of original system, *e.g.* time, force, speed, age represented in terms of distance in graphs.

(c) **Mathematical model:** It employes a set of mathematical symbols (letter, number, etc.) to represent the decision variables of the system. It consists of set of equations to describe the behaviour of the system.

(*ii*) According to Purpose

(*a*) **Descriptive model:** It describes some aspects of the problem based on available data, *e.g.* the result of an opinion poll.

(*b*) **Normative model:** It is used to find optimal solution in repetitive problems, *e.g.* linear programming.

(*iii*) According to Nature

(*a*) **Deterministic model:** In this, parameters are completely defined and outcomes are certain, *e.g.* linear programming.

(*b*) **Probabilistic model:** In this, outcome of decisions cannot be predicted with certainty.

(*iv*) According to Behaviour

(*a*) **Static model:** In this, effects of changes taking place with time are not considered.

(*b*) **Dynamic model:** Effects of changes with time during planning are considered.

1.5. METHODOLOGY OF OPERATIONS RESEARCH

The procedure for operation research study involves following steps:

(*i*) **Formulation and defining a problem:** This is done by collecting information in following way:

(*a*) Identify objectives and alternative courses of action,

(*b*) evaluate each alternative course of action.

(*ii*) **Constructing a mathematical model:** It requires identification of static and dynamic elements. It should include–decision variables and parameters, restrictions and objective function.

(*iii*) **Deriving solutions from the model**: Determine value of decision variables that optimize the given objective function.

(*iv*) **Testing the model:** This model is then tested against real world for any deviation.

(*v*) **Control over solution:** Controls over solution with desired degree of satisfaction are made so that it is adaptable to changing environment.

(*vi*) **Validation of model:** Validity of model is ascertained by comparing its performance with the past data available for actual system.

(*vii*) **Implementing the solution:** Final solution is put to use and observed.

1.6. APPLICATIONS OF OPERATIONS RESEARCH

Operation research is useful in following important fields:

(*i*) **Agriculture:** Used in optimum allocation of land to various crops, optimum distribution of water etc.

(*ii*) **Finance:** Used to maximise per capita income, to find out profit plan etc.

(*iii*) **Industry:** Operation research is useful in deciding optimum allocation of various limited resources (man, machines, material, money, etc.).

(*iv*) **Marketing:** With the help of operation research, marketing manager can decide minimum per unit sale price, where to distribute the product, how to select advertisement media, etc.

(*v*) **Production:** Production manager can use operation research in scheduling and sequencing, in calculating optimum product mix etc.

1.7. ADVANTAGES OF OPERATIONS RESEARCH

(*i*) Operation research is helpful in maintaining better coordination among different departments.

(*ii*) Operation research is used to establish a new system (facility) in a better way.

(*iii*) In complicated situations, operation research with the help of computer analysis, helps to take better decisions.

(*iv*) Operation research helps in better control of supervision of employees.

1.8. OPERATIONS RESEARCH TECHNIQUES

Different techniques of operation research which can be applied to any decision making problem are listed below:

(*i*) Linear programming

(*ii*) Decision models

(*iii*) Network theory–PERT, CPM

(*iv*) Queuing theory

(*v*) Sequencing

(*vi*) Game theory–modelling techniques

(*vii*) Simulation–Process of designing an experiment representing real situation

(*viii*) Replacement theory

(*ix*) Reliability

(*x*) Non-linear programming

(*xi*) Dynamic programming

1.9. LIMITATIONS

Limitations of operation research are mainly related with time and money.

(*i*) Operation research can be used to find optimal solution only when all parameters of problem can be quantified *e.g.*, human behaviour cannot be quantified.

(*ii*) There is resistance from employees used to with conventional thinking, to the suggestions made from operation research expert.

(*iii*) Most of the executives in different organisations do not have basic knowledge required for operation research techniques therefore there is always some distance between operation research expert and executives.

(*iv*) Present day problems depend on various factors which are interdependent and requires quantification. This increases the magnitude of computation work.

(*v*) Due to dynamic nature of problems, frequent changes are to be incorporated. This adds on to cost and time.

(*vi*) Due to lack of awareness among executives, they do have faith in conventional approaches rather than in operation research techniques.

(*vii*) There is lack of commitment and involvement of top level management which discourages middle level managers.

EXERCISE

1.1. What is Operations Research?
1.2. Describe four applications of operations research.
1.3. What are the characteristics of operations research?
1.4. Explain different types of models used in operations research.
1.5. "Operations research is the art of winning war without actually fighting it". Discuss.
1.6. Discuss the importance of operations research in decision making process.
1.7. What is methodology of operations research?
1.8. Discuss advantages and limitations of operations research.
1.9. How can operations research models be classified?
1.10. What is the function of a model in decision making?
1.11. Distinguish between deterministic and probabilistic models.
1.12. Distinguish between static and dynamic models.

OBJECTIVE TYPE QUESTIONS

1.1. Operations research approach is
(*a*) scientific (*b*) multidisciplinary
(*c*) intuitive (*d*) all of the above

1.2. A model is
(*a*) selective representation of reality
(*b*) an abstraction (*c*) an approximation
(*d*) an idealization (*e*) all of the above

1.3. Decision variables are
(*a*) controllable (*b*) uncontrollable
(*c*) parameters (*d*) none of the above

1.4. A physical model is an example of
(*a*) A verbal model (*b*) An analogue model
(*c*) An iconic model (*d*) A mathematical model
(*e*) All of the above

1.5. A model
(*a*) refines crude data (*b*) is a tool for decision maker
(*c*) forces a manager to be explicit (*d*) all of the above

1.6. A photograph is an example of

(*a*) Iconic model
(*b*) Analogue model
(*c*) Symbolic model
(*d*) All of the above

1.7. Example of a deterministic model is

(*a*) Linear programming model
(*b*) queuing theory model
(*c*) both
(*d*) none

Answers

1.1. (*b*) 1.2. (*e*) 1.3. (*a*) 1.4. (*c*) 1.5. (*d*) 1.6. (*a*) 1.7. (*a*)

Chapter 2

Linear Programming-Graphical Method

2.1. INTRODUCTION

Linear programming deals with the optimization (maximization or minimization) of a function of variables known as objective functions. It is subject to a set of linear equalities and/or inequalities known as constraints. Linear programming is a mathematical technique, which involves the allocation of limited resources in an optimal manner, on the basis of a given criterion of optimality.

Linear programming is a technique for determining an optimum schedule of interdependent activities in view of the available resources. The term programming means scheduling. Programming is just another word for 'Planning' and refers to the process of determining a particular plan of action from amongst several alternatives. The word 'linear' indicates that all relationships involved in a particular problem are linear.

***George B. Dantzing** is the father of linear programming.*

According to **Samuelson** "Linear programming is the analysis of problems in which linear function of a number of variables is to be maximized (or minimized) when those variables are subject to a number of restraints in the form of linear inequalities."

2.2. FORMULATION OF LINEAR PROGRAMMING PROBLEMS

It consists of the following steps:

1. Write down the decision variables of the problem.
2. Formulate the objective function to be optimized (maximized or minimized) as a linear function of the decision variables.
3. Formulate the other conditions of the problem such as resource limitations, market constraints, interrelations between variables etc., as linear inequations or equations in terms of the decision variables.
4. Mention the non-negativity constraints associated with decision variables since the negative values of the decision variables do not have any valid physical interpretation.

2.3. APPLICATIONS OF LINEAR PROGRAMMING

1. To determine the most economical diet for patients.
2. Development of a production schedule by minimizing total production and inventory costs to satisfy future demand for a product.
3. It is used to determine allocation of land for different crops so as to maximize the profit.

4. Selection of a product mix which maximizes the profit of the firm and makes best use of available resources.
5. To determine the most economic pattern and timing for flights.
6. Selection of different blends of raw materials in feed mills.
7. Allocation of a limited advertising budget among various advertising media like TV, radio and newspaper in order to maximize advertising effectiveness.
8. Selection of an investment portfolio from variety of stocks and bonds available in such a way as to maximize the return of investment or minimization of risk.

2.4. REQUIREMENTS OF LINEAR PROGRAMMING PROBLEM

1. Decision variables should be interrelated linearly in terms of resources and need simultaneous solution.
2. A linear programming problem must have well defined objective function to optimize (maximize or minimize) available resources. It is expressed as linear function of decision variables.

   Maximize $Z = C_1 x_1 + C_1 x_2$
3. Limited resources like men, machines, materials, space, markets etc. are to be allocated among various competing ends. These constraints must be capable of being expressed as linear functions of decision variables.
4. There must be alternative courses of actions.
5. All decision variables must have non-negative values, as negative values have no physical interpretation.

   x_1 and $x_2 \geq 0$

2.5. ASSUMPTIONS OF LINEAR PROGRAMMING MODEL

Since all the requirements of linear programming problem are rarely met. Following assumptions are made for the analysis purpose:

1. The objective function and constraints are expressed as linear equations or inequations. These can be graphically represented as a straight line.
2. All the decision parameters (coefficients of objective function and constraints) are known with certainty and are constant.
3. The value of objective function is equal to sum of contributions made by each decision variable.
4. The total sum of resources is equal to sum of resources used by each decision variable.
5. The decision variables will take only no-negative values. Fractional values are allowed but some time not desirable.

2.6. MATHEMATICAL MODEL OF LINEAR PROGRAMMING

$$\text{Maximize } Z = C_1x_1 + C_2x_2 + \ldots + C_nx_n \qquad \ldots(1)$$

Subject to constraints

$$\left.\begin{array}{l} a_{11}x_1 + a_{12}x_2 + \quad + a_{1n}x_n \leq b_1 \\ a_{21}x_1 + a_{22}x_2 + \quad + a_{2n}x_n \leq b_2 \\ \\ a_{m_1}x_1 + a_{m_2}x_2 + \quad + a_{mn}x_n \leq b_m \end{array}\right] \qquad \ldots(2)$$

and $\quad x_1, x_2 \ldots, x_n \geq 0 \qquad \ldots(3)$

Here,

1. Z is the objective function (1)
2. $x_1, x_2, ... x_n$ are decision variables
3. Inequalities (2) are called constraints
4. $C_1, C_2 ... C_n$ represents the per unit contribution of decision variables to the objective function.
5. a_{ij} are substitution coefficients and represent amount of resources consumed per unit of variables.
6. $b_1, b_2 .... b_m$ represent availability of constraint.

2.7. BASIC DEFINITIONS

1. **Solution:** The value of decision variables ($x_1, x_2 ...$), which satisfy the constraints of LPP, is called its solution.
2. **Feasible solution:** Any solution of LPP, which satisfies the non-negativity constraints (x_1 and $x_2 \geq 0$) of LPP, is called its feasible solution.
3. **Basic solution:** For a set of m simultaneous equations in n unknowns ($n > m$), a solution obtained by setting ($n - m$) of the variables equal to zero and solving the remaining **m** equations in **m** unknowns is called a basic solution. Zero valued variables ($n - m$) are called **non-basic** variables and remaining m variables are called **basic** variables and constitute a basic solution.
4. **Basic feasible solution:** A basic solution of LPP, which is feasible also (satisfies non-negativity constraints), is called a basic feasible solution.

   A basic feasible solution can be of two types:

   (*i*) **Non-degenerate solution:** A basic feasible solution, which has exactly m positive non-zero basic variables, is known as non-degenerate solution.

   (*ii*) **Degenerate solution:** A basic feasible solution is called degenerate if one or more of the basic variables become equal to zero.
5. **Optimal feasible solution:** Any basic feasible solution, which optimizes (maximizes or minimizes) the objective function of a LPP, is called its optimal feasible solution.
6. **Convex set:** Convex set is a polygon in which if any two points on the polygon are selected arbitrarily then a straight-line segment joining these two points lies completely within the polygon. The extreme points of the convex set are the basic solutions of LPP.

2.8. GRAPHICAL METHOD (SEARCH APPROACH METHOD) OF SOLVING LINEAR PROGRAMMING PROBLEM

Linear programming problem involving two decision variables can be solved by graphical method.

Main steps of solving a Linear programming problem by graphical method (search approach method) are described below:

1. Identify the problem - decision variables, the objective function and the restrictions.
2. Formulate the problem mathematically.
3. Consider each inequality constraint as equation.
4. Plot a graph representing each of the constraints of the problem and identify the feasible region (solution space). The feasible region is the intersection of

all the regions represented by the constraints of the problems and is restricted to the first quadrant only (non-negativity constraint).

If the inequality constraint is ≤ type then the region below the line in the first quadrant is shaded. For the inequality constraint ≥ type, the region above the line in the first quadrant is shaded. The common region thus obtained is known as feasible region.

5. The feasible region obtained in step 4 may be bounded or unbounded. Calculate the coordinates of all the corner points of the feasible region.
6. Determine the value of the objective function at each corner (solution) point obtained in step 5.
7. Select the corner point that optimizes (maximizes or minimizes) the value of the objective function. It gives the optimum feasible solution.

Example 2.1. A company produces 2 types of hats. Every hat A require twice as much labour time as the second hat B. If the company produces only hat B then it can produce a total of 500 hats a day. The market limits daily sales of the hat A and hat B to 150 and 250 hats. The profits on hat A and B are Rs. 8 and Rs. 5 respectively. Solve graphically to get the optimal solution.

Solution : Let x_1 and x_2 be number of units of type A and type B hats respectively.

$$\text{Max } Z = 8x_1 + 5x_2$$

$$\text{subject to } 2x_1 + x_2 \leq 500$$

$$x_1 \leq 150$$

$$x_2 \leq 250$$

$$x_1, x_2 \geq 0$$

First, write the inequality of the constraint into an equation and plot the lines in the graph.

$2x_1 + x_2 = 500$ passes through (0, 500) (250, 0)

$x_1 = 150$ passes through (150, 0)

$x_2 = 250$ passes through (0, 250)

We mark the region below the lines lying in the first quadrant as the inequality of the constraints are ≤.

The feasible region is OABCDO

Corner points	Value of $Z = 8x_1 + 5x_2$
O (0, 0)	0
A (150, 0)	1200
B (150, 200)	2200
C (125, 250)	2250 (Maximum Z = 2250)
D (0, 250)	1250

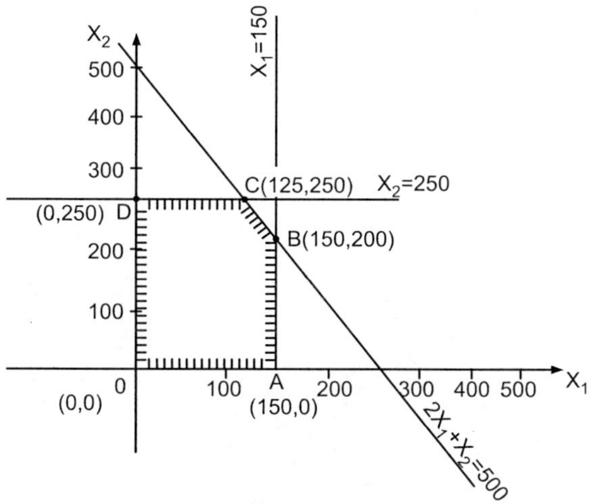

Fig. 2.1

The maximum value of Z is attained at C(125, 250)

∴ The optimal solution is $x_1 = 125, x_2 = 250$ *i.e.* The company should produce 125 hats of type A and 250 hats of type B in order to get the maximum profit of Rs. 2250.

2.9. ISO-COST OR ISO-PROFIT METHOD

1. Identify the problem- decision variables, the objective function and the restrictions.
2. Formulate the problem mathematically.
3. Consider each inequality constraint as equation.
4. Plot a graph representing all the constraints of the problem and identify the feasible region (solution space). The feasible region is the intersection of all the regions represented by the constraints of the problems and is restricted to the first quadrant only due to non-negativity constraints.
5. The feasible region obtained in step 4 may be bounded or unbounded.
6. Choose a convenient profit (or cost) and draw iso-profit (iso-cost) line so that it falls within the feasible region (assume zero profit / cost for convenience).
7. Move this iso-profit (iso-cost) line parallel to itself farther (closer) from (to) the origin till further movement brings the line outside the feasible region.
8. Identify the optimum solution as the coordinates of that point on the feasible region touched by the highest possible iso-profit line (or lowest possible iso-cost line).
9. Find the coordinates of the corner point of the feasible region selected in step 8.
10. Calculate the optimum feasible solution.

Example 2.2. Solve LPP by graphical method

Maximize $Z = 6\,x_1 + 8\,x_2$

subject to $5\,x_1 + 10\,x_2 \leq 60$

$5x_1 + 5\,x_2 \leq 40$

$x_1, x_2 \geq 0$

Solution: Converting inequalities into equations

$5x_1 + 10x_2 = 60$ passes through (12,0) & (0, 6)
$5x_1 + 5x_2 = 40$ passes through (8, 0) & (0, 8)
plotting these equations on graph.

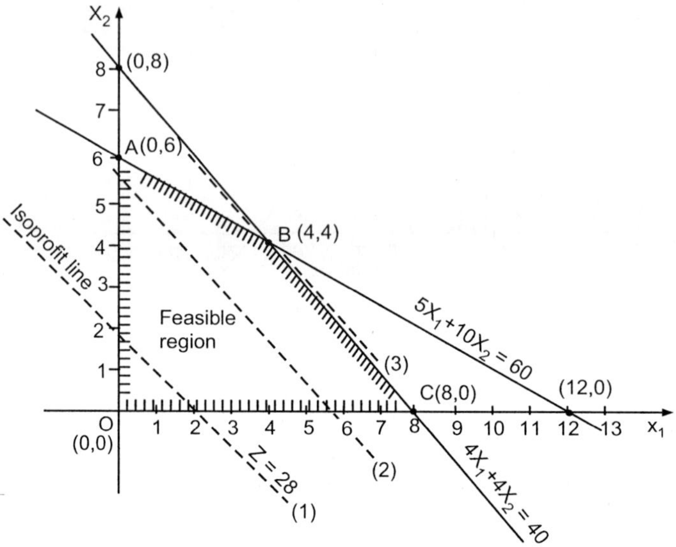

Fig. 2.2

On solving above equations, we get coordinates of corner point B – (4, 4), feasible region is OABC. Plot Z for some value say 28. Now moving isoprofit line parallel to itself away from origin till it touches extreme point (corner point B) coordinates of B are (4, 4). Therefore, optimal solution is $x_1 = 4, x_2 = 4, z = 56$.

2.10. SPECIAL CASES

1. **Infeasible solution:** When there is no value of the decision variables that satisfies all the constraints of LPP then it is said to have infeasible solution.

Example 2.3. Maximize $Z = 4x_1 + 2x_2$

subject to $2x_1 + 3x_2 \leq 18$

$x_1 + x_2 \geq 10$

$x_1, x_2 \geq 0$

Solution: Line AB represents

$$2x_1 + 3x_2 = 18$$

and line CD represents

$$x_1 + x_2 = 10$$

In the figure, there is no point which satisifies both the constraints in first quadrant. Hence, problem can not be solved or given L.P. problem has no feasible solution.

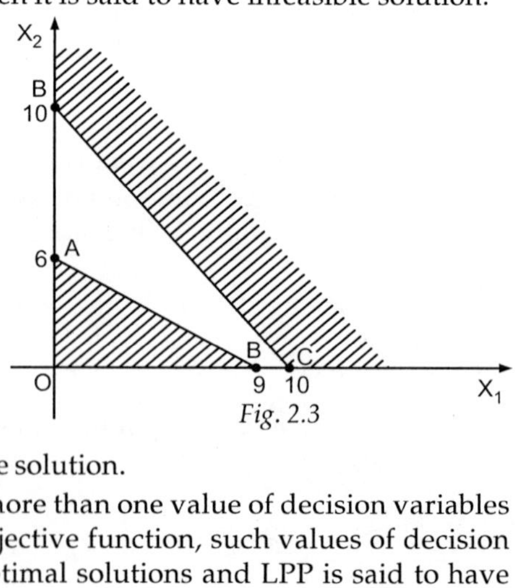

Fig. 2.3

2. **Multiple optimal solution:** When more than one value of decision variables yields the same optimal value of objective function, such values of decision variables are known as alternate optimal solutions and LPP is said to have multiple optimal solutions.

For an alternative solution to exist, following conditions should be satisfied.

1. Line representing objective function should be parallel to constraint line (same slop) that forms the boundary of feasible solution region.
2. The constraint line should form a boundary on the feasible region in the direction of movement of objective function line.

Example 2.4. Solve following LPP by graphical method.

Maximize $Z = 100x_1 + 40x_2$

subject to $5x_1 + 2x_2 \leq 1000$

$3x_1 + 2x_2 \leq 900$

$x_1 + 2x_2 \leq 500$

and $x_1, x_2 \geq 0$

Solution: The region shaded in figure is feasible region (OABC)

Corner Points	Value of $Z = 100x_1 + 40x_2$
O(0, 0)	O
A(200, 0)	20,000
B(125, 187.5)	20,000 (Max. value of Z)
C(0, 250)	10,000

Hence, the maximum value of Z is 20,000 and it occurs at two corner points A and B and this maximum value also occurs at all points on the line joining A and B. Therefore, there are infinite number of optimal solutions.

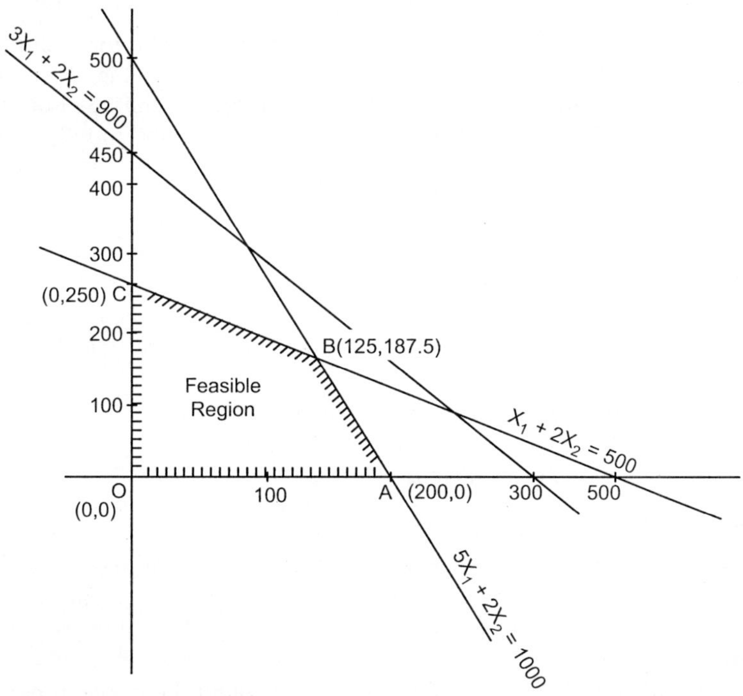

Fig. 2.4.

3. **Redundancy:** A constraint in the given LPP is said to be redundant if the feasible region of the problem remains unchanged on deleting that constraint.

4. **Unbounded solution:** When the value of objective function can be increased or decreased infinitely without any limitation, such solution of LPP is known as unbounded solution. In this situation, optimal solution may or may not exist. Optimal solution exists if the value of objective function in unbounded feasible region of a maximization problem is less than its value at vertex.

Example 2.5. Maximize $Z = 2x_1 + 2x_2$

Subject to $x_1 - x_2 \leq 1$

$x_1 + x_2 \geq 3$

$x_1, x_2 \geq 0$

Solution:

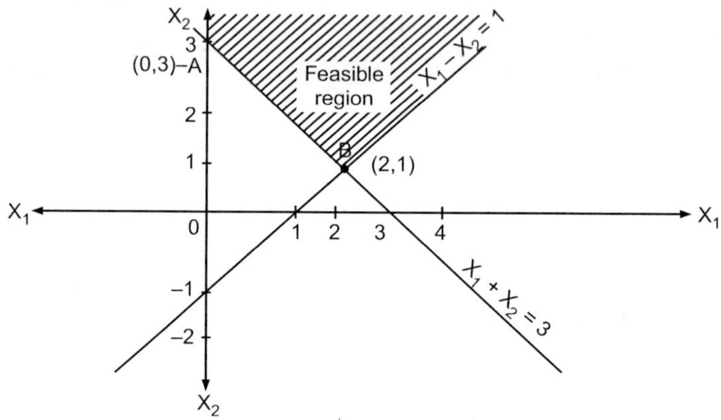

Fig. 2.5.

Here, the feasible region extends infinitely to the right. Therefore solution space is unbounded. The value of objective function at corner point A is, $Z(A) = 6$ and at B is, $Z(B) = 6$. But the value of objective function is greater than 6 at other points towards infinity of solution space. Hence the problem has unbounded solution.

2.11. ADVANTAGES OF LINEAR PROGRAMMING

1. Linear programming technique helps in making the optimum utilization of resources. It indicates how a decision maker can employ his productive factors most effectively.
2. The quality of decisions may also be improved by linear programming techniques. The user of this technique becomes more objective and less subjective.
3. Highlighting of bottlenecks in the production process.
4. It provides a clear picture of management problems.
5. It provides practically applicable solutions.

2.12. LIMITATIONS OF LINEAR PROGRAMMING

1. In some real life problems objective function and constraints are not linear.
2. In solution of LPP, fractional values of decision variables are allowed, which may not be feasible in practice. For example it is not possible to produce 2.5 buses.
3. In practical situations, it is not possible to have all the decision parameters (coefficients of objective function and constraints) known with certainty and constant. Some parameters may vary with time.

4. Many large real-world problems, complex in nature, cannot be solved by linear programming techniques.
5. Linear programming deals with single objective where as real life problems may have multi-objectives.

Example 2.6. Consider two different types of foodstuffs say F_1 and F_2. Assume these foodstuffs contain vitamin V_1 and V_2. Minimum daily requirement of V_1 is 40 units and V_2 is 50 units. Food F_1 contains 2 units of vitamin V_1 and 3 units of vitamin V_2. Further, F_2 contains 4 units of V_1 and 2 units of V_2. The cost of 1 unit of F_1 is Rs.3 and of F_2 is Rs. 2.5. Find minimum cost of diet that would supply the body atleast minimum requirements of each vitamin. Formulate L.P.P. and solve it graphically.

(UPTU: 2004-2005)

Solution: Let F_1 units of food F_1 and F_2 units of food F_2 constitute the diet.

Formulating the LPP

Objective function

Minimize $Z = 3F_1 + 2.5\,F_2$

subject to $2F_1 + 4F_2 \geq 40$ ..... (Requirement of V_1)

$3F_1 + 2\,F_2 \geq 50$ ..... (Requirement of V_2)

Writing equations for graphical solution

$$2F_1 + 4\,F_2 = 40 \quad \text{...(1)}$$

$$3\,F_1 + 2F_2 = 50 \quad \text{...(2)}$$

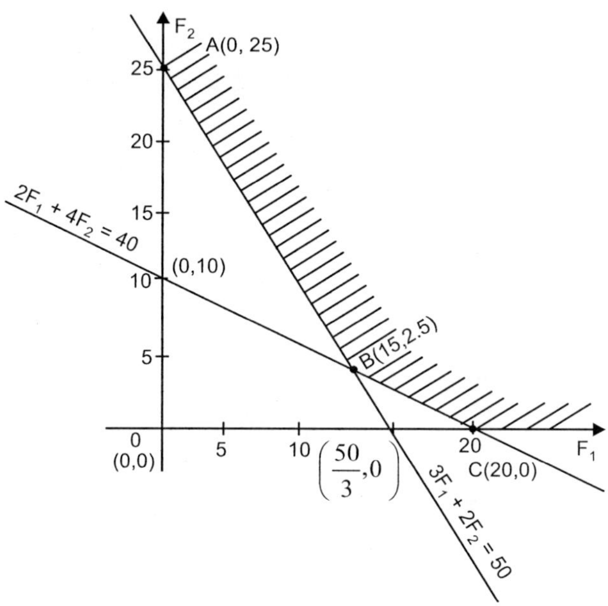

Fig. 2.6.

Line representing $2F_1 + 4\,F_2 = 40$ passes through (20,0) & (0,10) and line representing $3F_1 + 2F_2 = 50$ passes through $\left(\frac{50}{3}, 0\right)$ & (0,25). Feasible space is beyond ABC.

Solving equation (1) and (2) gives $F_1 = 15$, $F_2 = 2.5$

Hence, coordinates of point B are (15, 2.5).

Corner points	Value of $Z = 3F_1 + 2.5\,F_2$
A (0, 25)	62.5
B (15, 2.5)	51.25 (Minimum value)
C (20,0)	60

Therefore, the optimal solution is $F_1 = 15$ units, $F_2 = 2.5$ units

Minimum cost of diet = 51.25 Rupees.

Example 2.7. Solve the following LP problem using graphical method:

Maximize $Z = 6\,x_1 + 8\,x_2$ **(UPTU: 2004-2005)**

Subject to $5\,x_1 + 10\,x_2 \le 60$

$4\,x_1 + 4\,x_2 \le 40$

x_1 and $x_2 \ge 0$

Solution: Converting inequalities into equations

$$5\,x_1 + 10\,x_2 = 60 \quad ...(1)$$

passes through

A (0, 6) and D (12, 0)

$$4\,x_1 + 4\,x_2 = 40 \quad ...(2)$$

passes through

E(0, 10) and C(10, 0)

Feasible region is OABC

On Solving (1) and (2)

$$B = (8, 2)$$

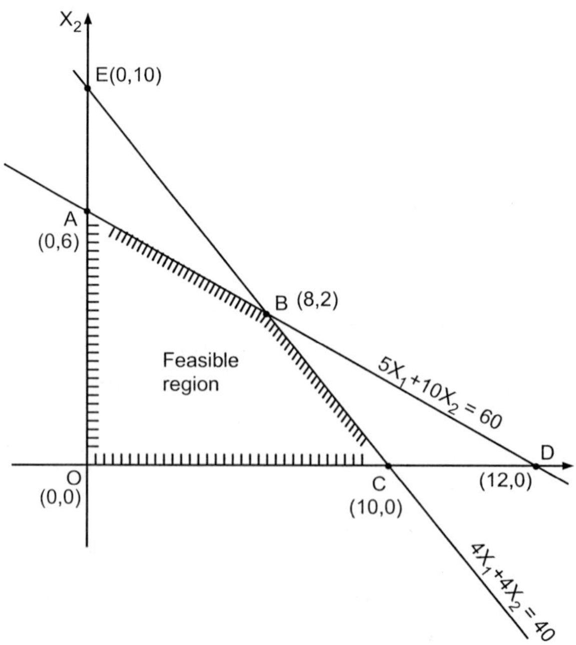

Fig. 2.7.

Corner points	Value of $Z = 6X_1 + 8X_2$
O(0, 0)	0
A(0, 6)	48
B(8, 2)	64 (Maximum value)
C(10, 0)	60

So, maximum value of Z occurs at $B(8, 2)$.

Optimal solution is $x_1 = 8, x_2 = 2$

and $Z_{max} = 64$

Example 2.8. Using graphical method or otherwise, **(UPTU: 2003-2004)**

Minimize $f = 3x_1 + 2x_2$,

subject to $8x_1 + x_2 \geq 8$

$2x_1 + x_2 \geq 6$

$x_1 + 3x_2 \geq 6$

$x_1 + 6x_2 \geq 8$

$x_1 \geq 0;\ x_2 \geq 0$

Solution : Converting inequalities into equations and ploting lines on graph.

$8X_1 + X_2 = 8$

$2X_1 + X_2 = 6$

$X_1 + 3X_2 = 6$

$X_1 + 6X_2 = 8$

Minimize

$f = 3x_1 + 2x_2$

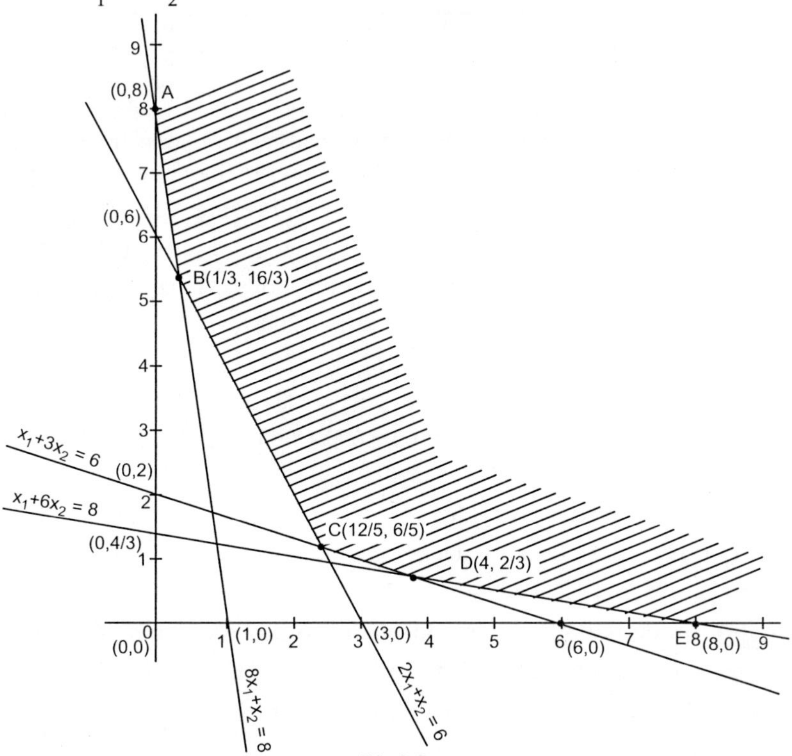

Fig.2.8.

Feasible region is ABCDE

Corner points	Value of $f = 3x_1 + 2x_2$
O(0, 8)	16
$B\left(\frac{1}{3}, \frac{16}{3}\right)$	$\frac{35}{3} = 11.67$
$C\left(\frac{12}{5}, \frac{6}{5}\right)$	$\frac{48}{5} = 9.6$ (Minimum value)
$D\left(4, \frac{2}{3}\right)$	13.33
E(8, 0)	24

Optimal solution is $x_1 = \frac{12}{5}$, $x_2 = \frac{6}{5}$

$$f = 9.6$$

Example 2.9. A manufacturing company can sell a maximum of 10,000 units of a products, which are quality approved. The company has 3 processing plants, each with a processing capacity of 5000 units, cost of processing and percentage of rejections varies for various plants.

Plant	**Cost of processing/Unit**	**% rejection**
A	100	5%
B	125	3%
C	150	2%

The only cost other than processing cost is raw material cost, which is Rs. 200 per unit. All quality approved goods are sold for Rs. 600 each and rejections are sold per Rs. 50 each. The company has policy of not having more than overall 3% rejections.

Formulate LPP to decide optimum production by various plants.

(UPTU. 2007-2008)

Solution. Let plants A, B and C produce x_1, x_2 and x_3 units of the product respectively.

Raw material cost $= 200\,x_1 + 200\,x_2 + 200\,x_3$

Processing cost $= 100\,x_1 + 125\,x_2 + 150\,x_3$

Total cost $= 300\,x_1 + 325\,x_2 + 350\,x_3$

Selling price $= (.95\,x_1 + .97\,x_2 + .98\,x_3) \times 600 + (.05\,x_1 + .03\,x_2 + .02\,x_3) \times 50$

$= 572\,x_1 + 583.5\,x_2 + 589\,x_3$

Profit = Total sale price – cost price

$= (572\,x_1 + 583.5\,x_2 + 589\,x_3) - 300\,x_1 + 325\,x_2 + 350\,x_3$

$= 272.5\,x_1 + 255.5\,x_2 + 239\,x_3$

Therefore, the objective function is to maximize total profit (z).

$$\text{Max. } z = 272.5\,x_1 + 255.5\,x_2 + 239\,x_3$$

Subject to, $x_1 + x_2 + x_3 \leq 10{,}000$ (selling capacity)

$x_1 \leq 5000$

$x_2 \leq 5000$

$x_3 \leq 5000$

$x_1, x_2, x_3 \geq 0$

EXERCISE

2.1. A firm produces three products A,B and C. Their profits per unit are Rs. 300, Rs. 200 and Rs. 400, respectively. The firm has two machines and the required processing time in minutes on each machine for each product is given in the following table:

		Product		
		A	B	C
Machine	1	4	3	5
	2	2	2	4

Machines 1 and 2 have 2000 and 2500 machine-minute, respectively. The upper limits for the production volumes of the product A, B and C are 100 units, 200 units and 50 units, respectively. But, the firm must produce a minimum of 50 units of the product A. Develop a LP model for this manufacturing situation to determine the production volume of each product such that the total profit is maximized. [**Ans.** $x_1 = 100, x_2 = 200, x_3 = 50, z = 90,000$]

2.2. Solve the following LP problem graphically:

Minimize $Z = 20\,x_1 + 10\,x_2$

subject to $x_1 + 2x_2 \leq 40$

$3x_1 + x_2 \geq 30$

$4x_1 + 3x_2 \geq 60$

x_1 and $x_2 \geq 0$ [**Ans.** $x_1 = 6, x_2 = 12, z = 240$]

2.3. Solve the following LP problem graphically :

Minimize $Z = 45\,x_1 + 55x_2$

subject to $x_1 + 2x_2 \leq 30$

$2x_1 + 3x_2 \leq 80$

$x_1 - x_2 \geq 8$

x_1 and $x_2 \geq 0$ [**Ans.** $x_1 = 8, x_2 = 0, z = 360$]

2.4. Solve the following LP problem using Graphical method :

Minimize $Z = x_1 + x_2$

subject to $2x_1 + x_2 \geq 2$

$x_1 + x_2 \geq 1$

x_1 and $x_2 \geq 0$ [**Ans.** $x_1 = 1, x_2 = 0, z = 1$]

2.5. Solve of the following LP problem graphically.

Maximize $Z = 5x_1 + 6x_2$

subject to $4x_1 + 7x_2 \leq 20$

$5x_1 + 2x_2 \leq 10$

$6x_1 + 8x_2 \leq 25$

x_1 and $x_2 \geq 0$

[**Ans.** $x_1 = \frac{10}{9}, x_2 = \frac{20}{9}, z = 18.89$]

2.6. What do you understand by a linear programming problem? What are its limitations?

2.7. Write the applications of linear programming in management.

2.8. Explain the graphical method of solving a linear programming problem involving two variables.

2.9. Define an optimization problem. What constitutes an objective function? Give the conditions; an optimization problem must satisfy so that it is called a Linear Programming Problem (LPP). **(UPTU 2003-2004)**

2.10. Define following terms:

(*i*) Objective function (*ii*) Decision variables
(*iii*) Unbounded solution (*iv*) Feasible solution
(*v*) Infeasible solution (*vi*) Optimum solution
(*vii*) Alternate optimum solution (*viii*) Degenerate solution
(*ix*) Basic variable

2.11. What are advantages and limitations of linear programming models?

2.12. Discuss the basic characteristics and assumptions of linear programming model.

2.13. What are implications of assuming linear objective function and constraints?

2.14. Illustrate graphically the following cases of linear programming problems:

(*i*) Multiple optimal solutions
(*ii*) Infeasible solution
(*iii*) Unbounded solution

2.15. Discuss similarities and differences between minimization and maximization problems using graphical method of linear programming.

2.16. Discuss 'Each linear programming problem that has a feasible region has an infinite number of solutions'.

OBJECTIVE TYPE QUESTIONS

2.1. Constraints may represent

(*a*) Limitations (*b*) Requirements
(*c*) Balanced conditions (*d*) All of the above

2.2. A constraint limits the values that

(*a*) Objective function can assume (*b*) The decision variables can assume
(*c*) Neither of the above (*d*) Both (*a*) and (*b*)

2.3. Linear Programming is

(*a*) Constraint optimization model(*b*) Constraint decision making model
(*c*) A mathematical programming model
(*d*) All of the above

2.4. Model formulation is important because

(*a*) It enables us to use Algebric techniques
(*b*) In a business context, most managers prefer to work with formal model
(*c*) It forces management to address a clearly defined problem
(*d*) It allows the manager to better communicate with the management

2.5. The non-negativity requirement is included in an LPP because
(*a*) It makes the model easier to solve
(*b*) It makes the model correspond more closely to real world problem
(*c*) Both (*a*) and (*b*) (*d*) Neither of above

2.6 An iso-profit line represents
(*a*) An infinite number of solutions all of which yield the same profit
(*b*) An infinite number of solutions all of which yield the same cost
(*c*) None of (*a*) and (*b*) (*d*) Both (a) and (*b*)

2.7. The distinguishing feature of an LP model is
(*a*) Relationship among all variables is linear.
(*b*) It has single objective function and constraints.
(*c*) Value of decision variables is non-negative.
(*d*) All of the above.

2.8. Which of the following is not a characteristic of LP model.
(*a*) Alternative courses of action.
(*b*) An objective function of maximization type.
(*c*) Limited amount of resources
(*d*) Non-negativity condition on the value of decision variables.

2.9. If an iso-profit line yielding the optimal solution coincides with a constraint line, then
(*a*) the solution is unbounded
(*b*) the solution is infeasible
(*c*) the constraint which coincides is redundant
(*d*) none of the above

2.10. A feasible solution to the linear programming problem should
(*a*) satisfy the problem constraints
(*b*) optimize the objective function
(*c*) satisfy the problem constraints and non-negativity restrictions
(*d*) satisfy the non-negativity restrictions

2.11. A variable which has no physical meaning, but is used to obtain an initial basic feasible solution to the linear programming problem is referred to as
(*a*) basic variable (*b*) non-basic variable
(*c*) artificial variable (*d*) basis

Answers

2.1. (*d*) 2.2. (*d*) 2.3. (*d*) 2.4. (*d*) 2.5. (*b*) 2.6. (*a*) 2.7. (*a*)
2.8. (*b*) 2.9. (*c*) 2.10. (*c*) 2.11. (*b*)

❑❑❑

Chapter 3

Linear Programming-Simplex Method

3.1. INTRODUCTION

The Simplex method was developed in 1957 by George B. Dentzig. The Simplex method is an iterative procedure for solving a linear programming problem in a finite number of steps. The Simplex method is based on following principle.

"The optimal solution to a Linear Programming Problem, if it exists, always occurs at one of the corner points of the feasible solution space."

It is a systematic and step by step procedure for finding out corner point solutions and taking them for optimality. The evaluation of corner points always starts at the point of origin (initial basic feasible solution), which is one of the corners of feasible solution space. This solution is then tested for optimality.

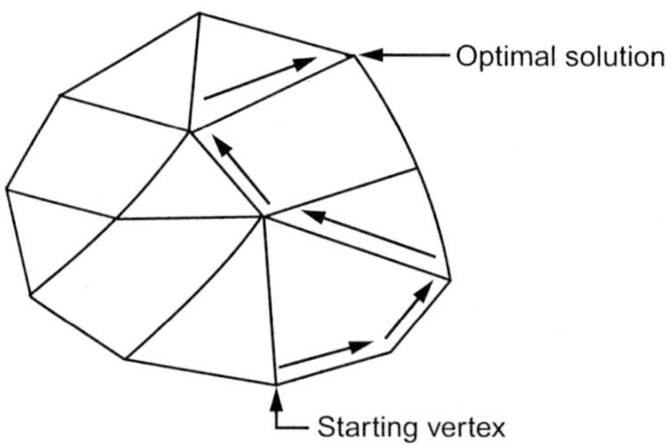

Fig. 3.1. Polytope.

A series of linear inequalities defines a polytope as a feasible region. The simplex method begins at a starting vertex and moves along the edges of the polytope until it reaches the vertex of the optimum solution.

3.2. DEFINITIONS OF BASIC TERMS

(*i*) **Standard form:** A linear problem in which all the constraints are written as equalities. Optimal solution of standard form is same as optimal solution of original linear programming problem.

(*ii*) **Slack variable:** A variable added to the left hand side of a less than or equal to constraint to convert it into an equality is called slack variable. This is amount of unused resources and contribution associated with this is zero.

Slack = Requirement - Production

e.g., $a_{11} x_1 + a_{12} x_2 \leq b_1$

Inequality can be removed by adding a slack variable.

$$a_{11} x_1 + a_{12} x_2 + s_1 = b_1$$

(*iii*) **Surplus variable:** a variable subtracted from the left hand side of the greater than or equal to constraint, to convert it into an equality is called surplus variable. This is the amount over and above the required minimum level.

Surplus = Production - Requirement

e.g., $a_{11} x_1 + a_{12} x_2 \geq b_1$

Here, inequality is removed by subtracting a surplus variable.

$$a_{11} x_1 + a_{12} x_2 - s_2 = b_1$$

(*iv*) **Basic solution:** For a system of m linear equations in n variables ($n > m$) a solution obtained by setting $n - m$ variables equal to zero and solving for remaining m variables is called basic solution.

(*v*) **Non-basic variables:** The variables ($n - m$) that are set equal to zero in any solution are called non-basic variables.

(*vi*) **Basic variables:** The variables (m) whose values are determined by solving **m** equations are known as basic variables.

(*vii*) **Basic feasible solution**: The basic solution, which is feasible also, is known as Basic feasible solution.

(*viii*) **Optimal solution**: A basic feasible solution that optimizes (maximizes or minimizes) the value of objective function is called an optimal solution.

(*ix*) **Simplex table:** The parameters of a linear programming problem are represented in a tabular form when simplex method is employed. This table is used to make calculations at each iteration.

(*x*) **Z_j row:** The numbers in this row under each variable represents the decrease in the value of objective function when one unit of non-basic variable is introduced in the basis in place of a basic variable.

(*xi*) **Index row ($C_j - Z_j$):** Each element in the index row represents the net contribution to objective function by introducing one unit of variable at the top of the column. Elements of this row are calculated by subtracting values of elements in **Z_j row** from the corresponding values of elements in per unit contribution row (C_j) for that column.

(*xii*) **Key column:** The column with largest positive number in the index row of maximization problem or largest negative number in the index row of minimization problem is known as key column. The variable of this column enters in the solution in next iteration.

(*xiii*) **Key row:** The row with smallest positive ratio found by dividing the values of quantity column by corresponding values in key column for each row is called as key row. The variable of this row leaves the basis in next iteration.

(*xiv*) **Key number:** The non-zero positive element at the intersection of key row and key column is called as key number.

3.3. STEPS OF SIMPLEX METHOD

3.3.1. Case 1: Maximization Type Problems (Constraints of ≤ Type)

1. Formulate the mathematical model of the real world problem. *e.g.*,
   Maximize $Z = C_1x_1 + C_2x_2 + ... + C_nx_n$
   subject to the constraints
   $$\begin{aligned} a_{11}x_1 + a_{12}x_2 + ... + a_{1n}\, x_n &\le b_1 \\ a_{21}x_1 + a_{22}\, x_2 + ... + a_{2n}x_n &\le b_2 \\ \vdots \qquad \vdots \qquad\qquad \vdots \quad \vdots \quad &\vdots \\ a_{m1}x_1 + a_{m2}x_2 + ... + a_{mn}\, x_n &\le b_m \\ x_1, x_2, ..., x_n &\ge 0 \end{aligned}$$
2. If the objective function is to be minimized then convert the given problem into a problem of maximization by Minimize Z = Maximize (- Z).
3. Check whether all b_i are positive; if any one of b_i is negative then multiply both sides of that constraint by -1 so as to get all b_i positive.
4. Write the mathematical model in the standard form by adding slack variables in the left hand side of constraints and assign a zero coefficient to these in objective function (Z).
   The given problem can be expressed in standard form as follows:
   Maximize $Z = C_1x_1 + C_2x_2 + ... C_nx_n + 0s_1 + 0s_2 + ... + 0s_m$
   subject to constraints
   $$\begin{aligned} a_{11}\, x_1 + a_{12}\, x_2 + ... + a_{1n}\, x_n + s_1 &= b_1 \\ a_{21}\, x_1 + a_{22}\, x_2 + ... + a_{2n}\, x_n + s_2 &= b_2 \\ \vdots \qquad \vdots \qquad\qquad \vdots \qquad &\vdots \\ a_{m1}\, x_1 + a_{m2}\, x_2 + ... + a_{mn}\, x_n + s_m &= b_m \end{aligned}$$
5. Find initial basic feasible solution by substituting the values of decision variables equal to zero.
   Substitute, $x_1 = x_2 = ... = x_n = 0$
   and obtain, $s_1 = b_1, s_2 = b_2 ... s_m = b_m$
6. Develop initial Simplex table.

C_j (Contribution Per unit)			C_1	C_2	....	C_n	0	0	....	0	
	Basic Variables	Value of Basic Variables	Coeff. Matrix				Identity Matrix				Minimum Ratio
C_B	B	$b(=X_b)$	X_1	X_2	....	X_n	S_1	S_2	....	S_m	
0	S_1	b_1	a_{11}	a_{12}	....	a_{1n}	1	0	....	0	b_1/a_{1j}
0	S_2	b_2	a_{21}	a_{22}	....	a_{2n}	0	1	....	0	
⋮	⋮	⋮	⋮	⋮		⋮	⋮	⋮		⋮	
0	S_m	b_m	a_{m1}	a_{m2}	....	a_{mn}	0	0	....	1	
	$Z = \Sigma C_{Bj}\, a_{ij}$		0	0		0	0	0		0	
Net Contribution	$C_j - Z_j$		C_1	C_2		C_n	0	0		0	

$C_j \rightarrow$ Objective row (coefficient of variables in objective function) remain unchanged during succeeding table.

$C_B \rightarrow$ Objective column (coefficient of current basic variables in objective function).

$B \rightarrow$ Basic variables (product mix) in basis. In initial table basic variables are slack variables.

$b \rightarrow$ solution values (quantity column) of basic variables—indicates resources.

Body Matrix—Coefficient of decision variables in constraints set – a_{ij} gives rate of substitution between s_i and x_j or if we produce a unit of x_j then a_{ij} units of s_i must be sacrificed. Identity matrix—coefficient of slack variables in the initial table.

7. Test the solution for optimality: Calculate the elements of index row $(C_j - Z_j)$; if all the elements in index row are negative then current solution is optimal basic feasible solution. If not, current solution can be improved by introducing one non-basic variable in place of basic variable. Go for next iteration.
8. Revision of current Simplex table.
   (*i*) Identify key column, with largest positive number in the index row. Non-basic variable at the top of this column is entering variable for the next iteration.
   (*ii*) Identify key row, corresponding to smallest non-negative ratio found by dividing the values of quantity column by corresponding values in key column for each row. Variable of this row is outgoing variable.
   (*iii*) Identify key element, the non-zero positive element at the intersection of key column and key row. Encircle the key element.
9. Construct new simplex table.
   (*i*) Calculate the new values for the key row (except those in objective column and variable column) by dividing every element of key row by key number.
   (*ii*) Entries in variable and objective column of key row are taken respectively from key column values of variable and objective row.
   (*iii*) The new values of the elements in the remaining rows (other than those in objective and variable column which remains the same as in previous table) are calculated by using following formula.

$$\text{New No.} = \text{Old No.} - \left[\frac{\text{Associated no. in key row} \times \text{Corresponding no. in key column}}{\text{key no.}}\right]$$

10. Go to step 7 and repeat the procedure until either an optimal solution is reached or there is an indication of unbounded solution.

Example 3.1. A firm manufactures and sells two products *A* and *B*. Each unit of A requires 1 hr of machining and 2 hr of skilled labour, where as each unit to B uses 2 hr of machining and 1 hr of labour. For the coming month the machining capacity is limited to 720 machining hrs and skilled labour is limited to 780 hrs. Not more than 320 units of *A* can be sold in the market during a month. Determine the optimal product-mix and the maximum contribution. Unit contribution from A is Rs 6 and from *B* is Rs 4. What will be incremental contribution per unit of machine hr, per unit of labour per unit of A saleable?

Solution: Maximize $Z = 6x_1 + 4x_2$

$$\left.\begin{aligned} \text{subject to constraint, } x_1 + 2x_2 &\le 720 \\ 2x_1 + 1x_2 &\le 780 \\ x_1 &\le 320 \end{aligned}\right] \text{ LP Model}$$

2. Equations, Maximize $Z = 6x_1 + 4x_2 + 0s_1 + 0s_2 + 0s_3$
subject to constraint, $x_1 + 2x_2 + s_1 = 720$

$$2x_1 + x_2 + s_2 = 780$$

$$x_1 + s_3 = 320$$

3. Initial BFS, $x_1 = 0, x_2 = 0$

$\Rightarrow$ $s_1 = 720, s_2 = 780, s_3 = 320$

4.

Profit/Unit C_B	Product Mix b	Quantity x_b	6 x_1	4 x_2	0 s_1	0 s_2	0 s_3	Minimum Ratio
0	s_1	720	1	2	1	0	0	720
0	s_2	780	2	1	0	1	0	390
0	s_3	320	(1)	0	0	0	1	320
	$Z_j = \Sigma C_{Bj} a_{ij}$		0	0	0	0	0	
Index row	$C_j - Z_j =$		6	4	0	0	0	

Profit = $\Sigma C_B X_b = 0$ Key no. = 1

5. 2nd Table

C_B	$C_j \longrightarrow$ b	x_b	6 x_1	4 x_2	0 s_1	0 s_2	0 s_3	Minimum Ratio
0	s_1	400	0	2	1	0	–1	200
0	s_2	140	0	(1)	0	1	–2	140
6	x_1	320	1	0	0	0	1	
	$Z_j = \Sigma C_{Bj} a_{ij}$		6	0	0	0	6	
	$C_j - Z_j =$		0	4	0	0	–6	

6. 3rd Table

C_B	$C_j \longrightarrow$ b	x_b	6 x_1	4 x_2	0 s_1	0 s_2	0 s_3	Minimum Ratio
0	s_1	120	0	0	1	–2	(3)	40
4	x_2	140	0	1	0	1	–2	—
6	x_1	320	1	0	0	0	1	320
	$Z_j = \Sigma C_{Bj} a_{ij}$		6	4	0	4	–2	
	$C_j - Z_j$		0	0	0	–4	2	

7. 4th Table

	$C_j \longrightarrow$		6	4	0	0	0	Minimum
C_B	b	x_b	x_1	x_2	s_1	s_2	s_3	Ratio
0	s_3	40	0	0	1/3	–2/3	1	
4	x_2	220	0	1	2/3	–1/3	0	
6	x_1	280	1	0	–1/3	2/3	0	
	Z_j	2560	6	4	2/3	8/3	0	
Index row	$C_j - Z_j$		0	0	–2/3	–8/3	0	

Optimal solution : $Z = 2560$, $x_1 = 280$, $x_2 = 220$, $s_3 = 40$
incremental values, $s_1 = 2/3$ per hr, $s_2 = 8/3$ per hr

3.3.2. Case 2: Minimization Type Problems (Constraints of ≥ Type)

1. Formulate the mathematical model of the given linear programming problem.
2. Write the standard form.

Minimize $Z = C_1x_1 + C_2x_2 + \dots C_nx_n + 0s_1 + 0s_2 + \dots 0s_n$

$$\begin{array}{l|l} a_{11}\, x_1 + a_{12}\, x_2 + \dots.\; a_{1n}\, x_n - s_1 = b_1 & x_1, x_2, x_3 \geq 0 \\ a_{21}\, x_1 + a_{22}\, x_2 + \dots\; a_{2n}\, x_n - s_2 = b_2 & s_1, s_2, s_3 \geq 0 \\ a_{m1}\, x_1 + a_{m2}\, x_2 + \dots.\; a_{mn}\, x_n - s_m = b_m & \end{array}$$

initial basic solution, by putting $x_1 = 0$, $x_2 = 0 = \dots\dots$

$$s_1 = -\,b_1,\ s_2 = -\,b_2,\ s_3 = -\,b_3 \dots$$

It violates non-negativity constraints ($s_j \geq 0$), hence it is not feasible. '*m*' new variables called Artificial variables ($A_1, A_2, A_3 \dots A_m$) are added to left hand side (one to each equation)

$$\begin{array}{l|l} \Rightarrow\ a_{11}x_1 + a_{12}\, x_2 \dots. + a_{1n}\, x_n - s_1 + A_1 = b_1 & x_i \geq 0 \\ a_{21}\, x_1 + a_{22}\, x_2 \dots. + a_{2n}\, x_n - s_2 + A_2 = b_2 & s_j \geq 0 \\ a_{m1}\, x_1 + a_{m2}\, x_2 \dots + a_{mn}\, x_n - s_m + A_m = b_m & A_j \geq 0 \end{array}$$

3. The Initial basic feasible solution can be obtained by any of the two methods (*i*) Big M (Penalty) method (*ii*) Two phase simplex method.

(*i*) In **Big M** method a very high penalty **M** is assigned to each artificial variables in the objective function. Therefore, it is known as **Big M** (or **Penalty**) method.

Initial basic feasible solution. It is obtained by setting $(n + 2m - m)$ variables equal to zero *i.e.*, $x_1, x_2, x_3 = 0$

$$s_1, s_2, s_3 \dots = 0,$$

$$\therefore \quad A_1 = b_1,\ A_2 = b_2,\ A_3 = b_3 \dots$$

Minimise $Z = C_1x_1 + C_2x_2 + \dots + C_nx_n + 0s_1 + 0s_2 + \dots + MA_1 + MA_2 + \dots$

(*i*) Artificial variable do not have any physical meaning.
(*ii*) (– *M*) is used for maximization problems and + M for minimization problems.
(*iii*) ≥ requires both surplus (s) and Artificial (A) variables.

(*iv*) = requires only *A* to be added.

(*v*) No solution exists if all Artificial variables can not be driven out in optimal big *M* table (resources are not sufficient to meet the demand)

4. Table

C_j →			C_1	C_2	C_3	C_m	0	0	0		0	M	M	M	Minimum Ratio
C_B	b	x_b	x_1	x_2	x_3	x_m	s_1	s_2	s_3	....	s_m	A_1	A_2	A_3	
M	A_1	b_1	a_{11}	a_{12}	a_{13}............	a_{1m}	–1	0	0	....	0	1	0	0	
M	A_2	b_2	a_{21}	a_{22}	a_{23}............	a_{2m}	0	–1	0	....	0	0	1	0	
M	A_3	b_3	a_{31}	a_{32}	a_{33}............	a_{3m}	0	0	–1	....	0	0	0	1	
	Z_j		0	0	0	0	0	0	0		0	0	0	0	
	$C_j - Z_j$		C_1	C_2	C_3	C_m	0	0	0		0	M	M	M	

5. **Test for optimality :** If all elements in index row are ≥ 0 then solution is optimal. If not, then select most negative number in index row-key column (incoming variable). Thus we choose the variable that decreases the cost most. Other steps are same.

(*i*) If at least one artificial variable is present in solution at zero level (*A* = 0) and in index row the coefficient of *M* in each is non-negative then solution of problem does not exist *i.e.,* current solution is degenerate.

(*ii*) If *A* = + ve in (*i*) then problem does not possess any optimal solution.

(*iii*) An Artificial variable, once driven, can never re enter, because of big *M* associated with it in objective function. Do not compute column of this artificial variable A in subsequent iteration.

Example 3.2. A diet for a patient must contain at least 4000 units of vitamins, 50 units of minerals and 1400 units of calories. Two fruits *A* and *B* are available at a cost of Rs 4 and Rs 3 per unit respectively. If one unit of *A* contains 200 units of vitamins, 1 unit of mineral and 40 units of calories and one unit of fruit *B* contain 100 units of vitamins, 2 units of minerals and 40 units of calories. Find, what combination of fruits be used to have least cost?

Solution : 1. Minimize $Z = 4x_1 + 3x_2$

subject to
$$200x_1 + 100x_2 \geq 4000$$
$$x_1 + 2x_2 \geq 50$$
$$40x_1 + 40x_2 \geq 1400$$
$$x_1, x_2 \geq 0$$

2. Minimize $Z = 4x_1 + 3x_2 + 0s_1 + 0s_2 + 0s_3 + MA_1 + MA_2 + MA_3$

subject to,
$$200x_1 + 100x_2 - s_1 + A_1 = 4000$$
$$x_1 + 2x_2 - s_2 + A_2 = 50$$
$$40x_1 + 40x_2 - s_3 + A_3 = 1400$$
$$x_1, x_2 \geq 0$$

3. Initial BFS, $x_1 = 0, x_2 = 0,$

$s_1 = s_2 = s_3 = 0$

⇒ $A_1 = 4000,$

$A_2 = 50$

$A_3 = 1400$

	$C_j \longrightarrow$		4	3	0	0	0	M	M	M	Minimum Ratio
C_B	b	x_b	x_1	x_2	s_1	s_2	s_3	A_1	A_1	A_3	
M	A_1	4000	(200)	100	–1	0	0	1	0	0	20
M	A_2	50	1	2	0	–1	0	0	1	0	50
M	A_3	1400	40	40	0	0	–1	0	0	1	35
	Z_j		241M	142 M	–M	–M	–M	M	M	M	
	$C_j - Z_j$		4–241M	3–142M	M	M	M	0	0	0	

4. Next table

	$C_j \longrightarrow$		4	3	0	0	0	M	M	M	θ
C_B	b	x_b	x_1	x_2	s_1	s_2	s_3	A_1	A_1	A_3	
4	X_1	20	1	1/2	–1/200	0	0		0	0	40
M	A_2	30	0	(3/2)	1/200	–1	0		1	0	20
M	A_3	600	0	20	1/5	0	–1		0	1	30
	Z_j		4	2+215M	–1/50+41M/200	–M	–M		M	M	
	$C_j - Z_j$		0	1–21.5M	1/50–41M/200	M	M		0	0	

5. Final table

	$C_j \longrightarrow$		4	3	0	0	0	M	θ
C_B	b	x_b	x_1	x_2	s_1	s_2	s_3	A_3	
4	X_1	10	1	0	–2/300	1/3	0	0	30
3	X_2	20	0	1	1/300	–2/3	0	0	—
M	A_3	200	0	0	2/15	40/3	–1	1	15
	Z_j		4	3	$\left(\frac{-5}{300}+\frac{2}{15}M\right)$	4M/3–2/3	–M		
	$C_j - Z_j$		0	0	$\left(\frac{-2M}{15}+\frac{5}{300}\right)$	–4M/3 + 2/3	M		

6. Current solution is not optimal solution. Select key column (most negative index row element)—column headed by s_2. Key row—Third row (outgoing variable A_3). Develope new table.

C_B	b	x_b	4 x_1	3 x_2	0 s_1	0 s_2	0 s_2	θ
4	x_1	50	1	0	–1/100	0	1/40	
3	x_2	30	0	1	1/100	0	–1/20	
0	s_2	15	0	0	1/100	1	–3/40	
	Z_j		4	3	–1/100	0	–1/20	
	$C_j - Z_j$		0	0	1/100	0	1/20	

Since, all elements in index row are either zero or positive, current solution is optimal solution.

∴ $x_1 = 5, x_2 = 30$

Minimum value of Z = 4 × 5 + 3 × 30 = Rs. 110

3.3.3. Case 3: Minimization Type Problems (Mixed Constraints)

Example 3.3. A company must produce 200 kg of a mixture consisting of ingredient x_1 and x_2 daily. x_1 costs Rs 3 per kg and x_2 Rs 8 per kg. No more than 80 kg of x_1 can be used and at least 60 kg of x_2 must be used. Find how much of each ingredient should be used if company wants to minimize cost.

Solution : 1. Minimize $Z = 3x_1 + 8x_2$

subject to $x_1 + x_2 = 200$

$x_1 \le 80,$

$x_2 \ge 60,$

$x_1, x_2 \ge 0$

2. $Z = 3x_1 + 8x_2 + 0s_1 + 0s_2 + MA_1 + MA_2$

subject to $x_1 + x_2 + A_1 = 200$

$x_1 + s_1 = 80,$

$x_2 - s_2 + A_2 = 60$

3. IBFS, $x_1, x_2 = 0,$

$A_1 = 200, s_1 = 80,$

$A_2 = 60\ (s_2 = 0)$

4.

C_B	b	x_b	3 x_1	8 x_2	0 s_1	0 s_2	M A_1	M A_2	θ
M	A_1	200	1	1	0	0	1	0	200
O	S_1	80	1	0	1	0	0	0	–
M	A_2	60	0	(1)	0	–1	0	1	60
	Z_j		M	2M	0	–M	M	M	
	$C_j - Z_j$		3–M	8–2M	0	M	0	0	

5. Next table

C_B	C_j b	x_b	3 x_1	8 x_2	0 s_1	0 s_2	M A_1	θ
M	A_1	140	1	0	0	1	1	140
O	s_1	80	(1)	0	1	0	0	80
8	x_2	60	0	1	0	–1	0	–
	Z_j		M	8	0	M–8	M	
	$C_j - Z_j$		3–M	+0	0	8–M	0	

6. Above solution is not optimal, hence develop new table.

C_B	C_j b	x_b	3 x_1	8 x_2	0 s_1	0 s_2	M A_1	θ
M	A_1	60	0	0	–1	(1)	1	60
3	x_1	80	1	0	1	0	0	–
8	x_2	60	0	1	0	–1	0	–
	Z_j		3	8	–M+3	M–8	M	
	$C_j - Z_j$		0	0	M–3	8–M	0	

7. Current solution is still not optimal, go for next iteration.

C_B	C_j b	x_b	3 x_1	8 x_2	0 s_1	0 s_2	θ
0	s_2	60	0	0	–1	1	
3	x_1	80	1	0	1	0	
8	x_2	120	0	1	–1	0	
	Z_j	(1200)	3	8	–5	0	
	$C_j - Z_j$		0	0	5	0	

In above table all elements of index row are either zero or positive, hence, solution is optimal solution.

$$x_1 = 80, x_2 = 120,$$
$$Z = 3 \times 80 + 8 \times 120 = 1200$$

3.3.4. Case 4: Maximization Type Problems (Mixed Constraints)

Example 3.4. Maximise $Z = 4x_1 + 5x_2 - 3x_3$

subject to, $x_1 + x_2 + x_3 = 10$

$$x_1 - x_2 \geq 1$$
$$2x_1 + 3x_2 + x_3 \leq 30$$

1. Maximise $Z = 4x_1 + 5x_2 - 3x_3 + 0s_1 + 0s_2 - MA_1 - MA_2$
subject to, $x_1 + x_2 + x_3 + A_1 = 10$
$x_1 - x_2 - s_1 + A_2 = 1$
$2x_1 + 3x_2 + x_3 + s_2 = 30$

2. IBFS, $x_1, x_2, x_3 = 0,$
$A_1 = 10, A_2 = 1, s_2 = 30$

3. Develope simplex table

C_B	C_j / b	x_b	4 / x_1	5 / x_2	−3 / x_3	0 / s_1	0 / s_2	−M / A_1	−M / A_2	θ
−M	A_1	10	1	1	1	0	0	1	0	10
−M	A_2	1	(1)	−1	0	−1	0	0	1	1
O	S_2	30	2	3	1	0	1	0	0	15
	Z_j		−2M	0	−M	+M	0	−M	−M	
	$C_j - Z_j$		4+2M	5	−3+M	−M	0	0	0	

Since, all elements of index row are not negative, solution is not optimal.

4. Next table

C_B	C_j ⟶ / b	x_b	4 / x_1	5 / x_2	−3 / x_3	0 / s_1	0 / s_2	−M / A_1	θ
−M	A_1	9	0	(2)	1	1	0	1	9/2=4.5
4	X_1	1	1	−1	0	−1	0	0	–
O	S_2	28	0	5	1	2	1	0	28/5=5.
	Z_j		4	−2M−4	−M	−M−4	0	−M	
	$C_j - Z_j$		0	9+2M	−3+M	M+4	0	0	

In this table also, 2 elements are positive, therefore, solution is not optimal. Go for next iteration.
key column → column headed by x_2
key row → first row (outgoing variable is A_1)
key number = 2

5. Final table

	C_j		4	5	–3	0	0	θ
C_B	b	x_b	x_1	x_2	x_3	s_1	s_2	
5	x_2	9/2	0	1	1/2	1/2	0	
4	x_1	5.5	1	0	1/2	–1/2	0	
O	s_2	5.5	0	0	–1.5	–.5	1	
	Z_j		4	5	4.5	.5	0	
	$C_j - Z_j$		0	0	–7.5	–.5	0	

Optimal solution $x_1 = 5.5,\ x_2 = 4.5,$

$s_2 = 5.5,$

$Z = 89/2 = 44.5$

3.4. SPECIAL CASES

3.4.1. Unbounded Solution

It occurs when one or more of decision variables can be increased indefinitely without violating any restrictions. It occurs in maximization problem when at any stage, all the entries in minimum ratio column are either negative or infinite. The variable in key column can be increased indefinitely.

Example 3.5. Solve following LPP by simplex method.

Maximize $Z = 4x_1 + x_2 + 3x_3 + 5x_4$

subject to $4x_1 - 6x_2 - 5x_3 - 4x_4 \geq -20$

$-3x_1 - 2x_2 + 4x_3 + x_4 \leq 10$

$-8x_1 - 3x_2 + 3x_3 + 2x_4 \leq 20$

$x_1, x_2, x_3, x_4 \geq 0$

Solution : 1. Standard form—multiply first constaint by – 1 and write standard form

$Z = 4x_1 + x_2 + 3x_3 + 5x_4 + 0s_1 + 0s_2 + 0s_3$

$-4x_1 + 6x_2 + 5x_3 + 4x_4 + s_1 = 20$

$-3x_1 - 2x_2 + 4x_3 + x_4 + s_2 = 10$

$-8x_1 - 3x_2 + 3x_3 + 2x_4 + s_3 = 20$

2. Initial solution

Put $x_1 = x_2 = x_3$

$= x_4 = 0$

$s_1 = 20,$

$s_2 = 10,\ s_3 = 20$

and $Z = 0$

3. Simplex table

C_B	B	x_B	4 x_1	1 x_2	3 x_3	5 x_4	0 s_1	0 s_2	0 s_3	θ
0	s_1	20	–4	6	5	(4)	1	0	0	20/4 = 5
0	s_2	10	–3	–2	4	1	0	1	0	10/1 = 10
0	s_3	20	–8	–3	3	2	0	0	1	20/2 = 10
	Z_j		0	0	0	0	0	0	0	
Index row ($C_j - Z_j$)			4	1	3	5	0	0	0	

4. This solution is not optimal as index row has positive elements.
key column–column of x_4 → entering variable
key row–1st row, s_1 → leaving variable, key element = 4

5. Develop new Table.

C_B	B	x_B	4 x_1	1 x_2	3 x_3	5 x_4	0 s_1	0 s_2	0 s_3	θ
5	x_4	5	–1	3/2	5/4	1	1/4	0	0	—
0	s_2	5	–2	–7/2	11/4	0	–1/4	1	0	—
0	s_3	10	–6	–6	1/2	0	–1/2	0	1	—
	Z_j		–5	15/2	25/4	5	5/4	0	0	
Index row ($C_j - Z_j$)			9	–13/2	–13/4	0	–5/4	0	0	

↑ Key column

6. Current solution is not optimal solution because positive element (9) appears in index row. Column corresponding to this is key column and x_1 is incoming variable. But all the minimum ratios (θ) are negative. That means x_1 can be increased indefinitely without violating a constraint. Solution is therefore unbounded one.

3.4.2. Multiple Optimal Solution

It occurs when there can be infinite number of solutions possible for a given problem. This is possible when one of the non-basic variables in index row of optimal solution table has a zero value. Considering column with zero value in index row as key column, another optimal solution can be obtained.

Example 3.6. Solve following LPP by simplex .

Maximize $Z = 6x_1 + 4x_2$

subject to $-2x_1 + x_2 \leq 2$

$x_1 - x_2 \leq 2$

$3x_1 + 2x_2 \leq 9$

$x_1, x_2 \geq 0$

Solution: 1. Standard form

$$Z = 6x_1 + 4x_2 + 0s_1 + 0s_2 + 0s_3$$
$$-2x_1 + x_2 + s_1 = 2$$
$$x_1 - x_2 + s_2 = 2$$
$$3x_1 + 2x_2 + s_3 = 9$$

2. Initial basic solution

$$x_1 = x_2 = 0$$
$$s_1 = 2, s_2 = 2,$$
$$s_3 = 9, Z = 0$$

3. Simplex table

C_B	B	$C_j \rightarrow$ X_B	6 x_1	4 x_2	0 s_1	0 s_2	0 s_3	θ
0	s_1	2	–2	1	1	0	0	–
0	s_2	2	(1)	–1	0	1	0	2
0	s_3	9	3	2	0	0	1	3
	Z_j		0	0	0	0	0	
	$C_j - Z_j$		6	4	0	0	0	

↑ Key column

4. This solution is not optimal as index row contains positive elements.

key column—column headed by x_1, key row—2nd row, s_2 outgoing variable, key element = 1.

5. New table

C_B	B	$C_j \rightarrow$ X_b	6 x_1	4 x_2	0 s_1	0 s_2	0 s_3	θ
0	s_1	6	0	–1	1	2	0	–
6	x_1	2	1	–1	0	1	0	–
0	s_3	3	0	(5)	0	–3	1	3/5
	Z_j		6	–6	0	6	0	
Index row	$C_j - Z_j$		0	10	0	–6	0	

↑ Key column

Present solution is not optimal.

key column—column headed by x_2, x_2 is next entering variable, key row—3rd row, s_3 is leaving variable, key element = 5.

6. Next table

$C_j \rightarrow$			6	4	0	0	0	θ
C_B	B	X_B	x_1	x_2	s_1	s_2	s_3	
0	S_1	33/5	0	0	1	7/5	1/5	–
6	X_1	13/5	1	0	0	2/5	1/5	–
4	X_2	3/5	0	1	0	–3/5	1/5	
	Z_j		6	4	0	0	2	
Index row $(C_j - Z_j)$			0	0	0	0	–2	

Since, all the elements are zero or negative; optimal solution is reached.

$$x_1 = \frac{13}{5},\ x_2 = \frac{3}{5}$$

$$Z = 6 \times \frac{13}{5} + 4 \times \frac{3}{5} = \frac{78}{5} + \frac{12}{5} = \frac{90}{5} = 18$$

Now, In above table, index row elements of basic variable x_1 and x_2 are zero and index row element of one of the nonbasic variable s_2 is zero. Hence multiple optimal solutions are possible. s_1 can be replaced by s_2

$C_j \rightarrow$			6	4	0	0	0	θ
C_B	B	X_B	x_1	x_2	s_1	s_2	s_3	
0	S_2	33/7	0	0	5/7	1	1/7	
6	X_1	5/7	1	0	–2/7	0	1/7	
4	X_2	24/7	0	1	3/7	0	2/7	
	Z_j		6	4	0	0	2	
Index row $(C_j - Z_j)$			0	0	0	0	–2	

This solution is also optimal and gives values

$$x_1 = \frac{5}{7},\ x_2 = \frac{24}{7}$$

$$Z = 6 \times \frac{5}{7} + 4 \times \frac{24}{7} = \frac{126}{7} = 18$$

Thus value of objective function remains same.

3.4.3. Non-Feasible Solution

When the constraints of a LPP are inconsistent, no feasible solution to the problem exists. Also, a solution is non-feasible if an artificial variable appears in basis of optimal table.

Example 3.7. Maximize $Z = 3x_1 + 2x_2$

subject to
$$2x_1 + x_2 \le 2$$
$$3x_1 + 4x_2 \ge 12$$
$$x_1, x_2 \ge 0$$

Solution : 1. Standard form

$$Z = 3x_1 + 2x_2 + 0s_1 + 0s_2 - MA_1$$
$$2x_1 + x_2 + s_1 = 2$$
$$3x_1 + 4x_2 + A_1 - s_2 = 12$$

2. Initial feasible solution

$$x_1 = x_2 = 0$$
$$s_1 = 2, A_1 = 12, Z = -\ 12\ M$$

3. Simplex table

$C_j \rightarrow$			3	2	0	0	–M	θ
C_B	B	x_B	x_1	x_2	s_1	s_2	A_4	
0	S_1	2	2	(1)	1	0	0	2/1 = 2
–M	A_1	12	3	4	0	–1	1	12/4 = 3
	Z_j		–3M	–4M	0	M	–M	
Index row ($C_j - Z_j$)			(3+3M)	(2+4M)	0	–M	0	

This solution is not optimal

key column—headed by $x_2 \rightarrow$ entering variable. Key row—1st row, s_1 is leaving variable, key element = 1.

4. New table

$C_j \rightarrow$			3	2	0	0	–M	θ
C_B	B	x_B	x_1	x_2	s_1	s_2	x_4	
2	x_2	2	2	1	1	0	0	
–M	A_1	4	–5	0	–4	–1	1	
	Z_j		(4+5M)	2	(2+4M)	M	–M	
Index row ($C_j - Z_j$)			(–1–5M)	0	(–2–4M)	–M	0	

Above is the optimum table but artificial variable A_1 still exists as a basic variable with value (+4). Hence, problem is infeasible.

3.4.4. Degeneracy

Degeneracy in a linear programming problem may arise (*i*) at initial stage or (*ii*) at any subsequent iteration stage. In case (*i*), it is revealed when a basic variable acquires a zero value in initial basic feasible solution. At subsequent stage of the solution, either the number of basic variables is not equal to the number of constraints or number of zero variables does not equal the number of decision variables. It is preceded by a tie for existing variable (θ same for two or more rows) and an arbitrary selection for it is done. If a proper key row is selected, degeneracy can be resolved (avoided). If minimum ratio (θ) is zero then cycle is repeated indefinitely without arriving at optimal solution (rare situation). To make selection of key row in case of tie following rules are adopted:

(*i*) Coefficient of slack variable in the table where degeneracy is detected, are divided by corresponding +ve numbers of key column in row starting from left to right.

(*ii*) The row which contains the smallest ratio comparing from left to right column wise becomes key row.

(*iii*) In case of tie between slack and artificial variables to leave the basis, artificial variable shall leave the basis.

From the theoretical point of view, degeneracy has two implications (*i*) cycling—repeating a sequence of iterations, never improving the objective value. (*ii*)Different iterations give identical values for all the variables and objective function (z), although differing in basic-non basic categorisation of the variables.

From the practical point of view, it is observed that one of the constraints is redundant.

Example 3.8. Maximize $Z = 3x_1 + 9x_2$

subject to $x_1 + 4x_2 \leq 8$

$x_1 + 2x_2 \leq 4$

$x_1, x_2 \geq 0$

Solution : 1. Standard form

$Z = 3x_1 + 9x_2 + 0s_1 + 0s_2$

$x_1 + 4x_2 + s_1 = 8$

$x_1 + 2x_2 + s_2 = 4$

2. Initial basic solution

$x_1 = x_2 = 0$

$s_1 = 8, s_2 = 4$

$Z = 0$

3. Simplex table

C_B	B	$C_j \rightarrow$ X_B	3 x_1	9 x_2	0 s_1	0 s_2	θ
0	S_1	8	1	4	1	0	8/4 = 2
0	S_2	4	1	(2)	0	1	4/2 = 2
	Z_j		0	0	0	0	
Index row ($C_j - Z_j$)			3	9	0	0	

↑ Key column

There is tie in minimum ratio 'θ'. This is condition of degeneracy. To resolve it, coefficient of slack variables of rows which have tie are divided by corresponding elements of key column.

s_1	s_2
1/4	0/4
0/2 = 0	½

Row with minimum resulting ratio comparing in column from left to right is selected as key row (2nd row).

4. New table

C_B	B ($C_j \longrightarrow$)	x_B	3 x_1	9 x_2	0 s_1	0 s_2	θ
0	s_1	0	–1	0	1	–2	
9	x_2	2	1/2	1	0	1/2	
	Z_j		9/2	9	0	9/2	
Index row $(C_j - Z_j)$			–3/2	0	0	–9/2	

All the elements in index row are either zero or negative, hence current solution is optimal

$$x_2 = 2, x_1 = 0$$
$$Z = 3 \times 0 + 9 \times 2 = 18$$

3.5. Dual Problem

Associated with every LP problem there is another intimately related LP problem that is based upon the same data and having the same solution. The original LP problem is called primal problem and associated problem is called dual problem.

3.5.1. Duality Theorem

For every maximization (or minimization) problem in LP, there is unique similar problem of minimization (or maximization) involving the same data, which describe original problem.

Either of two problems can be treated as primal and other its dual.

Maximise $Z_x = C_1x_1 + C_2x_2 + ... + C_n x_n$

subject to, $a_{11} x_1 + a_{12}x_2 + ... + a_{1n} x_n \leq b_1$ Primal

$a_{21} x_1 + a_{22} x_2 + ... + a_{2n} x_n \leq b_2$

$\vdots$

$a_{m1} x_1 + a_{m2} x_2 + ... + a_{mn} x_n \leq b_m$

Dual: Minimise $Z_y = b_1y_1 + b_2y_2 + ... + b_m y_m$

subject to, $a_{11}y_1 + a_{21} y_2 + ... a_{m1} y_m \geq C_1$

$a_{12} y_1 + a_{22} y_2 + ... + a_{m2} y_m \geq C_2$

$\vdots$

$a_{1n} y_1 + a_{2n} y_2 + ... a_{mn} y_m \geq C_n$

		Primal x_1	x_2	......	x_n	Relation	Constraint
Dual Variables	y_1	a_{11}	a_{12}	......	a_{1n}	$\leqslant$	b_1
	y_2	a_{21}	a_{22}	......	a_{2n}	$\leqslant$	b_2
	⋮	⋮					⋮
	y_m	a_{m1}	a_{m2}	......	a_{mn}	$\leqslant$	b_m
Relation		$\geqslant$	$\geqslant$		$\geqslant$		mini Z_y
Constraint		C_1	C_2		C_n	max Z_x	

3.5.2. Rules

1. Number of variables in DP is equal to number of constraints in PP.
2. Number of constraints in DP is equal to number of variables in PP.
3. Coefficients of objective function in DP come from right hand side of PP.
4. If PP is maximum model, DP is a minimum model and vice-versa.
5. Coefficient of first constraint function for DP is coefficient of first variable in PP.
6. Right hand side of DP constraints comes from coefficients of objective function in PP.
7. i^{th} dual constraint is = if and only if i^{th} variable in in PP is unconstrained in sign.
8. If PP is maximization (minimization) model then after applying rule 7 assign to the remaining dual constraints a sense same as (opposite to) the corresponding variable in PP.
9. i^{th} variable in dual problem is unconstrained in sign if and only if the ith constraint in PP is equality.
10. If PP is maximum (minimum) model then after applying rule 9, assign to remaining dual variables a sense opposite to (same as) the corresponding constraint in PP.
11. Dual of dual is original problem.

3.5.3. Characteristics of Dual Problem

1. Dual of dual is a primal problem.
2. If either the primal or dual problem has finite optimal solution, then other one also possesses the same. Optimum values of objective functions of two problems are same.
3. If any of two problems has only an infeasible solution, then value of objective function of other is unbounded.
4. Value of objective function for any feasible solution of primal is less than the value of objective function for feasible solution of dual.
5. If primal has feasible solution but dual does not have, then primal will not have finite optimal solution and vice-versa.

3.5.4. Advantages

1. Calculations can be reduced if primal contains large number of rows and smaller number of variables (column).
2. Economic interpretation of dual helps management in making future decisions.
3. A linear programming problem in which initial solution is infeasible, can be solved by using concept of Duality.

   Write Dual of following primal problems.

Example 3.9. Minimize $Z_x = 6000x_1 + 4000x_2$

subject $4x_1 + x_2 \geq 12$

$$9x_1 + x_2 \geq 20$$
$$7x_1 + 3x_2 \geq 18$$
$$10x_1 + 40x_2 \geq 40$$

Solution : Dual, Max $Z_y = 12y_1 + 20y_2 + 18y_3 + 40y_4$

subject to $4y_1 + 9y_2 + 7y_3 + 10y_4 \le 6000$

$y_1 + y_2 + 3y_3 + 40y_4 \le 4000$

Example 3.10. Maximise $Z = 3x_1 + 4x_2$

subject to, $2x_1 + 3x_2 \le 16$

$5x_1 + 2x_2 \ge 20$

$x_1, x_2 \ge 0$

$-5x_1 - 2x_2 \le -20$

Solution : Dual, Min. $Z = 16y_1 - 20y_2$

subject to , $2y_1 - 5y_2 \ge 3$

$3y_1 - 2y_2 \ge 4$

Example 3.11. Min. $Z = 40x_1 + 120x_2$

subject to, $x_1 - 2x_2 \le 8$

$3x_1 + 5x_2 = 90$

$15x_1 + 44x_2 \le 660$

Solution: $-x_1 + 2x_2 \ge -8$

$3x_1 + 5x_2 \ge 90$

$-3x_1 - 5x_2 \ge -90$

$-15x_1 - 44x_2 \ge -660$

Maximum $Z = -8y_1 + 90y_2 - 90y_3 - 660y_4 = -8y_1 + 90y - 660y_4$

$-y_1 + 3y_2 - 3y_3 - 15y_4 \le 40$

$2y_1 + 5y_2 - 5y_3 - 44y_4 \le 120$

$-y_1 + 3y - 15y_4 \le 40$

$2y_1 + 5y - 44y_4 \le 120$

$y_1 \ge 0, y_4 \ge 0$

y un-restricted, $y_2 - y_3 = y$

Example 3.12. Max $Z = 20x_1 + 15x_2 + 18x_3 + 10(x_5 - x_6)$

subject to, $4x_1 - 3x_2 + 10x_3 + 4x_4 \le 60$ x_4 unrestricted in sign

$x_1 + x_2 + x_3 = 27$

$-x_2 + 4x_3 + 7x_4 \ge 35$

Solution: $4x_1 - 3x_2 + 10x_3 + 4(x_5 - x_6) \le 60$

$x_1 + x_2 + x_3 \le 27$

$-x_1 - x_2 - x_3 \le -27$

$x_2 - 4x_3 - 7(x_5 - x_6) \le -35$

Dual is $Z_y = 60y_1 + 27y - 35y_4$

$4y_1 + y \ge 20$

$-3y_1 + y + y_4 \ge 15$

$10y_1 + y - 4y_4 \ge 18$

$$4y_1 + 7y_4 = 10$$

y unrestricted in sign.

Example 3.13. Write the dual of following LPP. **(UPTU. 2007-08)**

minimize $Z = x_1 + \frac{1}{2} x_2$

subject to $6x_1 + 2x_2 \geq 24$

$3x_1 + 2x_2 \geq 18$

$x_1 + 3x_2 \geq 12$

$x_1, x_2 \geq 0$

Solution: Dual problem

maximize $Z_y = 24y_1 + 18y_2 + 12y_3$

subject to $4y_1 + 3y_2 + y_3 \leq 1$

$2y_1 + 2y_2 + 3y_3 \leq \frac{1}{2}$

$y_1, y_2, y_3 \geq 0.$

3.6. POST OPTIMALITY (OR SENSITIVITY) ANALYSIS

It is under taken to explore the effect of changes in LP parameters on the optimal solution. It is concerned with the extent of sensitivity of the optimum solution to an LPP for change in one or more of (*i*) RHS coefficient of constraints (*ii*) a new constraint is added (*iii*) Profit/cost coefficient of objective function (*iv*) LHS coefficient of variables in constraints.

3.6.1. Need for Analysis

(*i*) Above coefficients are estimates only and are liable to be inaccurate due to human error, limited sampling, in course of time deviations in actual values.

(*ii*) Management can manipulate these coefficients and relax constraints.

3.6.2 Changes Affecting Feasibility

Feasibility may be affected only if (*i*) RHS of constraints is changed (*ii*) new constraint is added. In both cases, infeasibility occurs when at least one element of RHS of optimal table becomes -ve i.e. one or more of current basic variable are –ve.

(*i*) If current solution remains optimal and feasible; no action is necessary.

(*ii*) If current solution becomes infeasible; use dual simplex to recover feasibility

(*iii*) If current solution becomes non-optimal; use primal simplex to recover optimality.

(*iv*) If current solution becomes both non-optimal and infeasible; use general simplex method to obtain new solution.

Example 3.14. If x_1, x_2, x_3 and x_4 represents the units of products A, B, C and D respectively

Maximize $Z = 4x_1 + 6x_2 + 3x_3 + x_4$

subject to $1.5x_1 + 2x_2 + 4x_3 + 3x_4 \leq 550$ (Machine I)

$4x_1 + x_2 + 2x_3 + x_4 \leq 700$ (Machine II, hours)

$$2x_1 + 3x_2 + x_3 + 2x_4 \leq 200 \text{ (Machine III, hours)}$$

(*i*) If machine I remains out of order for 24 hours, what will be the loss to the manufacturer?

Solution : 1. Standard form

$$Z = 4x_1 + 6x_2 + 3x_3 + x_4 + 0s_1 + 0s_2 + 0s_3$$
$$1.5x_1 + 2x_2 + 4x_3 + 3x_4 + s_1 = 550$$
$$4x_1 + x_2 + 2x_3 + x_4 + s_2 = 700$$
$$2x_1 + 3x_2 + x_3 + 2x_4 + s_3 = 200$$

2. Initial solution

$$x_1 = x_2 = x_3 = x_4 = 0$$
$$s_1 = 550,\ s_2 = 700,$$
$$s_3 = 200,\ Z = 0$$

3. Simplex table

C_B	B	x_B	6 x_1	4 x_2	3 x_3	1 x_4	0 s_1	0 s_2	0 s_3	θ
0	s_1	550	1.5	2	4	3	1	0	0	275
0	s_2	700	4	1	2	1	0	1	0	700
0	s_3	200	2	(3)	1	2	0	0	1	200/3=66.6
	Z_j		0	0	0	0	0	0	0	
	$(C_j - Z_j)$		4	6	3	1	0	0	0	

↑ Key column (under x_2)

Solutions is not optimal, key row—3rd row, s_3 is leaving variable, x_2 is next entering variable, key elements = 3.

4. Next table

C_B	B	x_B	4 x_1	6 x_2	3 x_3	1 x_4	0 s_1	0 s_2	0 s_3	θ
0	s_1	1250/3	1/6	0	(10/3)	5/3	1	0	–2/3	125
0	s_2	1900/3	10/3	0	5/3	1/3	0	1	–1/3	240
6	x_2	200/3	2/3	1	1/3	2/3	0	0	1/3	200
	Z_j		4	6	2	4	0	0	2	
Index row $(C_j - Z_j)$			0	0	1	–3	0	0	–2	

↑ Key column (under x_3)

This table contains one positive element (1) solution is not optimal.

Key row—Ist row, s_1 is leaving variable, key element = 10/3.

5. Next table

C_B	B	x_B	$C_j \to$ 6 x_1	4 x_2	3 x_3	1 x_4	0 s_1	0 s_2	0 s_3	θ
3	x_3	125	1/20	0	1	1/2	(3/10)	0	–2/10	
0	s_2	425	39/12	0	0	–1/2	–1/2	1	0	
6	x_2	25	13/20	1	0	1/2	(–1/10)	0	2/5	
	Z_j		4.05	6	3	4.5	3/10	0	18/10	
Index row $(C_j - Z_j)$			–.05	0	0	–3.5	–3/10	0	—18/10	

This solution is optimal solution.

$$x_3 = 125,\ x_2 = 25,\ s_2 = 573/3$$
$$Z_{max} = 4 \times 0 + 6 \times 25 + 3 \times 125 + 0 = 525$$

(*i*) If machine I remains out of order for 1 hour, it will decrease production of $C(x_3)$ by 3/10 units and increase the production of $B(x_2)$ by 1/10 units.

When machine *I* is out of order for 24 hours, x_3 becomes $125 - \frac{3}{10} \times 24 = 117.8$ and, x_2 becomes $x_2 = 25 + \frac{1}{10} \times 24 = 27.4$ units

$$\text{Net loss} = \frac{-3}{10} \times 24 = -7.2 = \text{Rs } 7.2$$

Example 3.15. Maximize $f = 240x_1 + 104x_2 + 60x_3 + 19x_4$

Subject to
$$20x_1 + 9x_2 + 6x_3 + x_4 \le 20$$
$$10x_1 + 4x_2 + 2x_3 + x_4 \le 10$$
$$x_i = 0\ ;\ i = 1 \text{ to } 4$$

Using Simplex method, find the solution of the above problem.

(UPTU: 2003-2004)

Solution : 1. Standard form

Maximize $f = 240x_1 + 104x_2 + 60x_3 + 19x_4 + 0s_1 + 0s_2$

Subject to $20x_1 + 9x_2 + 6x_3 + x_4 + s_1 = 20$

$10x_1 + 4x_2 + 2x_3 + x_4 + s_2 = 10$

2. Initial feasible solution

Put $x_1 = x_2 = x_3 = x_4 = 0$

$s_1 = 20,\ s_2 = 10$

$f = 0$

3. Simplex table

C_B	B	x_B	$C_j \to$ 240 x_1	104 x_2	60 x_3	19 x_4	0 s_1	0 s_2	θ
0	s_1	20	20	9	6	1	1	0	20/20 = 1
0	s_2	10	(10)	4	2	1	0	1	10/10 = 1
	Z_j		0	0	0	0	0	0	
Index row $(C_j - Z_j)$			240	104	60	19	0	0	

↑ Key column

4. This solution is not optimal as elements of index row are positive or zero.

Key column—column with x_1 at top.

There is tie in minimum ratio (θ). This reveals condition of degeneracy. To resolve it, coefficients of slack variables of rows which have tie are divided by corresponding element of key column . Row with minimum resulting ratio comparing column by column, left to right, is selected as key row.

s_1	s_2
1/20	0/20 = 0
0/10 = 0	1/10

Hence 2nd row is key row.

Key element = 10.

5. New table

C_j →			240	104	60	19	0	0	θ
C_B	B	x_B	x_1	x_2	x_3	x_4	s_1	s_2	
0	s_1	0	0	1	(2)	–1	1	–2	0
240	x_1	1	1	2/5	1/5	1/10	0	1/10	5
	Z_j		240	96	48	24	0	24	
Index row ($C_j - Z_j$)			0	8	12	–5	0	–24	

↑ Key column

6. Current solution is not optimal.

Key column—x_3 at head (largest value 12)

Key row—1st row

Key element —2

7. Next table

C_j			240	104	60	19	0	0	θ
C_B	B	x_B	x_1	x_2	x_3	x_4	s_1	s_2	
60	x_3	0	0	(1/2)	1	–1/2	1/2	–1	0
240	x_1	1	1	3/10	0	1/5	–1/10	3/10	10/3
	Z_j		240	102	60	18	6	12	
	($C_j - Z_j$)		0	2	0	1	–6	–12	

↑ Key column

Key row— 1st row, key element = $\frac{1}{2}$

8. New table

$C_j \longrightarrow$			240	104	60	19	0	0	θ
C_B	B	x_B	x_1	x_2	x_3	x_4	s_1	s_2	
104	x_2	0	0	1	2	–1	1	–2	–
240	x_1	1	1	0	–3/5	(1/2)	–2/5	9/10	2
		Z_j	240	104	64	16	8	8	
Index row $(C_j - Z_j)$			0	0	–4	3	–8	–8	

↑ Key column

Key row— 2nd row, key element = $\frac{1}{2}$

9. New table

$C_j \longrightarrow$			240	104	60	19	0	0	θ
C_B	B	x_B	x_1	x_2	x_3	x_4	s_1	s_2	
104	x_2	2	2	1	4/5	0	1/5	–1/5	
19	x_4	2	2	0	–6/5	1	–4/5	9/5	
		Z_j	246	104	60.4	19	32/5	67/5	
		$(C_j - Z_j)$	–6	0	–.4	0	–32/5	–67/5	

10. This solution is optimal solution hence,

$$x_2 = 2,\ x_4 = 2,$$
$$f = 104 \times 2 + 19 \times 2 = 246$$

Example 3.16. A scrap metal dealer has received an order from a customer for at least 2000 kg of scrap metal. The customer requires that at least 1000 kg of the shipment of metal must be of a high quality metal called **Alpha** that can be melted down and used to produce metal tubings. Further more, the customer will not accept delivery of the order if it contains more than 175 kg of metal that he deems unfit for commercial use, *i.e.*, metal that contains an excessive amount of impurities and can not be melted down and refined profitably. The dealer can purchase scrap metal from two different suppliers in unlimited quantities with the following % (by weight) of Alpha and unfit scrap.

	Supplier A	Supplier B
Alpha	25%	75%
Unfit Scrap	5%	10%

The costs per kg of metal purchased from supplier A and supplier B are Rs 1 and Rs 4 respectively. Using the simplex method determine the optimal quantities of metal for the dealer to purchase from each of the two suppliers.

Solution: 1. $x_1 + x_2 \geq 2000$

$$\frac{1}{4}x_1 + \frac{3}{4}x_2 \geq 1000$$

$$\frac{1}{20}x_1 + \frac{1}{10}x_2 \le 175$$

Minimize $Z = 1x_1 + 4x_2$

2. Minimize $Z = 1x_1 + 4x_2 + 0s_1 + 0s_2 + 0s_3 + MA_1 + MA_2$

$$x_1 + x_2 - s_1 + A_1 = 2000$$
$$A_1 = 2000$$
$$\frac{1}{4}x_1 + \frac{3}{4}x_2 - s_2 + A_2 = 1000$$
$$A_2 = 1000$$
$$\frac{1}{20}x_1 + \frac{1}{10}x_2 + s_3 = 175$$
$$s_3 = 175$$

3. Simplex table

C_B	b ($C_j \rightarrow$)	x_b	1 x_1	4 x_2	0 s_1	0 s_2	0 s_3	M A_1	M A_2	θ
M	A_1	2000	1	1	–1	0	0	1	0	200
M	A_2	1000	1/4	3/4	0	–1	0	0	1	4000/3
0	S_3	175	1/20	1/10	0	0	1	0	0	1750
	Z_j		1.25M	1.75M	–M	–M	0	M	M	
	$(C_j - Z_j)$		1–1.25M	4–1.75M	M	M	0	0	0	

↑ Key column

4. Next table

C_B	b ($C_j \rightarrow$)	x_b	1 x_1	4 x_2	0 s_1	0 s_2	0 s_3	M A_1	θ
M	A_1	2000/3	2/3	0	–1	4/3	0	1	500
4	x_2	4000/3	1/3	1	0	–4/3	0	0	
0	s_3	125/3	1/60	0	0	4/30	1	0	1250/4
	Z_j		.66M+4/3	4	–M	4/3M–16/3	0	M	
	$(C_j - Z_j)$		–1/3–2/3M	0	M	16/3–4/3M	0	0	

↑ Key column

5. Next table

C_B	b	x_b	1 x_1	4 x_2	0 s_1	0 s_2	0 s_3	M A_1	θ
M	A_1	250	(1/2)	0	–1	0	–10	1	500
4	x_2	1750	1/2	1	0	0	10	0	3500
0	s_2	1250/4	1/8	0	0	1	30/4	0	2500
	Z_j		M/2+2	4	–M	0	–10M+40	M	
	$(C_j – Z_j)$		–1–M/2	0	M	0	10M–40	0	

Key column (x_1)

6. Next table

C_B	b	x_b	1 x_1	4 x_2	0 s_1	0 s_2	0 s_3	θ
1	x_1	500	1	0	–2	0	–20	
4	x_2	1500	0	1	1	0	20	75
0	s_2	250	0	0	1/4	1	[10]	25
	Z_j		1	4	+2	0	60	
	$(C_j – Z_j)$		0	0	–2	0	–60	

Key column (s_3)

7. Next table

C_B	b	x_b	1 x_1	4 x_2	0 s_1	0 s_2	0 s_3	θ
1	x_1	1000	1	0	–3/2	0	0	
4	x_2	1000	0	1	1/2	–2	0	2000
0	s_3	25	0	0	[1/40]	1/10	1	1000
	Z_j		1	4	1/2	–6	0	
	$(C_j – Z_j)$		0	0	–1/2	6	0	

Key column (s_1)

8. Final table

C_B	$C_j \rightarrow$ b	x_b	1 x_1	4 x_2	0 s_1	0 s_2	0 s_3	θ
1	x_1	2500	1	0	0	8	60	
4	x_2	500	0	1	0	–4	–20	
0	s_1	1000	0	0	1	4	40	
	Z_j		1	4	0	–8	–20	
	$(C_j - Z_j)$		0	0	0	8	20	

Since all index row elements are positive solution is reached.

$$x_1 = 2500, x_2 = 500, Z = 4500$$

Example 3.17. Solve following linear programming problem using simplex method

Maximize $Z = 12x_1 + 15x_2 + 14x_3$

subject to, $-x_1 + x_2 \leq 0$

$-x_2 + 2x_3 \leq 0$

$x_1 + x_2 + x_3 \leq 10$

$x_1, x_2, x_3 \geq 0$

Solution: 1. Maxmise $Z = 12x_1 + 15x_2 + 14x_3 + 0s_1 + 0s_2 + 0s_3$

subject to, $-x_1 + x_2 + s_1 = \in$

$-x_2 + 2x_3 + s_2 = \in$

$x_1 + x_2 + x_3 + s_3 = 100$

$x_1, x_2, x_3 \geq 0$

2. Initial solution, $x_1, x_2, x_3 = 0,$

$Z = 0,$

$s_1 = \in, s_2 = \in, s_3 = 100$

3. Simplex table

C_B	C_j b	x_b	12 x_1	15 x_2	14 x_3	0 s_1	0 s_2	0 s_3	Minimum Ratio
0	s_1	ε	–1	(1)	0	1	0	0	∈
0	s_2	ε	0	–1	2	0	1	1	
0	s_3	100	1	1	1	0	0	0	100
	Z_j		0	0	0	0	0	0	
	$(C_j - Z_j)$		12	15	14	0	0	0	

4. Next table

C_B	b	x_b	12 x_1	15 x_2	14 x_3	0 s_1	0 s_2	0 s_3	Minimum Ratio
15	x_2	ε	–1	1	0	1	0	0	—
0	s_2	2ε	–1	0	2	1	1	0	—
0	s_3	100–ε	(2)	0	1	–1	0	1	$\frac{(100-\varepsilon)}{2}$
	Z_j		–15	15	0	15	0	0	
	$(C_j - Z_j)$		27	0	14	–15	0	0	

5. Next table

C_B	b	x_b	12 x_1	15 x_2	14 x_3	0 s_1	0 s_2	0 s_3	Minimum Ratio
15	x_2	100+ε/2	0	1	1/2	1/2	0	1/2	100+ε
0	s_2	100+3ε/2	0	0	(5/2)	1/2	1	1/2	100+3ε/5
12	x_1	100–ε/2	1	0	1/2	–1/2	0	1/2	100–ε
	Z_j		12	15	27/2	3/2	0	27/2	
	$(C_j - Z_j)$		0	0	1/2	–3/2	0	–27/2	

6. Final table

C_B	b	x_b	12 x_1	15 x_2	14 x_3	0 s_1	0 s_2	0 s_3	
15	x_2	400+2ε/10	0	1	0	4/10	–1/5	4/10	
14	x_3	100+3ε/5	0	0	1	1/5	2/5	1/5	
12	x_1	(400–8ε)/10	1	0	0	–6/10	–1/5	4/10	
	Z_j		12	15	14	16/10	1/5	136/10	
	$C_j - Z_j$		0	0	0	–16/10	–1/5	–136/10	

Since all elements of index row are negative or zero, solution is optimal.

$$x_1 = 40,\ x_2 = 40,\ x_3 = 20$$
$$Z = 12 \times 40 + 15 \times 40 + 14 \times 20 = 1360$$

Example 13.18. Maximise $Z' = 10x + 6y + 4z$

subject to $x + y + z \leq 100$

$$10x + 4y + 5z \leq 600$$
$$2x + 2y + 6z \leq 300$$
$$x, y, z \geq 0$$

Solution : Initial solution $x = 0, y = 0, z = 0$

$\Rightarrow \quad Z' = 0$

Introducing slack variables, s_1, s_2, s_3

$$Z' = 10x + 6y + 4z + 0s_1 + 0s_2 + 0s_3 \quad \text{...(A)}$$

$$x + y + z + s_1 + 0s_2 + 0s_3 = 100 \quad \text{...(1)}$$

$$10x + 4y + 5z + 0s_1 + s_2 + 0s_3 = 600 \quad \text{...(2)}$$

$$2x + 2y + 6z + 0s_1 + 0s_2 + s_3 = 300 \quad \text{...(3)}$$

at $x = y = z = 0$,

$s_1 = 100,\ s_2 = 600,\ s_3 = 300$

Construction of linear model

(*i*)

C_B	b	C_j / x_b	10 x	6 y	4 z	0 s_1	0 s_2	0 s_3
0	s_1	100	1	1	1	1	0	0
0	s_2	600	(10)	4	5	0	1	0 ← Key row
0	s_3	300	2	2	6	0	0	1
	Z_j		0	0	0	0	0	0
	$(C_j - Z_j)$		+10	6	4	0	0	0
			↑ Key column					

key no = 10

(*ii*) As an the element in the index row are not zero or – ve, linear model is again constructed

C_B	b	C_j / x_b	10 x	6 y	4 z	0 s_1	0 s_2	0 s_3
0	S_1	40	0	(3/5)	1/2	1	–1/10	0 ← Key row
10	X	60	1	2/5	1/2	0	1/10	0
0	S_3	180	0	6/5	5	0	–1/5	1
	Z_j	= 600	10	4	5	0	1	0
	$(C_j - Z_j)$		0	+2	–1	0	–1	0
				↑ Key column				

key no. $= \dfrac{3}{5}$

(*iii*) Again constructing the linear model :

C_B	b	C_j / x_b	10 x	6 y	4 z	0 s_1	0 s_2	0 s_3
6	y	200/3	0	1	5/6	5/3	–1/6	0
10	x	100/3	1	0	1/6	–2/3	1/6	0
0	s_3	100	0	0	4	–2	0	1
	Z_j	= 2200/3	10	6	40/6	10/3	4/6	0
	$(C_j - Z_j)$		0	0	–16/6	–10/3	–4/6	0

All the element in the index are either zero or – ve no. hence we get the optimal solution here.

$$y = \frac{200}{3},\ x = \frac{100}{3},\ Z = 0$$

$$\therefore \quad Z'\text{max} = 10x + 6y + 4z = \frac{10 \times 100}{3} + 6 \times \frac{200}{3} + 4 \times 0$$

$$Z'_{\text{max}} = \frac{1000}{3} + \frac{1200}{3} = \frac{2200}{3}$$

$$\boxed{Z'_{\text{max}} = \frac{2200}{3}}$$ **Ans.**

Example 3.19. Maximize $Z = 2x_1 + 4x_2 + x_3 + x_4$

subject to constraints $x_1 + 3x_2 + x_4 \le 4$

$2x_1 + x_2 \le 3$

$x_2 + 4x_3 + x_4 \le 3$

Solution : 1. Introducing slack variables (standard form)

$Z = 2x_1 + 4x_2 + x_3 + x_4 + 0s_1 + 0s_2 + 0s_3$

$$x_1 + 3x_2 + 0x_3 + x_4 + s_1 + 0s_2 + 0s_3 = 4 \quad \text{...(1)}$$

$$2x_1 + x_2 + 0x_3 + 0x_4 + 0s_1 + s_2 + 0s_3 = 3 \quad \text{...(2)}$$

$$0x_1 + x_2 + 4x_3 + x_4 + 0s_1 + 0s_2 + s_3 = 3 \quad \text{...(3)}$$

2. Initial feasible solution

at $\quad x_1 = x_2 = x_3 = x_4 = 0$

$Z = 0$

and $\quad s_1 = 4,\ s_2 = 3,\ s_3 = 3$

Constructing the linear model

3. (*i*)

C_B	b	C_j / x_b	2 x_1	4 x_2	1 x_3	1 x_4	0 s_1	0 s_2	0 s_3	θ
0	s_1	4	1	(3)	0	1	1	0	0	4/3 ← Key row
0	s_2	3	2	1	0	0	0	1	0	3
0	s_3	3	0	1	4	1	0	0	1	3
	Z_j		0	0	0	0	0	0	0	
	$(C_j - Z_j)$		+2	+4	+1	+1	0	0	0	

↑ Key column (x_2)

key no. =3

(*ii*) New table

		C_j	2	4	1	1	0	0	0	θ
C_B	b	x_b	x_1	x_2	x_3	x_4	s_1	s_2	s_3	
4	x_1	4/3	1/3	1	0	1/3	1/3	0	0	—
0	s_2	5/3	5/3	0	0	–1/3	–1/3	1	0	—
0	s_3	5/3	–1/3	0	(4)	2/3	–1/3	0	1	5/12
	Z_j	16/3	4/3	4	0	4/3	4/3	0	0	
	$(C_j - Z_j)$		+2/3	0	+1	–1/3	–4/3	0	0	

Key row

Key column

Key no. = 4

(*iii*) New table

		C_j	2	4	1	1	0	0	0	θ
C_B	b	x_b	x_1	x_2	x_3	x_4	s_1	s_2	s_3	
4	x_2	4/3	1/3	1	0	1/3	1/3	0	0	4
0	s_2	5/3	(5/3)	0	0	–1/3	–1/3	1	0	1
1	x_3	5/12	–1/12	0	1	1/6	–1/12	0	1/4	—
		69/12	+3/4	0	0	–1/2	–5/4	0	–1/4	

Key row

Key column

(*iv*) Again constructing new table

			2	4	1	1	0	0	0	θ
C_B		0	x_1	x_2	x_3	x_4	s_1	s_2	s_3	
4	x_2	1	0	1	0	6/15	6/15	–1/5	0	
2	x_1	1	1	0	0	–1/5	–1/5	3/5	0	
1	x_3	1/2	0	0	1	3/20	–1/10	1/20	1/4	
Index row		13/2	0	0	0	–21/60	–11/10	–9/20	–1/4	

4. All elements in the index row are either zero or – ve, hence the optimum solution is

$$x_2 = 1,\ x_1 = 1,\ x_3 = \frac{1}{2},\ x_4 = 0$$

$$Z_{max} = 2x_1 + 4x_2 + x_3 + x_4 = 2 \times 1 + 4 \times 1 + \frac{1}{2} + 0$$

$$\boxed{Z_{max} = \frac{13}{2}} \quad \textbf{Ans.}$$

Example 3.20. Maximize $Z = 95A + 41B + 84C + 76D$

subject to $4A + 2B + 3C + 3D \le 800$

$$10A + 6B + 8C + 7D \le 1200$$
$$10A + 8B + 8C + 8D \le 800$$
$$A, B, C, D \ge 0$$

Solution : 1. Introducing slack variables s_1, s_2, s_3

$$Z = 95A + 41B + 84C + 76D + 0s_1 + 0s_2 + 0s_3$$
$$4A + 2B + 3C + 3D + s_1 + 0s_2 + 0s_3 = 800$$
$$10A + 6B + 8C + 7D + 0s_1 + s_2 + 0s_3 = 1200$$
$$10A + 8B + 8C + 8D + 0s_1 + 0s_2 + s_3 = 800$$

2. Initial solution at $A = B = C = D = 0, Z = 0,$

$$s_1 = 800, s_2 = 1200 \text{ and } s_3 = 800$$

3. Linear model

(*i*)

C_B	b	C_j / x_b	95 A	41 B	84 C	76 D	0 s_1	0 s_2	0 s_3	θ
0	s_1	800	4	2	3	3	1	0	0	200
0	s_2	1200	10	6	8	7	0	1	0	120
0	s_3	800	(10)	8	8	8	0	0	1	80 ← Key row
Index Row		0	+95	+41	+84	+76	0	0	0	

↑ Key column (A)

Key no. = 10

(*ii*)

C_B	b	C_j / x_b	95 A	41 B	84 C	76 D	0 s_1	0 s_2	0 s_3	θ
0	s_1	480	0	–6/5	–1/5	–1/5	1	0	–2/5	—
0	s_2	400	0	–2	0	–1	0	1	–1	—
95	A	80	1	8/10	(8/10)	8/10	0	0	1/10	100 ← Key row
Index Row		7600	0	–35	+8	0	0	0	–95/10	

↑ Key column (C)

$$\text{Key no.} = \frac{8}{10} = \frac{4}{5}$$

(*iii*)

C_B	b	C_j / x_b	95 A	41 B	84 C	76 D	0 s_1	0 s_2	0 s_3
0	s_1	500	1/4	–1	0	0	1	0	–3/8
0	s_2	400	0	–2	0	–1	0	1	–1
84	C	100	5/4	1	1	1	0	0	1/8
Index Row		8400	–10	–43	0	–8	0	0	–21/2

4. All the elements in the index row = zero or – ve

∴ Optimum solution is $C = 100, A, B, D = 0$

$$\therefore \quad Z = 95A + 41B + 84C + 76D$$
$$= 84 \times 100 = 8400$$

$\boxed{Z_{max} = 8400}$ **Ans.**

Example 3.21. Maximize $Z = 3x_1 - x_2$

Solution : subject to $2x_1 + x_2 \leq 2$

$x_1 + 3x_2 \geq 3$

$x_2 \leq 4$

and $x_1, x_2 \geq 0$

1. Introducing slack variables.

$Z = 3x_1 - x_2 - MA_1 + 0s_1 + 0s_2 + 0s_3$

$2x_1 + x_2 + s_1 = 2$ ...(1)

$x_1 + 3x_2 + A_1 - s_2 = 3$ ...(2)

$x_2 + s_3 = 4$ ...(3)

2. Initial solution $x_1 = x_2 = 0, s_1 = 2, s_3 = 4$

$A_1 = 3$

$Z = 3x_1 - x_2 - MA_1 + 0s_1 + 0s_2 + 0s_3$

$= 0 - 0 - M \times 3 + 0 + 0 + 0 = -3M$

3. Table

C_B	b	C_j / x_b	3 / x_1	–1 / x_1	0 / s_1	0 / s_2	0 / s_3	–M / A_1	θ
0	s_1	2	2	1	1	0	0	0	2
–M	A_1	3	1	(3)	0	–1	0	1	1 ← Key row
0	s_3	4	0	1	0	0	1	0	4
Index row		–3M	+M+3	+3M–1	0	–M	0	0	
				↑ Key column					

key member = 3

4. New table

C_B	b	C_j / x_b	3 / x_1	–1 / x_2	0 / s_1	0 / s_2	0 / s_3	θ
0	s_1	1	(5/3)	0	1	1/3	0	3/5 ← Key row
–1	x_2	1	1/3	1	0	–1/3	0	3
0	s_3	3	–1/3	0	0	1/3	1	—
Index row		–1	+10/3	0	0	–1/3	0	
			↑ Key column					

key no $= \frac{5}{3}$

5. New table

C_B	b	C_j / x_b	3 / x_1	−1 / x_2	0 / s_1	0 / s_2	0 / s_3	θ
3	x_1	3/5	1	0	3/5	1/5	0	
−1	x_2	4/5	0	1	−1/5	−6/15	0	
0	s_3	16/5	0	0	1/5	2/5	1	
Index row		1	0	0	−2	−1	0	

Since all index row elements are positive hence it is final solution

$$x_1 = \frac{3}{5}$$

$$x_2 = \frac{4}{5}$$

$$Z_{max} = 3x_1 - x_2$$

$$= 3\times\frac{3}{5} - \frac{4}{5} = \frac{9}{5} - \frac{4}{5}$$

$$= \frac{9-4}{5} = \frac{5}{5} = 1$$

$Z_{max} = 1$ **Ans.**

Example 3.22. Solve following linear programming problem using simplex method.

Machine Centre	Hours per unit		Hours Available
	Product A	Product B	
J	3	2	42
K	2	2	30
L	2	4	48
Profit Contribution Per Unit	12	8	

Solution : 1. Mathematical model of above problem can be written as

Objective function Maximize $Z = 12A + 8B$

where A = number of units produced of A

B = number of units produced of B

subject to constraints $3A + 2B \leq 42$

$$2A + 2B \leq 30$$

$$2A + 4B \leq 48$$

2. Convert inequalities into equalities by introducing non-negative slack variables

$$3A + 2B + s_1 = 42$$

$$2A + 2B + s_2 = 30$$

$$2A + 4B + s_3 = 48$$

If we use full capacity, $s_1 = s_2 = s_3 = 0$

Objective function Z becomes

$$Z = 12A + 8B + 0s_1 + 0s_2 + 0s_3$$

3. Find initial solution by substituting $A = 0$ and $B = 0$ (Non-basic variables)

This gives, $s_1 = 42,\ s_2 = 30,\ s_3 = 48$

as the values of basic variables

4. Develop simplex table

C_B	C_j / b	x_b	12 / A	8 / B	0 / s_1	0 / s_2	0 / s_3	Minimum Ratio
0	s_1	42	(3)	2	1	0	0	42/3=14
0	s_2	30	2	2	0	1	0	30/2=15
0	s_3	48	2	4	0	0	1	48/2=24
	Z_j		0	0	0	0	0	
Index row	$(C_j - Z_j)$		12	8	0	0	0	

Here, C_B is objective column which shows the coefficients of basic variables.

b is variable column

x_B shows solution values

C_j is objective row which shows coefficients of variables in objective function.

Thus initial solution gives zero profit.

$$Z = 12 \times 0 + 8 \times 0 + 0 \times 42 + 0 \times 30 + 0 \times 48 = 0$$

5. Optimality test: since all elements in the index row are positive, solution is not optimal.

6. Identify key column, key row and key number
   (*i*) Column with largest positive element in index row is selected as key column (here it is 12)
   (*ii*) Row with minimum ratio 'θ' is selected as key row (1st row is key row)
   (*iii*) Key number is 3 (intersection of key row and key column).

7. Develope second simplex table
   (*i*) Divide key row elements by key number except objective and variable column elements.
   (*ii*) Calculate other elements by using formula.

$$\text{New number} = \text{Old number} - \left[\frac{\text{Associated no. in key row} \times \text{Corresponding no. in key column}}{\text{Key number}}\right]$$

C_B	C_j b	x_b	12 A	8 B	0 s_1	0 s_2	0 s_3	θ
12	A	14	1	2/3	1//3	0	0	
0	s_2	2	0	2/3	–2/3	1	0	
0	s_3	20	0	8/3	–2/3	0	1	
	Z_j		12	8	4	0	0	
Index row	$(C_j - Z_j)$		0	0	–4	0	0	

Since all the elements in the index row are negative or zero. Current solution is optimal solution.

∴ $A = 12,\ B = 0$

and $Z = 12 \times 12 + 8 \times 0 = 144$

Example 3.23. Solve following problem by simplex technique.

Maximize $Z = 3x_1 + 4x_2$

subject to constraints $x_1 + x_2 \leq 450$

$2x_1 + x_2 \leq 600$

$x_1, x_2 \geq 0$

Solution : 1. Writing standard form by introducing slack variables

$Z = 3x_1 + 4x_2 + 0s_1 + 0s_2$

subject to, $x_1 + x_2 + s_1 = 450$

$2x_1 + x_2 + s_2 = 600$

2. Initial feasible solution by assuming $x_1 = x_2 = 0$

$s_1 = 450$

$s_2 = 600$

∴ $Z = 0$

3. Simplex table

C_B	C_j b	x_b	3 x_1	4 x_2	0 s_1	0 s_2	θ
0	s_1	450	1	①	1	0	450/1=450
0	s_2	600	2	1	0	1	600/1=600
	Z_j		0	0	0	0	
Index row	$C_j - Z_j$		3	4	0	0	

Key row (arrow pointing to s_1 row)

Key column (largest positive index row element)

4. Since the elements in the index row are zero or positive, solution is not optimal.

5. Key row and key column—shaded in table

Key number—1

6. Develop new table

$C_j \longrightarrow$			3	4	0	0	θ
C_B	B	x_b	x_1	x_2	s_1	s_2	
4	x_2	450	1	1	1	0	
0	s_2	150	1	0	–1	1	
	Z_j		4	4	4	0	
Index row ($C_j - Z_j$)			–1	0	–4	0	

7. All the elements in index row of New table are negative or zero, hence current solution is optimal.

$$x_1 = 0,\ x_2 = 450$$

$$Z_{max} = 3 \times 0 + 4 \times 450 = 1800$$

Example 3.24. A company makes 3 products x_1, x_2, x_3. The contribution to profit per unit is given as under — $x_1 = 2$, $x_2 = 4$, $x_3 = 3$. Each of these products pass through 3 machine centres. Time required in each centre to produce one unit of product is given below.

Machine Centres	x_1	x_2	x_1	Time Available
I	3	2	1	60
II	4	1	3	40
III	2	2	2	80

Find the product mix which maximizes the profit.

Solution : 1. Objective function, maximize $Z = 2x_1 + 4x_2 + 3x_3$

constraints, $3x_1 + 2x_2 + x_3 \leq 60$

$$4x_1 + x_2 + 3x_3 \leq 40$$

$$2x_1 + 2x_2 + 2x_3 \leq 80$$

$$x_1, x_2, x_3 \geq 0$$

2. Standard form

$$Z = 2x_1 + 4x_2 + 3x_3 + 0s_1 + 0s_2 + 0s_3$$

$$3x_1 + 2x_2 + x_3 + s_1 = 60$$

$$4x_1 + x_2 + 3x_3 + s_2 = 40$$

$$2x_1 + 2x_2 + 2x_3 + s_3 = 80$$

3. Initial solution

Put $\quad x_1 = x_2 = x_3 = 0$

Find $\quad s_1 = 60,\ s_2 = 40,\ s_3 = 80$

Objective function $Z = 0$

4. Simplex table

C_B	B	x_b	$C_j \to$ 2 x_1	4 x_2	3 x_3	0 s_1	0 s_2	0 s_3	θ
0	s_1	60	3	(2)	1	1	0	0	60/2 = 30 ← Key row
0	s_2	40	4	1	3	0	1	0	40/1 = 40
0	s_3	80	2	2	2	0	0	1	80/2 = 40
	Z_j		0	0	0	0	0	0	
	$(C_j - Z_j)$		2	4	3	0	0	0	

↑ Key column (Largest Positive element)

5. This solution is not optimal solution. Key column—x_2 at top—entering variable, key row—1st row—s_1 is leaving variable, key number—2.

6. New table

C_B	B	x_b	$C_j \to$ 2 x_1	4 x_2	3 x_3	0 s_1	0 s_2	0 s_3	θ
4	x_2	30	3/2	1	1/2	1/2	0	0	60
0	s_2	10	5/2	0	(5/2)	–1/2	1	0	4
0	s_3	20	–1	0	1	–1	0	1	20
	Z_j		6	4	2	2	0	0	
	$(C_j - Z_j)$		–4	0	1	–2	0	0	

7. One element is still positive in index row 1. Solution is not optimal. Column corresponding to 1, x_3 at top is key column. x_3 is next entering variable and 2nd row is key row hence s_2 is leaving variable.

$$\text{Key element} = \frac{5}{2}$$

8. Develop another simplex table

C_B	B	x_b	$C_j \to$ 2 x_1	4 x_2	3 x_3	0 s_1	0 s_2	0 s_3	θ
4	x_2	28	1	1	0	6/10	–1/5	0	
3	x_3	4	1	0	1	–1/5	2/5	0	
0	s_3	16	–2	0	0	–4/5	–2/5	1	
	Z_j		7	4	3	9/5	2/5	0	
Index row $(C_j - Z_j)$			–5	0	0	–9/5	–2/5	0	

9. Now, all the elements in the index row are either zero or negative. Hence, current solution is optimal solution.

Solution values are

$$x_2 = 28, x_3 = 4, x_1 = 0$$
$$Z = 2 \times 0 + 4 \times 28 + 3 \times 4$$
$$= 112 + 12 = 124$$
$$Z = 124$$

Example 3.25. Solve following LPP by simplex

Maximize $Z = 3x_1 - x_2$

subject to $2x_1 + x_2 \geq 2$

$$x_1 + 3x_2 \leq 3$$
$$x_2 \leq 4$$
$$x_1, x_2 \geq 0$$

Solution : 1. Standard form

$$Z = 3x_1 - x_2 - MA_1 + 0s_1 + 0s_2 + 0s_3$$
$$2x_1 + x_2 + A_1 - s_1 = 2$$
$$x_1 + 3x_2 + s_2 = 3$$
$$x_2 + s_3 = 4$$

Here, A_1 is artificial variable and s_1 is surplus variable. Artificial variable is assigned a large penalty – M in the objective function as the problem is of maximization type.

2. Initial basic feasible solution

$$x_1 = x_2 = 0,$$
$$s_1 = 0$$
$$\Rightarrow \quad A_1 = 2,$$
$$s_2 = 3, s_3 = 4$$

3. Simplex table

$C_j \rightarrow$			3	–1	0	0	0	–M	
C_B	B	x_b	x_1	x_2	s_1	s_2	s_3	A_1	θ
–M	A_1	2	(2)	1	–1	0	0	1	2/2 = 1 ← Key row
0	s_2	3	1	3	0	1	0	0	3/1 = 3
0	s_3	4	0	1	0	0	1	0	4/0 = ∞
	Z_j		–2M	–M	M	0	0	–M	
Index row ($C_j - Z_j$)			(2M–3)	(M–1)	–M	0	0	0	
			↑ Key column						

4. entering variable is x_1— key column; leaving variable is $A_1 \rightarrow$ key row (1st row); key element = 2

5. Develop New table—Artificial variable A_1 is leaving the basic therefore all its entries in column of variables are omitted from the table.

$C_j \rightarrow$			3	–1	0	0	0	θ
C_B	B	x_b	x_1	x_2	s_1	s_2	s_3	
3	x_1	1	1	1/2	–1/2	0	0	–
0	s_2	2	0	5/2	(1/2)	1	0	4 ← Key row
0	s_3	4	0	1	0	0	1	∝
	Z_j		3	3/2	–3/2	0	0	
Index row ($C_j - Z_j$)			0	–5/2	3/2	0	0	
					↑ Key column			

6. This solutions is also not optimal as it contains 3/2 as positive element in its index row. Key column—s_1 at head, key row—second row, s_2 is leaving, key number is 1/2.

7. New table

$C_j \rightarrow$			3	–1	0	0	0	θ
C_B	B	x_b	x_1	x_2	s_1	s_2	s_3	
3	x_1	3	1	3	0	1	0	
0	s_1	4	0	5	1	2	0	
0	s_3	4	0	1	0	0	1	
	Z_j		3	9	0	3	0	
Index row ($C_j - Z_j$)			0	–10	0	–3	0	

All elements of index row are zero or negative, therefore, current solution is optimal

$$x_1 = 3,\ s_1 = 4,\ s_3 = 4,\ A_1 = 0$$
$$Z = 3 \times 3 - 1 \times 0 = 9$$

Example 3.26. Solve following problem by using simplex method.

Maximize $Z = 3x_1 + 2x_2 + 2x_3$

subject to $5x_1 + 7x_2 + 4x_3 \leq 7$

$$-4x_1 + 7x_2 + 5x_3 \geq -2$$
$$3x_1 + 4x_2 - 6x_3 \geq 29/7$$

Solution : 1. Standard form

$$Z = 3x_1 + 2x_2 + 2x_3 + 0s_1 + 0s_2 + 0s_3 - MA_1$$
$$5x_1 + 7x_2 + 4x_3 + s_1 = 7$$
$$4x_1 - 7x_2 - 5x_3 + s_2 = 2$$
$$3x_1 + 4x_2 - 6x_3 + A_1 - s_3 = 29/7$$

2. Initial feasible solution

$$x_1 = x_2 = x_3 = 0$$
$$s_1 = 7,\ s_2 = 2,\ A_1 = 29/7$$
$$Z = -29/7\ M$$

3. Simplex table

$C_j \longrightarrow$			3	2	2	0	0	0	–M	θ
C_B	B	x_b	x_1	x_2	x_3	s_1	s_2	s_3	A_1	
0	s_1	7	5	(7)	4	1	0	0	0	7/7 = 1
0	s_2	2	4	–7	–5	0	1	0	0	–
–M	A_1	29/7	3	+4	–6	0	0	–1	1	29/28
	Z_j		–3M	–4M	6M	0	0	M	–M	
Index row ($C_j - Z_j$)			3M+3	4M+2	2–6M	0	0	–M	0	

↑ Key column

4. Current solution is not optimal. Key column—x_2 at top, x_2 is entering variable; key row—1st row, s_1 is leaving variable; key variable—7.

5. New table

$C_j \longrightarrow$			3	2	2	0	0	0	–M	θ
C_B	B	x_b	x_1	x_2	x_3	s_1	s_2	s_3	A_1	
2	x_2	1	5/7	1	4/7	1/7	0	0	0	7/5
0	s_2	9	9	0	–1	1	1	0	0	1
–M	A_1	1/7	(1/7)	0	–58/7	–4/7	0	–1	1	1
	Z_j		(10/7–M/7)	2	(8/7+58/7M)	(2/7+4/7M)	0	M	–M	
Index row ($C_j - Z_j$)			(M/7+11/7)	0	(6/7–587M)	(–2/7–4/7M)	0	–M	0	

↑ Key column

6. This solution is not optimal as index row contain one positive element $\left(\frac{M}{7}+\frac{11}{7}\right)$ column corresponding to this is key column and -3rd row is selected as key row. In case of tie in minimum ratio 'θ', artificial variable leaves the basis.

$$\text{Key element} = \frac{1}{7}$$

7. New table

$C_j \longrightarrow$			3	2	2	0	0	0	θ
C_B	B	x_B	x_1	x_2	x_3	s_1	s_2	s_3	
2	x_2	2/7	0	1	42	3	0	5	2/294
0	s_2	0	0	0	(521)	37	1	63	0
3	x_1	1	1	0	–58	–4	0	–7	–
	Z_j		3	2	–90	–6	0	–11	
Index row ($C_j - Z_j$)			0	0	–92	6	0	11	

8. Current solution is also not optimal. key column—x_3 at top of column, key row—2nd row, s_2 is leaving variable; key element—521.

9. New table

$C_j \longrightarrow$			3	2	2	0	0	0	θ
C_B	B	x_B	x_1	x_2	x_3	s_1	s_2	s_3	
2	x_2	2/7	0	1	0	.017	(–42/521)	–.27	
2	x_3	0	0	0	1	0.07	1/521	.12	
3	x_1	1	1	0	0	.119	(58/521)	.013	
Z_j			3	2	2	.46	17/521	.139	
Index row $(C_j - Z_j)$			0	0	0	(–.46)	(–17/521)	(–.139)	

10. Since all elements of index row are negative, current solution is optimal

$$x_2 = \frac{2}{7}, \quad x_3 = 0,$$

$$x_1 = 1$$

$$Z = 3 \times 1 + 2 \times \frac{2}{7} + 2 \times 0$$

$$= \frac{25}{7}$$

EXERCISE

3.1. Define the following terms:

(*i*) Feasible solution
(*ii*) Degenerate solution
(*iii*) Unbounded solution
(*iv*) Alternate optimum solution
(*v*) Slack variables
(*vi*) Surplus variables
(*vii*) Artificial variables
(*viii*) Basic variables
(*ix*) Objective function
(x) Non-negativity constraints
(*xi*) Unique solution

3.2. Write and explain the assumptions of linear programming problem.

3.3. Explain the procedure of Simplex method to solve a linear programming problem.

3.4. What do you understand from sensitivity analysis in a LPP? What is the advantage of carrying out sensitivity analysis? Give examples.

(UPTU: 2003-2004)

3.5. Explain the general structure of a linear programming model.

(UPTU: 2004-2005)

3.6. What is the role of key element in simplex table?

3.7. The set of all feasible solutions of a linear programming problem forms a convex set; discuss.

3.8. Explain the use of artificial variables in linear programming problem.

3.9. Explain the Big-M method for solving a linear programming problem.

3.10. What is sensitivity analysis? Explain with suitable examples.

3.11. Explain the concept of degeneracy in linear programming.

3.12. How do the maximization and minimization problems differ when solving with simplex method.

3.13. When does the unbounded solution occur?

3.14. Explain duality theory used in linear programming. Show that a dual of a dual is the primal problem.

3.15. What are advantages of duality? State the fundamental properties of duality.

3.16. Solve the following LPP by simplex method

Maximize $Z = 3x_1 + 2x_2$

subject to $4x_1 + 3x_2 \leq 12$

$4x_1 + x_2 \leq 8$

$4x_1 - x_2 \leq 8$

$x_1, x_2 \geq 0$ [**Ans.** $x_1 = 3/2$, $x_2 = 2$, Max. $Z = 17/2$]

3.17. Use simplex method to solve the following LPP.

Minimize $Z = x_2 - 3x_3 + 2x_5$

subject to $3x_2 - x_3 + 2x_5 \leq 7$

$-2x_2 + 4x_3 \leq 12$

$-4x_2 + 3x_3 + 8x_5 \leq 10$

$x_2, x_3, x_5 \geq 0$ [**Ans.** $x_2 = 4$, $x_3 = 5$, $x_5 = 0$, $Z = -11$]

3.18. Solve following LPP by simplex method

Maximize $Z = x_1 + 2x_2 + 3x_3 - x_4$

subject to $x_1 + 2x_2 + 3x_3 = 15$

$2x_1 + x_2 + 5x_3 = 20$

$x_1 + 2x_2 + x_3 + x_4 = 10$ [**Ans.** $x_1, x_2, x_3 = 5/2$, $x_4 = 0$, $Z = 15$]

3.19. Solve the following LP problem using simplex method :

Maximize $Z = 5x_1 + 3x_2 + 7x_3$

subject to $x_1 + x_2 + 2x_3 \leq 22$

$3x_1 + 2x_2 + x_3 \leq 26$

$x_1 + x_2 + x_3 \leq 18$

x_1 and $x_2 \geq 0$ [**Ans.** $x_1 = 6, x_2 = 0, x_3 = 8, z = 86$]

3.20. A firm has available 240, 370 and 180 kg of wood, plastic and steel respectively. The firm produces two products *A* and *B*. Each unit of *A* requires 1, 3 and 2 kg of wood, plastic and steel respectively. The corresponding requirements for each unit of *B* are 3, 4 and 1 respectively. If *A* sells for Rs 4 and *B* for Rs 6, determine how many units of *A* and *B* should be produced in order in obtain the maximum income. Solve using simplex method.

[**Ans.** $X_1 = 30, X_2 = 70, Z = 540$]

3.21. A company makes two products *A* and *B*. Each unit of product *A* requires twice as much labour time as the product *B*. If all the units are of product *B*, the company can produce a total of 500 units a day. The market limits daily sales of products A and B to 150 and 250 units respectively. The profit per unit of A is Rs 8 and per unit of B is Rs 5. Determine optimal product mix to maximize the net profit of company.[**Ans.** $A = 1245, B = 250, Z = 2250$]

3.22. Use penalty method to

Minimize $Z = 2x_1 + x_2$

subject to $3x_1 + x_2 = 3$

$$4x_1 + 3x_2 \geq 6$$
$$x_1 + 2x_2 \leq 3$$
$$x_1, x_2 \geq 0$$

[**Ans.** $x_1 = 3/5, x_2 = 6/5, Z = 12/5$]

3.23. Solve the following problem by simplex method.

Maximize $15x_1 + 6x_2 + 9x_3 + 2x_4$

subject to $2x_1 + x_2 + 5x_3 + 0.6x_4 \leq 10$

$$3x_1 + x_2 + 3x_3 + 0.25x_4 \leq 12$$
$$7x_1 + x_4 \leq 35$$
$$x_1, x_2, x_3, x_4 \geq 0$$

[**Ans.** 66]

3.24. Two types of hens are available. Type A costs Rs 8 per hen and type B Rs 10 per hen. Type A lays on an average 4 eggs per week and type B, 5 eggs per week. Eggs sell at 50 paise each. The feeding cost of type A hen is Rs 1.5 per week and of type B is Rs 1.9 per week. [**Ans.** 66]

Maximum of Rs 180 is to be spent on purchasing hens. How many of each type are to be bought to have maximum profit? It is not possible to house more than 20 hens at a time. Solve using simplex method. [**Ans.** $A = 10, B = 10, Z = 11$]

3.25. Maximize $Z = 4x_1 + 5x_2 - 3x_3 + 50$

subject to $x_1 + x_2 + x_3 = 10$

$$x_1 - x_2 \geq 1$$
$$2x_1 + 3x_2 + x_3 \leq 40$$
$$x_1, x_2, x_3 \geq 0$$

[**Ans.** $x_1 = 11/2, x_2 = 9/2, Z = 189/2$]

3.26. Maximize $R = 22x + 3y + 25\,Z$

subject to $2x + 2y \leq 100$

$$2x + y + z \leq 100$$
$$x + 2y + 2z \leq 100$$
$$x, y, z \geq 0$$

[**Ans.** $x = 100/3, y = 50/3, Z = 50/3, R = 1650$]

3.27. Maximize $Z = 6x_1 - 2x_2$

subject to $2x_1 - x_2 \leq 2$

$$x_1 \leq 4$$
$$x_1, x_2 \geq 0$$

[**Ans.** $x_1 = 4, x_2 = 6$ and $Z = 12$]

3.28. Maximize $Z = 3x_1 + 2x_2 + 3x_3$

subject to $2x_1 + x_2 + x_3 \leq 2$

$$3x_1 + 4x_2 + 2x_3 \geq 8$$
$$x_1, x_2, x_3 \geq 0$$

[**Ans.** $x_2 = 2, Z = 4$]

OBJECTIVE TYPE QUESTIONS

3.1. In a maximization problem, the objective function coefficient of an artificial variable is
   (*a*) zero (*b*) + M
   (*c*) – M (*d*) none of the above

3.2. Key element is always found in
   (*a*) index row (*b*) key column
   (*c*) quantity column (*d*) none

3.3. A variable which does not appear in the basic variable (B) column of simplex table is
   (*a*) never equal to zero (*b*) always equal to zero
   (*c*) called a basic variable (*d*) none of the above

3.4. In the solution of linear programming problem by simplex method, for deciding the leaving variable
   (*a*) the maximum negative coefficient in the objective function row is selected.
   (*b*) the minimum positive ratio of the right hand side to the first decision variable is selected.
   (*c*) the maximum positive ratio of the right hand side to the coefficient in the key column is selected.
   (*d*) the minimum positive ratio of the right hand side to the coefficient in the key column is selected.

3.5. If a negative value appears in the solution values (x_B) column of the simplex table, then
   (*a*) the solution is infeasible (*b*) the solution is optimal
   (*c*) the solution is unbounded (*d*) all of the above

3.6. If for a given solution, a slack variable is equal to zero, then
   (*a*) the solution is infeasible
   (*b*) the solution is optimal
   (*c*) the entire amount of resource with the constraint in which the slack variable appears has been consumed.
   (*d*) All of the above

3.7. If an optimal solution is degenerate, then
   (*a*) there are alternative optimal solutions
   (*b*) the solution is of no use to the decision maker
   (*c*) the solution is infeasible
   (*d*) none of the above

Answers

3.1. (*c*) 3.2. (*b*) 3.3. (*b*) 3.4. (*d*) 3.5. (*c*) 3.6. (*c*) 3.7. (*a*).

❑❑❑

Unit II

Chapter 4

Transportation Problem

4.1. INTRODUCTION

Transportation problem is a special type of Linear programming problem in which the objective is to transport goods/product from various sources to different destinations in a way that transportation cost is minimized and warehouse (destination) requirements are met within the plant (source) capacity constraints. The solution of transportation problem requires the determination of how many units should be transported from each plant (source) to each demand destination in order to satisfy all destination demands while minimizing the total transportation cost.

It was originated in 1941 by two scientists, F.L. Hitchcock and T.C. koopmans.

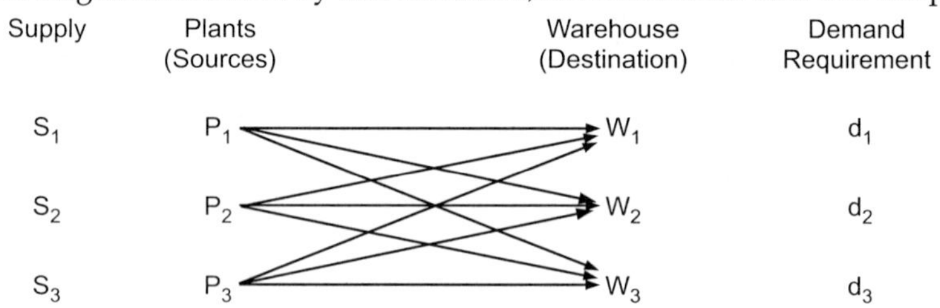

Fig. 4.1 Model of Transportation problems.

This model is used to decide where to locate a new facility, new warehouse, and new sales office.

4.1.1. Applications

(*i*) It is used to minimize transportation cost from plants to warehouses, warehouses to retailers.

(*ii*) It is used in distribution of army personnel and material to vessels and troop locations.

(*iii*) Determining lowest cost location for starting a new plant, sales office or warehouse.

4.1.2. Mathematical Model

Consider a transportation problem with 'm' sources and 'n' destinations. Let

s_i be the supply at source i

d_j be the demand at destination j

c_{ij} be the cost of transportation of one unit of product from i^{th} source to j^{th} destination.

Xij be the number of units to be transported from i^{th} source to j^{th} destination. This data can be expressed in the form of a table given below:

Table 4.1 General Transportation table

Plants (Origins)	Warehouse (Destinations)					Supply (Availab-ility)
	W_1	W_2	W_3	W_j	W_n	
P_1	X_{11} C_{11}	X_{12} C_{12}	X_{13} C_{13}	X_{1j} C_{1j}	X_{1n} C_{1n}	S_1
P_2	X_{21} C_{21}	X_{22} C_{22}	X_{23} C_{23}	X_{2j} C_{2j}	X_{2n} C_{2n}	S_2
P_i	X_{i1} C_{i1}	X_{i2} C_{i2}	X_{i3} C_{i3}	X_{ij} C_{ij}	X_{in} C_{in}	S_i
P_m	X_{m1} C_{m1}	X_{m2} C_{m2}	X_{m3} C_{m3}	X_{mj} C_{mj}	X_{mn} C_{mn}	S_m
Demand (Requirement)	d_1	d_2	d_3	d_j	d_n	$\sum_{i=1}^{m} S_i = \sum_{j=1}^{n} D_j$

Mathematically, transportation problem can be expressed as Linear programming problem.

Minimize total cost, $Z = x_{11}\, c_{11} + x_{12}\, c_{12} + ... + x_{21}\, c_{21} + ... + x_{mn}\, c_{mn}$

subject to constraints

$$x_{i1} + x_{i2} + ... + x_{in} = s_i \qquad ...(1)$$

$$x_{1j} + x_{2j} + ... + x_{mj} = d_j \qquad ...(2)$$

$$x_{ij} \geq 0 \text{ for all } i \text{ and } j \qquad ...(3)$$

or it can be written as

$$\text{Minimize } z = \sum_{i=1}^{m}\sum_{j=1}^{n} C_{ij}\, x_{ij}$$

$$\text{subject to constraints, } \sum_{j=1}^{n} x_{ij} = s_i \; (i = 1, 2, ... m)$$

$$\sum_{i=1}^{m} x_{ij} = d_j \qquad (j = 1, 2 .... n) \qquad \text{and } x_{ij} \geq 0$$

(*i*) If $\sum_{i=1}^{m} s_i = \sum_{J=1}^{n} d_j$, the given transportation problem is balanced. In this total supply is equal to total demand.

(*ii*) If $\sum_{i=1}^{m} s_i \neq \sum_{j=1}^{n} d_j$, the given problem is unbalanced.

4.2. BASIC DEFINITIONS

(*i*) **Feasible solution:** Non-negative values of x_{ij}, which satisfies the rim requirement (equation 1 & 2) is called the feasible solution. For feasible solution to exist it is necessary that $\sum_{i=1}^{m} s_i = \sum_{j=1}^{n} d_j$ (balanced transportation problem)

(*ii*) **Basic Feasible solution:** An initial feasible solution is called a basic feasible solution if number of non-negative allocations is equal to $m + n - 1$ where m is the number of rows and n the number of columns in a transportation table.

(iii) Non-degenerate Basic Feasible solution: A feasible solution is said to be non-degenerate, if it contains $m + n - 1$ occupied cells and each allocation is in independent positions.

Allocations are said to be in independent positions, if it is impossible to form a closed path. Closed path is formed by allowing horizontal and vertical lines through occupied cells.

(iv) Degenerate Basic Feasible solution: If a basic feasible solution contains less than $m + n - 1$ non-negative occupied cells, it is said to be degenerate.

(v) Optimal solution: A feasible solution is said to be optimal if it minimizes the total transportation cost.

(vi) Loops in transportation table: An ordered set of four or more cells in a transportation table is said to form a loop if:

(*i*) Any two adjacent cells of the ordered set lie either in the same row or in the same column.

(*ii*) An even number of at least four cells must form a closed loop.

(*iii*) Closed loop may or may not be square or rectangular in shape.

(*iv*) All cells that have a plus or minus sign, except the starting unoccupied cell, must be occupied cells.

(vii) Opportunity cost: The opportunity cost of an empty cell is the reduction in cost by including this cell in the solution.

4.3. SOLVING A TRANSPORTATION PROBLEM

Following step-by-step procedure is employed to solve a problem:

(*i*) Define the objective function to be minimized and the constraints imposed.

(*ii*) Develop the transportation table with rows (m) representing the sources and columns representing warehouses (n).

(*iii*) Determine an **initial basic feasible solution** of the problem.

(*iv*) Find whether the initial solution is feasible or not. Solution is said to be feasible if $m + n - 1$ cells are allocated with independent positions. Cells having allocations are called occupied cells and remaining cells are known as empty cells.

(*v*) Initial feasible solution obtained above is tested for optimality by calculating opportunity costs of empty cells. Positive opportunity costs for all empty cells means that solution is optimal.

(*vi*) If the solution is not optimal, modify the table by including that empty cell whose inclusion in the table results in the largest saving.

(*vii*) Repeat steps *v* & *vi* till an optimal solution is reached.

4.4. INITIAL BASIC FEASIBLE SOLUTION

For initial solution to be basic feasible solution, it must meet following conditions:

(*i*) Number of occupied cells must be equal to $m + n - 1$

(*ii*) Occupied cells must be independent positions.

(*iii*) Rim conditions must be satisfied - sum of quantities in occupied cells for rows and columns must be equal to corresponding capacity and requirement.

There are several methods to determine initial basic feasible solution of a transportation problem. Three methods are described below:

4.4.1. North West Corner Cell Method

It is simplest, efficient, reliable, and easy to calculate and understand, method of finding initial basic feasible solution. It is used for transportation where cost is not the important criteria. Various steps of this method are described below:

(*i*) Select the North-West (upper left-hand) corner cell of the transportation table and allocate as many units as possible equal to minimum between available supply and demand requirements *i.e.*, x_{11} = minimum of (s_1, d_1).

(*ii*) Adjust the quantity of supply and demand in the respective rows and columns allocation.

(*iii*) (*a*) If the supply (s_1) of the first row is exhausted then move down to the first cell of the second row and make the second allocation x_{21} = minimum of $(s_2, d_1 - x_{11})$.

(*b*) If the demand (d_1) for the first column is satisfied, then move horizontally to the next cell in the second column and first row and make the second allocation x_{12} = minimum of $(s_1 - x_{11}, d_2)$.

(*iv*) If for any cell, supply equals demand, then the next allocation can be made in cell either in the next row or next column.

(*v*) Continue the procedure moving down towards the lower right corner till the total available supply is fully allocated to the cells and all the requirements are met.

(*vi*) All allocations are circled to indicate the value of corresponding variable.

Example 4.1. Determine initial basic feasible solution of the following transportation problem using North-West corner method.

Solution:

Source	Destination			Capacity
	D_1	D_2	D_3	
S_1	(10) [2]	[2]	[3]	10
S_2	(10) [4]	(5) [1]	[2]	15
S_3	[1]	(10) [3]	(30) [1]	40
Demand	20	15	30	65 / 65

(*i*) Total demand = 20 + 15 + 30 = 65

Total capacity = 10 + 15 + 40 = 65

Therefore, problem is balanced problem.

(*ii*) Start from (S_1, D_1) cell and allocate the minimum of (s_1, d_1) = min. (10, 20) = 10. It completely exhausts the capacity of S_1 and still there is demand of 20 – 10 = 10 at D_1

(*iii*) Move to (S_2, D_1), allocate min. (15, 10) = 10. Now, demand of D_1 is exhausted with balance of 15 – 10 = 5 units at S_2. Move to (S_2, D_2) and allocate min. of (5, 15) = 5. Capacity is exhausted. Balance demand at D_2 = 15 – 5 = 10.

(*iv*) Move to (S_3, D_2), allocate min. of (40, 10) = 10

(*v*) Finally alocate 30 units in cell (S_3, D_3) to meet the requirements.

Initial solution is $x_{11} = 10, x_{21} = 10, x_{22} = 5, x_{32} = 10, x_{33} = 30$.

Total cost = 10 × 2 + 10 × 4 + 5 × 1 + 10 × 3 + 30 × 1 = 125

4.4.2. Least Cost Method

This method takes into account the minimum unit cost therefore reduces calculations and time to arrive at optimal solution. Various steps of this method are described below:

(*i*) Find the cell with the lowest transportation cost among all the rows and column. If the minimum cost is not unique, then select any cell with minimum cost arbitrarily. Allocate x_{ij} = minimum of (s_i, d_j) in cell with minimum c_{ij}.

(*ii*) Allocate as many units as possible to the cell determined in step i and cross off that row or column in which either supply is exhausted or demand is met.

If $x_{ij} = s_i$ cross off the ith row of the table and decrease d_j by s_i and go to step(*iii*)

If $x_{ij} = d_j$ cross off the jth column of the table and decrease s_i by d_j and go to step (*iii*)

If $x_{ij} = s_i = d_j$ cross off either the ith row or j^{th} column of the table but not both.

(*iii*) Repeat step (*i*) and (*ii*) for the reduced table till the total supply at different sources is exhausted to satisfy the demand at different warehouses.

Example 4.2. Solve example 4.1 using least cost method.

Solution:

Source	Destination			Capacity
	D_1	D_2	D_3	
S_1	(10) 2	2	3	~~10~~
S_2	4	(15) 1	2	~~15~~
S_3	(10) 1	3	(30) 1	~~40~~ −30 −10
Demand	~~20~~ ~~10~~	~~15~~	~~30~~	65 / 65

(*i*) Allocate maximum possible number of units to least cost cell. Since, it is not unique (1), choose (S_3, D_3) cell as highest no. of units can be allocated to this cell. Allocating S_3 units to (S_3, D_3) meets the requirement of D_3. Allocate remaining 10 units of s_3 to (S_3, D_1) cell. This exhausts the supply of S_3. Continue in this fashion to meet all requirements.

(*ii*) Final solution is : $x_{11} = 10$, $x_{22} = 15$,

$x_{31} = 10$, $x_{33} = 30$

Total cost $= 10 \times 2 + 15 \times 1 + 10 \times 1 + 30 \times 1 = 75$

4.4.3. Vogel's Approximation Method

It is also known as penalty or regret method. It gives initial solution, which is nearer to optimal solution or is the optimal solution itself. Allocation is made on the basis of opportunity cost (penalty) that would have incurred if allocation in cells with minimum cost were not made. Vogel's method has disadvantage of more calculation work. Various steps of this method are summarized below:

(*i*) Calculate penalty for each row and column by finding difference of smallest and next smallest cost in the same row (column). This difference is known as penalty.

(*ii*) Identify the row or column with the largest penalty and allocate maximum possible quantity to lowest cost cell in that row or column. If a tie occurs in penalties, select that row or column which has minimum cost. If there is tie in minimum cost also, select that row or column which will have maximum possible assignments.

(*iii*) Adjust supply and demand and cross out the satisfied row or column. If a row and column are satisfied simultaneously, only one is crossed out and other is assigned zero supply (demand). Row (column) with zero supply (demand) should not be used in calculating future penalties.

(*iv*) Recalculate the row and column differences for the reduced transportation table omitting the rows (columns) crossed out in previous step.

(*v*) Repeat step (*i*) to (*iii*) till an initial feasible solution is reached.

Example 4.3. Solve example 4.1 using VAM method

Solution:

Source	Destination			Capacity	P_1
	D_1	D_2	D_3		
S_1	2	2	3	10	0
S_2	4	1	2	15	1
S_3	1	3	1	40	0
Demand	20	15	30		
P_1	1	1	1 ↑		

(1) Calculated penalties of each row and column are shown in table.

(2) Identify row (column) with highest penalty (mark with arrow).

(3) Allocate 30 units (maximum possible) to lowest cost cell (S_3, D_3). Demand of D_3 is met. Delete this column and draw table again.

Source	Destination			Capacity	P_2	P_3
	D_1	D_2				
S_1	2	2		10	0	0
S_2	4	1		~~15~~	3 ←	
S_3	1	3		~~10~~	2	2 ←
Demand	20	15				
P_2	1	1				
P_3	1					

(4) Highest penalty (in P_2) is 3. Allocate 15 units to (S_2, D_2). Supply of S_2 is exhausted. Delete this row (S_2). Assign zero demand to D_2.

(5) Calculate penalties (P_3) again for remaining rows and columns. Now, highest penalty is 2. Allocate 10 units to lowest cost cell (S_3, D_1). Finally, allocate 10 units to (S_1, D_1). Initial solution is: $x_{11} = 10, x_{22} = 15.$

$$x_{31} = 10, x_{33} = 30$$

$$\text{Total cost} = 10 \times 2 + 15 \times 1 + 10 \times 1 + 30 \times 1 = 75$$

4.5. TEST FOR OPTIMALITY

After determining initial basic feasible solution, next step is to check whether this solution is optimal or not. An optimal solution is one in which there is no opportunity cost. Following two methods are generally used to test the optimality of solution:

(*i*) Stepping stone method (*ii*) Modified Distribution (MODI) method

4.5.1. Stepping Stone Method

Steps of this method are summarized below:

(*i*) Determine initial basic feasible solution using any method discussed above.

(*ii*) Ensure that number of occupied cells are exactly equal to $m + n - 1$.

(*iii*) Evaluate the unoccupied cells

(*a*) Select an unoccupied cell, where allocation can be made.

(*b*) Starting at this cell, trace a closed path using most direct route through at least three occupied cells and then back to original unoccupied cell with only horizontal and vertical moves. Cells at turning points are called stepping-stones. While tracing the path both unoccupied and occupied cells may be skipped over.

(*c*) Starting with + sign at unoccupied cell, assign + and - sign alternatively on each corner cell of closed path.

(*d*) Calculate net change in the cost along the closed path by adding unit costs of +ve cells and subtracting unit costs of -ve cells.

(*e*) Repeat (*a*) to (*d*) for all unoccupied cells.

(*iv*) Check the sign of each net changes. If all net changes are +ve, an optimal solution has been reached. If not it is possible to improve the current solution.

(*v*) Select the unoccupied cell with most negative net cost change and determine maximum number of units that can be assigned to cell marked with -ve sign. Add this number to unoccupied cell and to all other cells on the closed path marked +ve and subtract this number from cells on closed path marked with -ve sign.

(*vi*) Repeat procedure till optimal solution is reached.

Example 4.4. Find the optimal solution of the problem given below:

Source	Destination				Capacity
	D_1	D_2	D_3	D_4	
S_1	19	30	50	10	7
S_2	70	30	40	60	9
S_3	40	8	70	20	18
Demand	5	8	7	14	34 / 34

Solution: (1) Total capacity = Total demand = 34.

Therefore, problem is balanced.

(2) Obtain the initial solution of the above problem using least cost method.

Source	Destination				Capacity
	D_1	D_2	D_3	D_4	
S_1	+ 19	30	50	(7) − 10	~~7~~
S_2	(2) 70	30	(7) 40	60	9
S_3	(3) 40	(8) 8	70	(7) + 20	18
Demand	5	~~8~~	7	14	34 / 24

NO. of occupied cells = 6 (3 + 4 – 1 = 6)

Total cost = $7 \times 10 + 2 \times 70 + 7 \times 40 + 3 \times 40 + 8 \times 8 + 7 \times 20 = 814$

(3) Evaluate unoccupied cells – $d_{11}, d_{12}, d_{13}, d_{22}, d_{24}, d_{33}$.

Trace close path for d_{11} (shown dotted)

$(S_1, D_1) \rightarrow (S_1, D_4) \rightarrow (S_3, D_4) \rightarrow (S_3, D_1) \rightarrow (S_1, D_1)$

mark alternate + and – sign

Net cost change = 19 – 10 + 20 – 40 = **– 11**

Similarly tracing closed path for all unoccupied cells and calculating net cost change for each.

$d_{12} - (S_1, D_2) \rightarrow (S_1, D_4) \rightarrow (S_3, D_4) \rightarrow (S_3, D_1) \rightarrow (S_1, D_1)$

$d_{12} = 30 - 10 + 20 - 8 = 32$

$d_{13} - (S_1, D_3) \rightarrow (S_1, D_4) \rightarrow (S_3, D_4) \rightarrow (S_3, D_1) \rightarrow (S_2, D_1) \rightarrow (S_2, D_3) \rightarrow (S_1, D_3)$

$d_{13} = 50 - 10 + 20 - 40 + 70 - 40 = 50$

$d_{22} - (S_2, D_2) - (S_3, D_2) \rightarrow \rightarrow (S_3, D_1) \rightarrow (S_2, D_1) \rightarrow (S_2, D_2)$

$d_{22} = 30 - 8 + 40 - 70 =$ **– 8**

$d_{24} - (S_2, D_4) \rightarrow (S_3, D_4) \rightarrow (S_3, D_1) \rightarrow (S_2, D_1) \rightarrow (S_2, D_4)$

$d_{24} = 60 - 20 + 40 - 70 =$ **– 10**

$d_{33} - (S_3, D_3) \rightarrow (S_3, D_1) \rightarrow (S_2, D_1) \rightarrow (S_2, D_3) \rightarrow (S_3, D_3)$

$d_{33} = 70 - 40 + 70 - 40 = 60$

(4) Present solution is not optimal as net cost change for all unoccupied cells is not positive. Unoccupied cell with highest negative net cost change is selected for improvement (d_{11}).

(5) In the closed path of d_{11}, minimum quantity of cell with minus sign is 3. Add 3 to cells marked plus and subtract 3 from cells marked minus.

(6) New solution is shown below

Source	Destination				Capacity
	D_1	D_2	D_3	D_4	
S_1	(3) + [19]	[30]	[50]	(4) – [10]	7
S_2	(2) – [70]	+ [30]	(7) [40]	[60]	9
S_3	[40]	(8) [8]	[70]	(10) + [20]	18
Demand	5	8	7	14	

No. of occupied cells = 6 = (3 + 4 – 1)

Total cost of improved solution is

$$= 3 \times 19 + 4 \times 10 + 2 \times 70 + 7 \times 40 + 8 \times 8 + 10 \times 20 = 781$$

(7) Calculating net change in cost for all unoccupied cells again.

$d_{12} = 30 - 8 + 20 - 10 = 32$

$d_{13} = 50 - 19 + 70 - 40 = 61$

$d_{22} = 30 - 70 + 19 - 10 + 20 - 8 =$ **– 19**

$d_{24} = 60 - 10 + 19 - 70 =$ **– 1**

$d_{31} = 40 - 19 + 10 - 20 = 11$

$d_{33} = 70 - 20 + 10 - 19 + 70 - 40 = 71$

(8) This solution is also not optimal. d_{22} is selected for improvement. Minimum quantity (2) of cell with minus sign (d_{21}) is added in cells with plus sign and subtracted from cells with minus sign.

(9) New solution is

Source	Destination				Capacity
	D_1	D_2	D_3	D_4	
S_1	(5) 19	30	50	(2) 10	7
S_2	70	(2) 30	(7) 40	60	9
S_3	40	(6) 8	70	(12) 20	18
Demand	5	8	7	14	

Total cost = $5 \times 19 + 2 \times 10 + 2 \times 30 + 7 \times 40 + 6 \times 8 + 12 \times 20 = 743$

(10) Calculating net cost change for all unoccupied cells.

$$d_{12} = 30 - 10 + 20 - 8 = 32$$
$$d_{13} = 50 - 10 + 20 - 8 + 30 - 40 = 42$$
$$d_{21} = 70 - 19 + 10 - 20 + 8 - 30 = 19$$
$$d_{24} = 60 - 20 + 8 - 30 = 18$$
$$d_{31} = 40 - 20 + 10 - 19 = 11$$
$$d_{33} = 70 - 40 + 30 - 8 = 52$$

Since all values are positive, optimal solution is reached.

4.5.2. The Modified Distribution (MODI) Method

It evaluates each of unoccupied cells more effectively. In this, opportunity costs of all cells are calculated without having to trace their respective closed path. We trace only one closed path for most -ve unoccupied cell.

It involves following steps:

(*i*) Determine initial basic feasible solution consisting of $m + n - 1$ allocations in independent positions.

(*ii*) Determine a set of numbers for each row and column (set $c_{ij} = u_i + v_j$ for each of $m + n - 1$ occupied cells). Number of equations = $m + n - 1$, number of unknowns = $m + n$. Put one of the variable equal to zero.

(*iii*) Calculate opportunity cost using $\Delta_{ij} = c_{ij} - (u_i + v_j)$ for each unoccupied cells.

(*iv*) Check the sign of each opportunity cost. If opportunity costs of all unoccupied cells are +ve or zero then solution is optimal. If one or more cells have -ve opportunity costs, solution is not optimal. Go to step (*v*).

(*v*) Select the unoccupied cell with most -ve opportunity cost as the cell to be included in next solution.

(*vi*) Draw a closed path for the unoccupied cell selected in step (*v*).

(*vii*) Assign alternate +ve and -ve sign (starting from unoccupied cell with +ve sign).

(*viii*) Determine maximum number of units that should be allocated to this unoccupied cell (smallest stone with -ve position). This is added to cells with +ve sign and subtracted from those with -ve sign.

(*ix*) Repeat complete procedure till an optimal solution is attained.

(*x*) Obtain total transportation cost.

- For a given solution only one loop can be constructed for each unoccupied cell. Loop can be traced clockwise or anti-clockwise.
- Every row and column must have atleast one assignment.

- If two or more unoccupied cells have same largest - ve opportunity cost, the unoccupied cell by including which the decrease in the transportation cost is maximum is chosen.
- If optimal solution has zero opportunity cost for one or more unoccupied cells then there are alternative solutions.

Example 4.5. Consider the following transportation problem involving three sources and three destinations. The cell entries represent the cost of transportation per unit in rupees.

	Destination			Supply
	A	B	C	
Source 1	6	8	4	14
Source 2	4	9	3	12
Source 3	1	2	6	5
Demand	6	10	15	

(UPTU: 2004-2005)

Obtain the optimum cost of transportation.

Solution: (1) Total demand = 6 + 10 + 15 = 31

Total supply = 14 + 12 + 5 = 31

Hence, problem is balanced.

(2) Initial solution by VAM

Source	Destination A_1	B_2	C_3	Capacity	P_1	P_2	P_3
S_1	6	8	(14) 4	14	2	2 ←	
S_2	(6) 4	(5) 9	(1) 3	12	1	1	1
S_3	1	(5) 2	6	5	1		
Demand	6	10	15				
P_1	3	6	1				
P_2	2	1	1				
P_3							

Total cost = $14 \times 4 + 6 \times 4 + 5 \times 9 + 1 \times 3 + 5 \times 2 = 138$

(3) No. of occupied cells = 5 (3 + 3 – 1 = 5)

Source	Destination A	B	C	Supply	Row Number
S_1	6	+ 8	– (14) 4	14	u_1
S_2	(6) 4	(5) 9	(1) 3	12	u_2
S_3	1	(5) 2	6	5	u_3
Demand	6	10	15		
Column number	v_1	v_2	v_3		

That means non-degenerate solution exists. Using MODI method to find optimal solution. Introducing row number u_1, u_2, u_3 and column number v_1, v_2, v_3.

For occupied cells, $c_{ij} = u_i + v_j$

$$C_{1C} = u_1 + v_3 = 4, \qquad C_{2A} = u_2 + v_1 = 4$$
$$C_{2B} = u_2 + v_2 = 9, \qquad C_{2C} = u_2 + v_3 = 3$$
$$C_{3B} = u_3 + v_2 = 2$$

We have six unknowns and five equations. Let $u_2 = 0$, find other variables

$$v_1 = 4, v_2 = 9, v_3 = 3, u_1 = 1, u_3 = -7$$

(4) Calculate cost change for all unoccupied cells using $\Delta_{ij} = C_{ij} - (u_i + v_j)$

Unoccupied Cell	$\Delta_{ij} = C_{ij} - C(u_i + v_j)$	Net Cost Change
1 – A	$\Delta_{1A} = 6 - (1 + 4) = 6 - 5 = 1$	+1
1 – B	$\Delta_{1B} = 8 - (1 + 9) = 8 - 10 = -2$	–2
3 – A	$\Delta_{3A} = 1 - (-7 + 4) = 4$	+4
3 – C	$\Delta_{3C} = 6 - (-7 + 3) = 10$	+10

Since net cost change for 1 – *B* is negative, further improvement is possible.

(5) Trace close path of 1 – *B*. It is (1 – *B*) → (1 – *c*) → (2 – *c*) → (2 – *B*) → (1 – *B*)

Minimum quantity of cell with minus sign is added to cells with plus sign and subtracted from cells with minus sign in closed loop.

(6) Construct new table and calculate total cost

$$= 5 \times 8 + 9 \times 4 + 6 \times 3 + 6 \times 4 + 5 \times 2 = 128$$

Source	Destination			Supply	Row Number
	A	B	C		
1	6	(5) 8	(9) 4	14	$u_1 = 0$
2	(6) 4	9	(6) 3	12	$u_2 = -1$
3	1	(5) 2	6	5	$u_3 = -6$
Demand	6	10	15		
Column number	$v_1 = 5$	$v_2 = 8$	$v_3 = 4$		

(7) Test this solution for further improvement

$$u_1 + v_2 = 8, u_1 + v_3 = 4,$$
$$u_2 + v_1 = 4$$
$$u_2 + v_3 = 3,$$
$$u_3 + v_2 = 2$$

Let $u_1 = 0$, then $v_2 = 8, v_3 = 4,$

$$u_2 = -1$$
$$u_3 = -6, v_1 = 5$$

(8) Calculate cost change for unoccupied cells

Unoccupied Cell	$\Delta_{ij} = C_{ij} - C(u_i + v_j)$	Net Cost Change
1 – A	$\Delta_{1A} = 6 - (0 + 5) = 1$	+1
1 – B	$\Delta_{2B} = 9 - (-1 + 8) = 2$	+2
3 – A	$\Delta_{3A} = 1 - (-6 + 5) = 2$	+2
3 – C	$\Delta_{3C} = 6 - (-6 + 4) = 8$	+8

All net cost change values are positive therefore current solution is optimal. Hence, optimum cost of transportation is 128.

4.6. DEGENERACY

If number of occupied cells is less than $m + n - 1$ at any stage, problem would degenerate. Degeneracy can occur at two stages (*i*) at initial solution (*ii*) during testing of optimal solution

(*i*) **At initial solution:** To resolve degeneracy at initial solution, an artificial quantity ε is used in one or more of the unoccupied cells so that number of occupied cells become $(m + n - 1)$. ε is so small that it does not affect the demand & supply constraint. ε = 0 when actually used in movements of goods. ε is generally placed in the unoccupied cell with lowest transportation cost. Use of ε is confined to evaluation of unoccupied cell. ε remains in solution till degeneracy is removed/solution is reached.

(*ii*) **Degeneracy occurs during testing of optimal solution:** Degeneracy occurs when inclusion of unoccupied cell with most - ve opportunity cost results in vacating of two or more occupied cells simultaneously. It is resolved by allocating ε to one or more of recently vacated cells to have $m + n - 1$ occupied cells.

Example 4.6. Solve following transportation problem.

Solution: (*i*) Initial solution using least cost method.

Source	Destinations A	B	C	Capacity	u_i
I	(50) [1]	[2]	[3]	50	$u_1 = 0$
II	[3]	[2]	(80) [1]	80	$u_2 = 1$
III	(70) [4]	[5]	[6]	70	$u_3 = 3$
IV	[3]	(80) [1]	(20) [2]	100	$u_4 = 2$
Demand	120	80	100		
v_j	$v_1 = 1$	$v_2 = -1$	$v_3 = 0$		

No. of allocations = 5 < (4 + 3 – 1)

Hence solution is degenerate (Type – 1)

(*ii*) Assign small quantity ε to independent cell (4, 1), use MODI method to find optimal solution. For occupied cells, $c_{ij} = u_i + v_j$

$c_{11} = u_1 + v_1, 1 = 0 + v_1 \quad \Rightarrow \quad v_1 = 1$ (Let $u_1 = 0$)

$c_{31} = u_3 + v_1, 4 = u_3 + 1 \quad \Rightarrow \quad u_3 = 3$

Similarly, calculate other numbers $u_4 = 2, v_2 = -1, v_3 = 0, u_2 = 1$

(*iii*) Calculate net cost change values for unoccupied cells

$\Delta_{12} = 2 - (0 - 1) = 3,$ $\Delta_{13} = 3 - (0 + 0) = 3$

$\Delta_{21} = 3 - (1 + 1) = 1,$ $\Delta_{22} = 2 - (1 - 1) = 2$

$\Delta_{32} = 5 - (3 - 1) = 3,$ $\Delta_{33} = 6 - (3 + 0) = 3$

$\Delta_{41} = 3 - (2 + 1) = 0$

All Δij values are positive, therefore, this solution is optimal. Finally put $\varepsilon = 0$

Optimal transportation cost $= 50 \times 1 + 80 \times 1 + 70 \times 4 + 80 \times 1 + 20 \times 2 =$ Rs. 530

4.7. PROHIBITED ROUTES

When some of the routes are restricted for some reasons like flood, snow fall, strike, road repairing etc., then assign a very high cost *M* (or ∞) to prohibited routes (cells) so that these routes are not included in optimal solution.

Example 4.7. Find the optimum solution of the following problem in which it is not possible to transport any quantity from factory *B* to Godown 5.

Solution: (*i*) We assign a very high cost *M* or ∞ to cell (2, 5), since this route is restricted. Obtain initial solution using VAM.

Factory	Godowns 1	2	3	4	5	6	Stock	u_i
A	7 (+)	(20) 5	7	7	(ε) 5 (−)	(40) 3	60	$u_1 = 0$
B	(10) 9 (−)	11	(10) 6 (+)	11	M	5	20	$u_2 = -3$
C	11	10	(30) 6 (−)	(20) 2	(40) 2 (+)	8	90	$u_3 = -3$
D	(50) 9	10	9	6	9	12	50	$u_4 = -3$
Demand	60	20	40	20	40	40	220 / 220	
v_j	$v_1 = 12$	$v_2 = 5$	$v_3 = 9$	$v_4 = 5$	$v_5 = 5$	$v_6 = 3$		

No. of occupied cells = 8 < (4 + 6 – 1). There is degeneracy in initial solution. Assign a small quantity ε in cell (1, 5) → unoccupied cell with lowest transportation cost.

(*ii*) Test optimality of initial solution by MODI method.

$\Delta_{11} = 7 - (0 + 12) = -5,$ $\Delta_{13} = 7 - (0 + 9) = -2$

$\Delta_{14} = 7 - (0 + 5) = 2,$ $\Delta_{22} = 11 - (-3 + 5) = 9$

$\Delta_{24} = 11 - (-3 + 5) = 9,$ $\Delta_{25} = M - (-3 + 5) = M - 2$

$\Delta_{26} = 5 - (-3 + 3) = 5,$ $\Delta_{31} = 11 - (-3 + 12) = 2$

$\Delta_{32} = 10 - (-3 + 5) = 8,$ $\Delta_{36} = 8 - (-3 + 3) = 8$

$\Delta_{42} = 10 - (-3 + 5) = 8,$ $\Delta_{43} = 9 - (-3 + 9) = 3$

$\Delta_{44} = 6 - (-3 + 5) = 4,$ $\Delta_{45} = 9 - (-3 + 5) = 7, \Delta_{46} = 12 - (-3 + 3) = 12$

Solution is not optimal as Δ_{11} and Δ_{13} are negative. Trace close loop for Δ_{11} (highest negative).

Minimum quantity of cell with – sign is ε. Add ε to cells with plus sign and subtract ε from cells with minus sign. Redraw table and check for optimality.

Factory	Godowns						Stock	u_i
	1	2	3	4	5	6		
A	(ε) 7	(20) 5	7	7	5	(40) 3	60	$u_1 = 0$
B	9 (10–ε)	11	6 (10+ε)	11	M	5	20	$u_2 = 2$
C	11	10	6 (30–ε)	(20) 2	2 (40+ε)	8	90	$u_3 = 2$
D	(50) 9	10	9	6	9	12	50	$u_4 = 2$
Demand	60	20	40	20	40	40		
v_j	$v_1 = 7$	$v_2 = 5$	$v_3 = 4$	$v_4 = 0$	$v_5 = 0$	$v_6 = 3$		

$\Delta_{13} = 7 - (0 + 4) = 3,$ $\quad \Delta_{14} = 7 - (0 + 0) = 7, \Delta_{15} = 5 - (0 + 0) = 5$

$\Delta_{22} = 11 - (2 + 5) = 4,$ $\quad \Delta_{24} = 11 - (2 + 0) = 9$

$\Delta_{25} = M - (2 + 0) = M - 2,$ $\quad \Delta_{26} = 5 - (2 + 3) = 0$

$\Delta_{31} = 11 - (2 + 7) = 2,$ $\quad \Delta_{32} = 10 - (2 + 5) = 3$

$\Delta_{36} = 8 - (2 + 3) = 3,$ $\quad \Delta_{42} = 10 - (2 + 5) = 3$

$\Delta_{43} = 9 - (2 + 4) = 3,$ $\quad \Delta_{44} = 6 - (2 + 0) = 4$

$\Delta_{45} = 9 - (2 + 0) = 7,$ $\quad \Delta_{46} = 12 - (2 + 3) = 7$

All Δij are positive, therefore, solution is optimal but $\Delta_{26} = 0$ that means alternative optimal solution exists.

Minimum transportation cost ($\varepsilon = 0$)

$= (20 \times 5) + (40 \times 3) + (10 \times 9) + (10 \times 6) + (30 \times 6) + (20 \times 2) + (40 \times 2) + (50 \times 9)$

= Rs. 1120.

4.8. UNBALANCED TRANSPORTATION PROBLEM

The total demand must be equal to total supply for feasible solution of a problem to exist. However, when total supply is not equal to total demand, problem is said to be unbalanced. (*i*) When total supply is greater than total demand, a dummy destination is introduced in the table. The cost of transporting from any source to dummy destination is assumed to be zero. The demand of this dummy destination is equal to the difference between total supply and total demand. (*ii*) When total demand is greater than total supply, a dummy source is introduced in the table. The cost of transporting from this dummy source to any destination is assumed to be zero. The availability of this dummy source is equal to the difference between total demand and total supply.

Example 4.8. A product is produced by four factories A, B, C and D. The unit production cost in them are Rs 2, 3, 1 and 5 respectively. Their production capacities are factory A = 50 units, B = 70 units, c = 30 units and D = 50 units. These factories supply the product to four stores, demand of which are 25, 35, 105 and 20 units respectively. Unit transportation cost in rupees from each factory to each store is given in the table below. Determine the extent of deliveries from each of the factories to each of the stores so that the total production and transportation cost is the minimum.

	1	2	3	4
A	2	4	6	11
B	10	8	7	5
C	13	3	9	12
D	4	6	8	3

(UPTU: 2003-2004)

Solution:

	Stores				
Factories	1	2	3	4	Supply
A	2	4	6	11	50
B	10	8	7	5	70
C	13	3	9	12	30
D	4	6	8	3	50
Demand	25	35	105	20	185 / 200

As the demand ≠ supply, therefore adding a dummy store having demand 15 to increase the total demand to 200. Also adding the unit production cost of Rs 2 for each element of factory A, Rs. 3 for each element of factory B, Rs. 1 to factory C and Rs. 5 to factory D (to get the total production and transportation cost).

	Stories				
Factories	1	2	3	4	Supply
A	2+2	4+2	6+2	11+2	50
B	10+3	8+3	7+3	5+3	70
C	13+1	3+1	9+1	12+1	30
D	4+5	6+5	8+5	3+5	50
Demand	25	35	105	20	185 / 200

(2) Initial solution using VAM is given below:

	Stores					Penalty						
Factories	1	2	3	4	Dummy	Supply	P_1	P_2	P_3	P_4	P_5	P_6
A	(25) 4	(5) 6	(20) 8	13	0	~~50~~ ~~25~~ 20	4	2	2	2	← 5	–
B	13	11	(55) 10	8	(15) 0	~~70~~ 55	← 8	2	2	2	2	2
C	14	(30) 4	10	13	0	30	4	← 6	–	–	–	–
D	9	11	(30) 13	(20) 8	0	~~50~~ 30	8	1	1	3	5	← 5
Demand	25	~~35~~ 5	~~105~~ 85	~~20~~	~~15~~	200 / 200						
Penalty												
(P_1)	5	2	0	0	0							
(P_2)	5	2	0	0	–							
(P_3)	5 ↑	5	2	0	–							
(P_4)	–	↑ 5	2	0	–							
(P_5)	–	–	2	0	–							
(P_6)	–	–	3	0	–							

Total transportation cost + production cost = 25 × 4 + 5 × 6 + 30 × 4 + 20 × 8 + 55 × 10 + 30 × 13 20 × 8 + 15 × 0 = Rs. 1510

(3) Using stepping stone method

Factories	STORES					Supply
	1	2	3	4	Dummy	
A	(25) 4	(5) 6	(20) 8	13	0	50
B	13	11	(55) 10	8	(15) 0	70
C	14	(30) 4	10	13	0	30
D	9	11	(30) 13	(20) 8	0	50
Demand	25	35	105	20	15	185 / 200

No. of occupied cells = 8 = (5 + 4 – 1)

Therefore non-degenerate solution exists.

Evaluate unoccupied cells – d_{A4}, d_{Ad}, d_{B1}, d_{B2} d_{B4}, d_{C1}, d_{C3}, d_{C4}, d_{Cd}, d_{D1} d_{D2}, d_{Dd}.

$d_{A4} = 13 - 8 + 13 - 8 = 10$

$d_{Ad} = 0 - 0 + 10 - 8 = 2$

$d_{B1} = 13 - 10 + 8 - 4 = 7$

$d_{B2} = 11 - 6 + 8 - 10 = 3$

$d_{B4} = 8 - 8 + 13 - 10 = 3$

$d_{C1} = 14 - 4 + 6 - 4 = 12$

$d_{C3} = 10 - 8 + 6 - 4 = 4$

$d_{C4} = 13 - 8 + 13 - 8 + 6 - 4 = 12$

$d_{Cd} = 0 - 4 + 6 - 8 + 10 - 0 = 4$

$d_{D1} = 9 - 4 + 8 - 13 = 0$

$d_{D2} = 11 - 6 + 8 - 13 = 0$

$d_{Dd} = 0 - 13 + 10 - 0 = -3$

(4) Present solution is not optimal as net cost change for d_{Dd} is negative. Therefore, further improvement is possible in this cell. In closed path of d_{Dd}, minimum quantity of cell with minus sign is 15. Add 15 to cells with + sign and subtract 15 from cells with – sign. New solution is shown in table.

(5)

Factories	STORES					Supply
	1	2	3	4	Dummy	
A	(25) 4	(5) 6	(20) 8	13	0	50
B	13	11	(70) 10	8	0	70
C	14	(30) 4	10	13	0	30
D	9	11	(15) 13	(20) 8	(15) 0	50
Demand	25	35	105	20	15	

No. of occupied cells = 8 = (4 + 5 – 1) → Non-degenerate

Total cost = 25 × 4 + 5 × 6 + 20 × 8 + 70 × 10 + 30 × 4 + 15 × 13 + 20 × 8 + 15 × 0
= 1465.

(6) Calculating net cost change for unoccupied cells again.

$d_{A4} = 13 - 8 + 13 - 8 = 10$

$d_{Ad} = 0 - 0 + 13 - 8 = 5$

$d_{B1} = 13 - 10 + 8 - 4 = 7$

$d_{B2} = 11 - 6 + 8 - 10 = 3$

$d_{B4} = 8 - 8 + 13 - 10 = 3$

dBd $= 0 - 0 + 13 - 10 = 3$

$d_{C1} = 14 - 4 + 6 - 4 = 12$

$d_{C3} = 10 - 8 + 6 - 4 = 4$

$d_{C4} = 13 - 8 + 13 - 8 + 6 - 4 = 12$

$d_{Cd} = 0 - 0 + 13 - 8 + 6 - 4 = 7$

$d_{D1} = 9 - 13 + 8 - 4 = 0$

$d_{D2} = 11 - 13 + 8 - 6 = 0$

Since all net cost change values are positive, therefore, this solution is optimal. Total minimum cost = **1465**.

4.9. MAXIMIZATION TRANSPORTATION PROBLEM

In this the objective is to maximize the total profit when the profit associated with each route is given. Here, the profits are transformed into relative costs by subtracting the profit of each cell from the highest profit in the matrix. Now, the objective is to minimize these relative costs, which can be done in usual manner. After an optimal solution of minimization problem is obtained, relative costs in occupied cells are written in profit function and total profit is calculated.

Example 4.9. Determine the most profitable production and distribution schedule and the associated profit. The deficit production should be taken to yield zero profit.

Source		Factories				Per unit sale price	Requirement
		F_1	F_2	F_3	F_4		
Production Cost/Unit		15	18	14	13		
Raw material Cost/Unit		10	9	12	9		
Transportation Cost/Unit	S_1	3	9	5	4	34	80
	S_2	1	7	4	5	32	120
	S_3	5	8	3	6	31	150
Supply		10	150	50	100		310 / 350

Solution: (*i*) Develope profit matrix using

Profit = sale price – (Production cost + Raw material cost + Transportation cost)

Total requirement = 80 + 120 + 150 = 350

Total supply = 10 + 150 + 50 + 100 = 310

A dummy factory is added to produce 350 – 310 = 40 units.

Profit table

Factory	Sale Depot			Availability
	S_1	S_2	S_3	
F_1	6	6	1	10
F_2	–2	–2	–4	150
F_3	3	2	2	50
F_4	8	5	3	100
Requirement	80	120	150	350 / 310

Profit table after adding dummy factory

Factory	Sale Depot			Availibility
	S_1	S_2	S_3	
F_1	6	6	1	10
F_2	–2	–2	–4	150
F_3	3	2	2	50
F_4	8	5	3	100
F_D	0	0	0	40
Requirement	80	120	150	350 / 350

(*ii*) Develope loss matrix using – (To convert problem into minimization type)
Relative loss = highest profit – profit of the cell

Factory	Sale Depot			Availibility	u_i
	S_1	S_2	S_3		
F_1	8–6=2	(10) 2	7	10	$u_1 = 2$
F_2	10	(40) + 10	– (110) 12	150	$u_2 = 10$
F_3	5	(50) – 6	+ 6	50	$u_3 = 6$
F_4	(80) 0	(20) 3	5	100	$u_4 = 3$
F_D	8	8	(40) 8	40	$u_5 = 0$
Requirement	80	120	150		
v_j	$v_1 = -3$	$v_2 = 0$	$v_3 = 2$		

Obtain initial solution using VAM in above table.

(*iii*) Initial solution is tested for optimality using MODI method. Calculate net cost change values for all unoccupied cells.

$$\Delta_{11} = 2 - (2-3) = 3,\ \Delta_{13} = 7 - (2+2) = 3$$
$$\Delta_{21} = 10 - (10-3) = 3,\ \Delta_{31} = 5 - (6-3) = 2$$
$$\Delta_{33} = 6 - (6+2) = -2,\ \Delta_{43} = 5 - (3+2) = 0$$
$$\Delta_{51} = 8 - (0-3) = 11,\ \Delta_{52} = 8 - (0+0) = 8$$

Since, Δ_{33} is negative, current solution is not optimal. Trace close loop. Minimum quantity 50 in cell with minus sign is added in cells with plus and subtracted in cells with minus sign.

(*iv*) Revised solution

Factory	Sale Depot			Availiability	u_i
	S_1	S_2	S_3		
F_1	2	(10) 2	7	10	$u_1 = 4$
F_2	10	(90) 10	– (60) 12	150	$u_2 = 12$
F_3	5	– 6	(50) 6	50	$u_3 = 6$
F_4	(80) 0	(20) 3	5	100	$u_4 = 5$
F_D	8	8	(40) 8	40	$u_5 = 8$
Requirement	80	120	150		
Vj	$V_1 = -5$	$V_2 = -2$	$V_3 = 0$		

Net cost change values for unoccupied cells are

$\Delta_{11} = 2 - (4 - 5) = 3,$ $\quad\Delta_{13} = 7 - (4 + 0) = 3$

$\Delta_{21} = 10 - (12 - 5) = 3,$ $\quad\Delta_{31} = 5 - (6 - 5) = 4$

$\Delta_{32} = 6 - (6 - 2) = 2,$ $\quad\Delta_{43} = 5 - (5 + 0) = 0$

$\Delta_{51} = 8 - (8 - 5) = 5,$ $\quad\Delta_{52} = 8 - (8 - 2) = 2$

All Δij values are positive or zero, therefore current solution is optimal.

Optimum profit = $(80 \times 8) + (10 \times 6) + \{90 \times (-2)\} + (20 \times 5) + [60 \times (-4)] + (50 \times 2) + 40 \times 0 =$ Rs. 480.

4.10. TRANSHIPMENT PROBLEM

A transportation problem in which available items frequently moves from one source to another or to destination and finally to actual receipt destination is called transhipment problem. Transhipment problem is an enlarged transportation problem with $m + n$ sources and destinations. Cost of transportation from s_i to s_i and d_j to d_j is equal to zero.

Example 4.10. A firm has two factories and wishes to ship its products to two stores. The number of units available at F_1 and F_2 are are 5 and 25 respectively while the demand of stores S_1 and S_2 are 20 and 10 respectively. The firm, instead of shipping directly from sources to destinations, decides to investigate the possibility of transhipment. The unit transporation costs are given in the table.

	Factory		Store		Available
	F_1	F_2	S_1	S_2	
F_1	0	2	3	4	5
F_2	2	0	2	4	25
S_1	3	2	0	1	
S_2	4	4	1	0	
Demand ⟶			20	10	

Solution: (1) In transhipment problem each of factory and store represents a point of supply and demand. The total supply may be concentrated at any of these factories or stores at any stage. Since all the demand may be concentrated at any one location, the fictitious demand and supply at each location (called buffer stocks) is assumed. Buffer stock must, at least be equal to the total supply in original problem.

Here, buffer stock = 30 (20 + 10).
Convert the given problem into standard form.
(2) Obtain the initial solution of extended transportation problem using VAM.

Source	Destinations F_1	F_2	S_1	S_2	Supply	u_i
F_1	(30) 0	1 \ 2	0 \ 3	(5) 4	5+30=35	$u_1 = 0$
F_2	3 \ 2	(30) 0	(25) 2	1 \ 4	25+30=55	$u_2 = -1$
S_1	6 \ 3	4 \ 2	(25) 0	(5) 1	30	$u_3 = -3$
S_2	8 \ 4	7 \ 4	2 \ 1	(30) 0	30	$u_4 = -4$
Demand	30	30	20+30=50	10+30=40	150 \ 150	
v_j	$v_1 = 0$	$v_2 = 1$	$v_3 = 3$	$v_4 = 4$		

Above solution is optimal solution also, as net cost change value for all unoccupied cells is positive or zero (shown in bottom left corners). The allocations in diagonal cells are ignored, since they do not have any physical meaning.

Example 4.11. Solve the following transportation problem.

		Destination				
		P	Q	R	S	Supply
Source	A	21	16	25	13	11
	B	17	18	14	23	13
	C	32	17	18	41	19
	Demand	6	10	12	15	43

Solution: Since $\Sigma s_i = \Sigma d_j$, the given TP is a balanced one. Therefore, there exists a feasible solution.

(*i*) Initial solution by VAM

Origin/Dest.	P	Q	R	S	Supply	P_1	P_2	P_3	P_4	P_5	P_6
A	21	16	25	(11) 13	11	3	–	–	–	–	–
B	(6) 17	18	(3) 14	(4) 23	13	4	4	4	4 ←	–	–
C	32	(10) 17	(9) 18	48	19	1	1	1	1	1	17
Demand	6	10	12	15	43						
(P_1)	4	1	4	10 ↑							
(P_2)	15	1	4	18 ↑							
(P_3)	15 ↑	1	4	–							
(P_4)	–	1	4	–							
(P_5)	–	17	18 ↑	–							
(P_6)	–	17 ↑	–	–							

(2) From this table we see that the number of non-negative independent allocation is $6 = m + n - 1 = 3 + 4 - 1$.

Hence, the solution is non-degenerate basic feasible.

The initial transportation cost

$$= 11 \times 13 + 3 \times 14 + 4 \times 23 + 6 \times 17 + 17 \times 10 + 18 \times 9 = \text{Rs. } 711$$

To find the optimal solution: Apply Modi method in order to determine the optimum solution. Determine a set of numbers u_i and v_j for each row and column, with $u_i + v_j = c_{ij}$ for each occupied cell. To start with take $u_2 = 0$ as the 2nd row has the maximum number of allocation.

$C_{21} = u_2 + v_1 \Rightarrow 17 = 0 + v_1 \Rightarrow v_1 = 17$
$C_{23} = v_2 + v_3 \Rightarrow 14 = 0 + v_3 \Rightarrow v_3 = 14$
$C_{24} = u_2 + v_4 \Rightarrow 23 = 0 + v_4 \Rightarrow v_4 = 23$
$C_{14} = u_1 + v_4 \Rightarrow 13 = u_1 + 23 \Rightarrow u_1 = -10$
$C_{33} = u_3 + v_3 \Rightarrow 18 = v_3 + 14 \Rightarrow u_3 = 4$
$C_{32} = u_3 + v_2 \Rightarrow 17 = 4 + v_2 \Rightarrow v_2 = 13$

(4) Calculate net cost change for all unoccupied cells

$\Delta_{11} = 21 - (-10 + 17) = 14,\quad \Delta_{12} = 16 - (-10 + 13) = 13$
$\Delta_{13} = 23 - (-10 + 14) = 19,\quad \Delta_{22} = 18 - (0 + 13) = 5$
$\Delta_{31} = 32 - (4 + 17) = 11,\quad \Delta_{34} = 41 - (4 + 23) = 14$

Since, all net cost change values are positive, solution is optimal and unique.

Source	Destination				
	P	Q	R	S	u_1
A	21	16	23	(11) 13	$u_1 = -10$
B	(6) 17	18	(3) 14	(4) 23	$u_2 = 0$
C	32	(10) 17	(9) 18	41	$u_3 = 4$
v_j	$v_1 = 17$	$v_2 = 13$	$v_3 = 14$	$v_4 = 23$	

The optimal solution is

$$x_1 = 11, x_{21} = 6, x_{23} = 3, x_{24} = 4, x_{32} = 10, x_{33} = 9$$

The minimum transportation cost

$$= 11 \times 13 + 17 \times 6 + 3 \times 14 + 4 \times 23 + 10 \times 17 + 9 \times 18 = \text{Rs. } 711.$$

Example 4.12. Determine the optimal production schedule?

Solution:

	1	2	3	D	Av
1	12(50)	14	16	0	50
2	16(10)	12(40)	14	0	50
3	20	16(60)	12	0	60
4	24	26(20)	28(10)	0	30
5	32	28	24(30)	0(20)	50
Req.	60	120	40	20	

(1) ΣAv. $= 50 + 50 + 60 + 30 + 50 = 240$

Σ Req $= 60 + 120 + 40 = 220$

$\Rightarrow \Sigma Av \neq \Sigma Req$

therefore one dummy colum with Req = (240 – 220) = 20 is added

(2) Basic feasible solution by North-West method:

$$Z = 12 \times 50 + 16 \times 10 + 12 \times 40 + 16 \times 60 + 26 \times 20 + 28 \times 10 + 24 \times 30 + 20 \times 0$$

$$= 600 + 160 + 480 + 960 + 520 + 280 + 720 + 0$$

$$\boxed{Z = 3720 \text{ units}}$$

(2) VAM

	1	2	3	D	Av	P_1	P_2	P_3	P_4	P_5	P_6
1	12(50)	14	16	0	50	12	2	2	2		
2	16	12(50)	14	0	50	12	2	4 ←			
3	20	16(20)	12(40)	0	~~60~~ 20	12	4 ←	4	4	4	4
4	24(10)	26	28	0(20)	~~30~~ 10	24 ←	2	2	2	2	2
5	32	28(50)	24	0	50	24	4	4	4	4	4
Req	~~60~~ 10	~~120~~ ~~70~~ 50	40	20							
P_1	4	2	2	0							
P_2	4	2	2								
P_3	4	2									
P_4	8↑	2									
P_5	4	10									
P_6	8↑	2↑									

$$z = 20 \times 0 + 12 \times 40 + 12 \times 50 + 12 \times 50 + 16 \times 20 + 24 \times 10 + 28 \times 50$$

$$= 0 + 480 + 600 + 600 + 320 + 240 + 1400$$

$$\boxed{z = 3640 \text{ units}}$$

(4) Stepping stone method: Optimality test

	1	2	3	D	Av
1	12(50)	14	16	0	50
2	16	12(50)	14	0	50
3	20	16(20)	12(40)	0	60
4	24(10)	26	28	0(20)	30
5	32	28(50)	24	0	50
Req.	60	120	40	20	

m = no. of rows = 4

n = no. of columns = 5

$m + n - 1 = 4 + 5 - 1 = 8$

no. of basic variables = 7 < 8 ($m + n - 1$ value)

Here problem is of **degenerate** transportation type; to solve this problem 'd' is added in all a_i's (Av) and 'md' is added to bn. Then perturbed problem is solved by earlier methods as a fresh problem from the begining. After the problem is solved, we put $d = 0$ to get optimal solution.

Example 4.13. Determine the optimal solution.

Solution: (1) $\Sigma ai = \Sigma b_j$; $20 + 30 + 15 + 13 = 40 + 6 + 8 + 18 + 6 \Rightarrow 78 = 78$
hence prob. is balanced.

	E	F	G	H	I	Av (a)
A	3(20)	4	6	8	9	20
B	2(20)	10(6)	1(4)	5	8	30
C	7	11	20(4)	40(11)	3	15
D	2	1	9	14(7)	16(6)	13
Demand (b)	40	6	8	18	6	78/78

(2) Basic feasible solution by **North-West** method

$$
\begin{aligned}
Z &= 3 \times 20 + 2 \times 20 + 10 \times 6 + 1 \times 4 + 20 \times 4 + 40 \times 11 + 14 \times 7 + 16 \times 6 \\
&= 60 + 40 + 60 + 4 + 80 + 440 + 98 + 96 = 878
\end{aligned}
$$

$$\boxed{Z = 878 \text{ units}}$$

(3) VAM

From table

$$
\begin{aligned}
Z &= 3 \times 20 + 2 \times 4 + 1 \times 8 + 5 \times 18 + 7 \times 9 + 3 \times 6 + 2 \times 7 + 1 \times 6 \\
&= 60 + 8 + 8 + 90 + 63 + 18 + 14 + 6 \\
Z &= 267 \text{ units}
\end{aligned}
$$

$$\boxed{Z = 267 \text{ units}}$$

	E	F	G	H	I	Av	P_1	P_2	P_3	P_4	P_5	P_6
A	3(20)	4	6	8	9	20	1	1	1	1	5	5 ←
B	2(4)	10	1(8)	5(18)	8	~~30~~ 22	1	3	3	3	3	3
C	7(9)	11	20	40	3(6)	~~15~~ 9	4	4	4 ←			
D	2(7)	1(6)	9	14	16	~~13~~ 7	1	1	1	1	12 ←	
Req	~~40~~ ~~31~~ 24	6	8	18	6							
P_1	0	3	5 ↑	3	5							
P_2	0	3		3	5 ↑							
P_3	0	3		3								
P_4	0	3 ↑		3								
P_5	0			3								
P_6	1			3								

(4) Stepping store method – Optimality test

	E	F	G	H	I	Av
A	3(20)	4	6	8	9	20
B	2(4)	10	1(8)	5(18)	8	30
C	7(9)	11	20	40	3(6)	15
D	2(7)	1(6)	9	14	16	13
Demand (b)	40	6	8	18	6	

$$d_{12} = 4 - 3 + 2 - 1 = 2$$
$$d_{13} = 6 + (-3) + 2 - 1 = 4$$
$$d_{14} = 8 - 3 + 2 - 5 = 2$$
$$d_{15} = 9 - 3 + 7 - 3 = 10$$
$$d_{21} = 10 - 2 + 2 - 1 = 9$$
$$d_{25} = 8 - 2 + 7 - 3 = 10$$
$$d_{32} = 11 - 7 + 2 - 1 = 5$$
$$d_{33} = 20 - 7 + 2 - 1 = 14$$
$$d_{34} = 40 - 7 + 2 - 5 = 30$$
$$d_{43} = 9 - 2 + 2 - 1 = 8$$
$$d_{44} = 14 - 2 + 2 - 5 = 9$$
$$d_{45} = 16 - 2 + 7 - 3 = 18$$

Since all empty cells are +ve, hence optimal

soln. is **Z = 267 units**

Example 4.14. (1) $\Sigma a_i = \Sigma b_j$ $7 + 9 + 18 = 5 + 8 + 7 + 14 \Rightarrow 34 = 34$

hence prob. is balanced.

Solution:

	E	F	G	H	Av
A	10(5)	30(2)	50	10	7
B	70	30(6)	40(3)	60	9
C	40	8	70(4)	20(14)	18
Demand (b)	5	8	7	14	34/34

(2) Basic feasible solution by **North-West method**:

$$Z = 10 \times 5 + 30 \times 2 + 30 \times 6 + 40 \times 3 + 70 \times 4 + 20 \times 14$$
$$= 50 \times 60 + 180 + 120 + 280 + 280$$

Z = 970 units

(3) VAM

	E	F	G	H	Av	P_1	P_2	P_3	P_4
A	10(5)	30	50	10(29)	~~7~~ 2	0	20	40	40
B	70	30	40(7)	60(2)	9	10	10	20	20
C	40	8(8)	70	20(10)	~~18~~ 10	12	12	50 ←	
Req	5	8	7	~~14~~ ~~4~~ 2					
P_1	30 ↑	22	10	10					
P_2		22 ↑	10	10					
P_3			10	10					
P_4			10	50 ↑					

$$Z = 10 \times 5 + 10 \times 2 + 40 \times 7 + 60 \times 2 + 8 \times 8 + 20 \times 10$$
$$= 50 + 20 + 280 + 120 + 64 + 200 = 734$$

$$\boxed{Z = 734 \text{ units}}$$

(4) Stepping stone method: Optimality test

Evaluation of empty cells

$$d_{12} = 30 - 10 + 20 - 8 = 32$$
$$d_{13} = 50 - 40 + 60 - 10 = 60$$
$$d_{21} = 70 - 10 + 10 - 60 = 10$$
$$d_{22} = 30 - 8 + 20 - 20 - 60 = -18$$
$$d_{31} = 40 - 10 + 10 - 20 = 20$$
$$d_{33} = 70 - 20 + 60 - 40 = 70$$

	E	F	G	H	Av
A	10(5)	30	50	10(2)	7
B	70	(+) 30	40(7)	(−) 60(2)	9
C	40	(−) 8(8)	70	(+) 20(10)	18
Demand (b)	5	8	7	14	34/34

hence there is only one cell – ve, which needs reallocation.

Previous cost for closed path $= 30 \times 0 + 60 \times 2 + 20 \times 10 + 8 \times 8 = 0 + 120 + 200 + 64 = 384$

New cost after reallocation $= 30 \times 2 + 20 \times 12 + 8 \times 6$
$= 60 + 240 + 68 = 348$

Cost reduction $= 384 - 348 = 36$

$Z = 734 - 36 = 698$

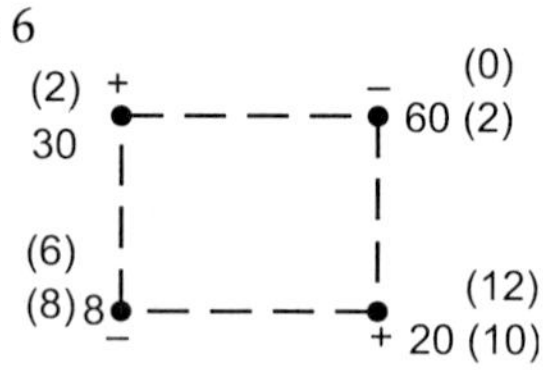

∴ Optimal solution is $\boxed{z = 698 \text{ units}}$

	W_1	W_2	W_3	W_4	Av
F_1	10(5)	30	50	10(2)	7
F_2	70	30	40(7)	60(2)	9
F_3	40	8(8)	70	20(10)	18
Req	5	8	7	14	

Example 4.15. Find optimal solution.

	S_1	S_2	S_3	S_4	S_5	S_6	S_7	D	Av
F_1	5(1000)	6(2000)	4(2500)	3(1500)	7	5	4	0	7000
F_2	9	4	3	4(3000)	3(1000)	2	1	0	4000
F_3	8	4	2	5	8(3000)	8(3500)	3(3000)	0(500)	10000
Req	1000	2000	2500	4500	4000	3500	3000	500	20500 +500 / 21000

Solution: (1) $\Sigma Av = 7000 + 4000 + 10{,}000 = 21{,}000$

Σ Demand $= 1000 + 2000 + 2500 + 4500 + 4000 + 3500 + 3000 = 20{,}500$

$\Sigma Av \neq \Sigma$ Demand $\quad \therefore$ problem is not balanced

To balance it, we add one Dummy column with demand = 21,000 – 20,500 = 500

(2) Basic feasible solution by North-West Method

$$Z = 5 \times 1000 + 6 \times 2000 + 4 \times 2500 + 3 \times 1500 + 4 \times 3000 + 3 \times 1000 + 4 \times 3000 + 8 \times 3500 + 3 \times 3000 + 0 \times 500$$

$$= 5000 + 12000 + 10{,}000 + 4500 + 12{,}000 + 3000 + 12000 + 28000 + 9000$$

$$\boxed{Z = 95{,}500 \text{ units}}$$

(3) VAM

	S_1	S_2	S_3	S_4	S_5	S_6	S_7	D	Av	P_1	P_2	P_3	P_4	P_5	P_6	P_7	P_8
F_1	5(1000)	6	4	3(4500)	7	5	4(1000)	0(500)	~~6500~~ 5500 ~~7000~~	3 ←	1	1	1	1	1	1	1
F_2	9	4	3	4	3	2(3500)	1(500)	0	500 ~~4000~~	1	1	1	2 ←				
F_3	8	4(2000)	2(2500)	5	8(4000)	8	3(1500)	0	1500 ~~4000~~ ~~6000~~ ~~10000~~	2	1	1	1	1	1	1	2
Req	1000	2000	2500	4500	4000	3500	~~3000~~ ~~2500~~ 1000	500									
P_1	3	0	1	1	1	3	2	0									
P_2	3 ↑	0	1	1	1	3	2										
P_3		0	1	1	1	3 ↑	2										
P_4		0	1	1	1		2										
P_5		2	2	2	3 ↑		1										
P_6		2 ↑	2	2			1										
P_7			2 ↑	2			1										
P_8				2			1										

$$z = 5 \times 1000 + 4 \times 2000 + 2 \times 2500 + 3 \times 4500 + 4 \times 4000 + 2 \times 3500 + 4 \times 1000 + 1 \times 500 + 3 \times 1500 + 0 \times 500 = 5000 + 8000 + 10{,}000 + 13500 + 16000 + 7000 + 4000 + 500 + 4500 + 0$$

$$\boxed{z = 68500}$$

(4) Stepping stone method : Optimality test

	S_1	S_2	S_3	S_4	S_5	S_6	S_7	Av	Dummy
F_1	5(1000)	6	4	3(4500)	7	5	4(4000)	7000	0
F_2	9	4	3	4	3	2(3500)	1(500)	4000	0
F_3	8	4(2000)	2(2500)	5	4(4000)	8	3(1500)	10000	0
Req	1000	2000	2500	4500	4000	3500	3000		500

Evaluation of empty cells

$d_{12} = 6 - 4 + 3 - 4 = 1,$	$d_{13} = 4 - 2 + 3 - 4 = 1$
$d_{15} = 7 - 4 + 3 - 4 = 2,$	$d_{16} = 5 - 2 + 1 - 4 = 0$
$d_{21} = 9 - 5 + 4 - 1 = 7,$	$d_{22} = 4 - 4 + 3 - 1 = 2$
$d_{23} = 3 - 2 + 3 - 1 = 3,$	$d_{24} = 4 - 3 + 4 - 1 = 4$
$d_{25} = 3 - 4 + 3 - 1 = 2,$	$d_{28} = 0 - 0 + 4 - 1 = 3$
$d_{31} = 8 - 5 + 4 - 3 = 4,$	$d_{34} = 5 - 3 + 4 - 3 = 3$
$d_{36} = 8 - 3 + 1 - 2 = 4,$	$d_{38} = 0 - 0 + 4 - 3 = 1$

Since all cells are + ve, hence result obtained from VAM is optimal

$\therefore \quad \boxed{z = 68{,}500}$

Example 4.16.

	E	F	G	H	I	Av
A	8(70)	10(30)	12	17	15	100
B	15	13(140)	18(10)	11	9	150
C	14	20	6(40)	10(120)	13	160
D	13	19	7	6(90)	12(190)	280
Req.	70	170	50	210	190	

Solution: (1) $\Sigma \text{Av} = 100 + 150 + 160 + 280 = 690$

$\Sigma \text{Req} = 70 + 170 + 50 + 210 + 190 = 690$

$\therefore \Sigma \text{Av} = \Sigma \text{ Req}$

hence problem is balanced.

(2) Basic feasible solution by North-West method:

$$z = 8 \times 70 + 10 \times 30 + 13 \times 140 + 18 \times 10 + 6 \times 40 + 10 \times 120 + 6 \times 90 + 12 \times 190$$

$$= 560 + 300 + 1820 + 180 + 240 + 1200 + 540 + 2280$$

$$\boxed{z = 7120}$$

(3) VAM : From table

$$Z = 8 \times 70 + 10 \times 30 + 13 \times 140 + 10 \times 9 + 6 \times 50 + 13 \times 110 + 6 \times 210 + 12 \times 70$$

$$= 560 + 300 + 1820 + 90 + 300 + 1430 + 1260 + 840$$

$$\boxed{z = 6600}$$

	E	F	G	H	J	Av	P_1	P_2	P_3	P_4	P_5	P_6
A	8(70)	10(30)	12	17	15	~~100~~ 30	2	2	2	5	5	5
B	15	13(140)	18	11	9(10)	~~150~~ 140	2	2	4	4	4	4
C	14	20	6(50)	10	13(110)	~~160~~ 110	4	4	7 ←	7 ←		
D	13	19	7	6(210)	12(70)	~~280~~ 70	1	1	5	7	7 ←	
Req	70	170	50	210	~~190~~ ~~80~~ 10							
P_1	5 ↑	3	1	4	3							
P_2		3	1	4 ↑	3							
P_3		3	1		3							
P_4		3			3							
P_5		3			3							
P_6		3			6 ↑							
P_7		3										

(4) Stepping stone method : Optimality test

	E	F	G	H	J	Av
A	− 8(70)	+ 10(30)	12	17	+ 15	100
B	15	− 13(140)	18	11	9(10)	150
C	+ 14	20	6(40)	10	− 13(110)	160
D	13	19	7	6(210)	12(70)	280
Req.	70	170	50	210	190	

Evaluation of empty cells:

$d_{13} = 12 - 6 + 13 - 9 + 13 - 10 = 13,$ $\quad d_{14} = 17 - 6 + 12 - 9 + 13 - 10 = 17$

$d_{15} = 15 - 9 + 13 - 10 = 9,$ $\quad d_{21} = 15 - 13 + 10 - 8 = 4$

$d_{23} = 18 - 6 + 13 - 9 = 16,$ $\quad d_{24} = 11 - 6 + 12 - 9 = 8$

$d_{31} = 14 - 8 + 10 - 13 + 9 - 13 = -1$ $\quad d_{32} = 20 - 13 + 9 - 13 = 3$

$d_{42} = 19 - 13 + 9 - 12 = 3,$ $\quad d_{43} = 7 - 6 + 13 - 12 = 2$

$d_{34} = 10 - 6 + 12 - 13 = 3,$ $\quad d_{41} = 13 - 8 + 10 - 13 + 9 - 12 = -1$

d_{31} and d_{41} can be reallocated to make solution optimal.

New route cost for cell $d_{31} = 14 \times 10 + 8 \times 60 + 10 \times 40 + 13 \times 130 + 9 \times 20 + 13 \times 100$

$= 140 + 480 + 400 + 1690 + 180 + 1300 =$ **4190**

Previous route cost for cell $d_{31} = (8 \times 70) + (10 \times 30) + (13 \times 140) + (9 \times 10) + (13 \times 110)$

$= 560 + 300 + 1820 + 90 + 1430$

$= 4200$

hence cost is reduced by 10 units from new allocation

* Previous cost = 3610, new cost = 3600 | For Cell d_{41}

$\therefore$ reduction in cost = 3610 − 3600 = 10

$\therefore$ We reallocate in d_{31} and make new table and calculate empty cells.

	E	F	G	H	J	Av
A	8(60)	10(40)	12	17	15	100
B	15	13(130)	18	11	9(20)	150
C	14(10)	20	6(50)	10	13(100)	160
D	13	19	7	6(210)	12(70)	280
Req.	70	170	50	210	190	

$d_{13} = 12 - 6 + 14 - 8 = 12,$

$d_{14} = 17 - 6 + 12 - 9 + 13 - 10 = 17$

$d_{15} = 15 - 9 + 13 - 10 = 9,$

$d_{21} = 15 - 13 + 10 - 8 = 4$

$d_{23} = 18 - 13 + 10 - 8 + 14 - 6 = 15,$

$d_{24} = 11 - 9 + 12 - 6 = 8$

$d_{32} = 20 - 10 + 8 - 14 = 4,$

$d_{34} = 10 - 13 + 12 - 6 = 3$

$d_{41} = 13 - 12 + 13 - 14 = 0,$

$d_{42} = 19 - 12 + 9 - 13 = 3$, d43 = 7– 12 + 13 – 6 = 2

all cells are + ve $\quad \therefore$ Optimal solution is

$$Z = 6600 - 10 = 6590 \text{ units}$$

Example 4.17. The ABC tool company has a sales force of 25 men who work out of three regional offices. The company produces four basic lines of hand tools. Mr. Jain, the sales manger feels that 6 salesman are needed to distribute product line 1, 10 salesmen to distribute product line 2, 4 salesmen to distribute product line 3 and 5 salesmen to distribute product line 4. The cost (in Rs) per day of assigning salesmen for each of the offices for selling each of the product lines are as follows:

		Product lines			
		1	2	3	4
	A	Rs 20	Rs 21	Rs 16	Rs 18
Regional	B	17	28	14	16
office	C	29	23	19	20

At the present time, 10 salesmen are allocated to office A, 9 salesmen to office *B* and 7 salesmen to office C. How many salesmen should be assigned from each office to selling each product line in order to minimize cost? Identify alternate optimal solutions, if any.

Solution : 1. Find initial solution using VAM

	1	2	3	4	d	Avail-ablity	RP1	RP2	RP3	RP4	RP5	RP6
A	20	(4) 21	(1) 16	(5) 18	0	10–5 –1	16	2	2	2	3	
B	(6) 17	28	(3) 14	16	0	9–6	14	2	2			
C	29	(6) 23	19	20	(1) 0	7–1	19	1	1	1	3	
Demand	6	10	4–3	5	1	26						
CP1	3	2	2	2	0							
CP2	3	2	2	2								
CP3		2	2	2								
CP4		2	3	2								
CP5		2		2								
CP6		2										

2. To test the optimality of initial solution, trace closed loop for all empty cells and calculate opportunity cost for each.

$$d_{A1} = 20 - 16 + 14 - 17 = 1$$
$$d_{Ad} = 0 - 0 + 23 - 21 = 2$$
$$d_{B2} = 28 - 14 + 16 - 21 = 9$$
$$d_{B4} = 16 - 18 + 16 - 14 = 0$$
$$d_{Bd} = 0 - 0 + 23 - 21 + 16 - 14 = 4$$
$$d_{C1} = 29 - 17 + 14 - 16 + 21 - 23 = 8$$
$$d_{C3} = 19 - 16 + 21 - 23 = 1$$
$$d_{C4} = 20 - 18 + 21 - 23 = 0$$

all + ve → Optimal solution

$$Z = 4 \times 21 + 16 + 18 \times 5 + 6 \times 17 + 3 \times 14 + 6 \times 23 = 472$$

Alternate optimal solution exists.

Example 4.18. Transportation Problem

	W_1	W_2	W_3	W_4	Av
F_1	10(5)	30(2)	50	10	7
F_2	70	30(6)	40(3)	60	9
F_3	40	8	70(4)	20(14)	18
Req	5	8	7	14	

Solution : 1. $\Sigma Av = 7 + 9 + 18 = 34$

$$\Sigma \text{Req} = 5 + 8 + 7 + 14 = 34$$
$$\Sigma Av = \Sigma \text{Req}$$

∴ Problem is balanced.

2. Basic feasible solution by N.W method.

$$Z = 10 \times 5 + 30 \times 2 + 30 \times 6 + 40 \times 3 + 70 \times 4 + 20 \times 14$$
$$= 50 + 60 + 180 + 120 + 280 + 280$$

$$\boxed{Z = 970 \text{ units}}$$

3. Solution by VAM from table

$$Z = 10 \times 5 + 8 \times 8 + 40 \times 7 + 10 \times 2 + 60 \times 2 + 20 \times 10$$
$$= 50 + 64 + 280 + 20 + 120 + 200$$

$$\boxed{Z = 734 \text{ units}}$$

VAM

	W_1	W_2	W_3	W_4	Av	P_1	P_2	P_3	P_4	P_5
F_1	10(5)	30	50	10(2)	7	0	20	40	40	
F_2	70	30	40(7)	60(2)	9	10	10	20	20	
F_3	40	8(8)	70	20(10)	~~18~~ 10	12	12	50		
Req	5	8	7	~~14~~ ~~4~~ 2						
P_1	30	22	10	10						
P_2	↑	22	10	10						
P_3		↑	10	10						
P_4			10	50						
P_5				↑						

Final table

	W_1	W_2	W_3	W_4	Av
F_1	10(5)	30	50	10(2)	7
F_2	70	30	40(7)	60(2)	9
F_3	40	8(8)	70	20(10)	18
Req	5	8	7	14	

4. Check for optimal solution. Evaluate empty cells.

$$d_{12} = 30 - 8 + 20 - 10 = 32$$
$$d_{13} = 50 - 40 + 60 - 10 = 60$$
$$d_{21} = 70 - 10 + 10 - 60 = 10$$
$$d_{22} = 30 - 8 + 20 - 60 = -18$$
$$d_{31} = 40 - 10 + 10 - 20 = 20$$
$$d_{33} = 70 - 20 + 60 - 40 = 70$$

Only one empty cell is – ve which needs reallocation

Previous cost $= 60 \times 2 + 20 \times 10 + 8 \times 8$
$= 120 + 200 + 64 = 384$

New cost (after reallocation) $= 30 \times 2 + 60 \times 0 + 20 \times 12 + 8 \times 6$
$= 60 + 0 + 240 + 48 = 348$

Cost reduction = 384 – 348 = 36 units

5. Final table

	W_1	W_2	W_3	W_4	Av
F_1	10(5)	30	50	10(2)	7
F_2	70	30(2)	40(7)	60	9
F_3	40	8(6)	70	20(12)	18
Req	5	8	7	14	

again evaluate empty cells

$$d_{12} = 30 - 8 + 20 - 10 = 32$$
$$d_{13} = 50 - 40 + 30 - 8 + 20 - 10 = 42$$
$$d_{21} = 70 - 10 + 10 - 20 + 8 - 30 = 28$$
$$d_{24} = 60 - 20 + 8 - 30 = 18$$
$$d_{31} = 40 - 20 + 10 - 10 = 20$$
$$d_{33} = 70 - 40 + 30 - 8 = 52$$

All cells are positive

∴ Solution is optimal

$$Z = 10 \times 5 + 30 \times 2 + 8 \times 6 + 40 \times 7 + 10 \times 2 + 20 \times 12$$
$$= 50 + 60 + 48 + 280 + 20 + 240 = 698$$
$$Z = 698 \quad \text{Units}$$

Cost reduction = 734 – 698 = 36 Units

Example 4.19. Determine the optimal solution of the transportation matrix and check the optimality. The matrix is cost matrix. **(UPTU. 2007-08)**

Requirement	Availability		30	30	30
	Market↓	Godown	*a*	*b*	*c*
28	x		7	9	3
50	*y*		2	5	7

Solution: Adding a dummy requirement to balance the problem : Finding IBFS using VAM.

		Market			Supply	P_1	P_2
		X	Y	D (Dummy)			
Godown	a	7	(30) 2	0	30	2	5 ←
	b	9	(18) 5	0 (12)	30	5 ←	4
	c	(28) 3	(2) 7	0	30	3	4
Demand		28	50	12			
	P_1	4	3	0			
	P_2	4	3	0			

Testing the optimality

	X	Y	D	Supply	
a	7	(30) 2	0	30	$u_1 = 0$
b	9	(18) + 5	– 0 (12)	30	$u_2 = 3$
c	(28) 3	(2) – 7	0 +	30	$u_3 = 5$
Demand	28	50	12		
	$v_1 = -2$	$v_2 = 2$	$v_3 = -3$		

$$\Delta_{ax} = 7 - (0 - 2) = 9$$
$$\Delta_{bx} = 9 - (3 - 2) = 8$$
$$\Delta_{aD} = 7 - (0 - 3) = 3$$
$$\Delta_{cD} = 0 - (5 - 3) = \mathbf{-2}$$

Since net cost change Δ_{cD} is negative, further improvement is possible; tracing the closed loop for cell C-D. Minimum quantity for negative cell (2) is added to +ve cells and substracted from the –ve cells.

	X	Y	D	Supply	
a	7	(30) 2	0	30	$u_1 = 0$
b	9	(20) 5	(10) 0	30	$u_2 = 3$
c	(28) 3	7	(2) 0	30	$u_3 = 3$
Demand	28	50	12		
	$v_1 = 0$	$v_2 = 2$	$v_3 = -3$		

Calculating net cost change

$$\Delta_{ax} = 7 - (0 + 0) = 7$$
$$\Delta_{bx} = 9 - (3 + 0) = 6$$
$$\Delta_{cY} = 7 - (3 + 2) = 2$$
$$\Delta_{aD} = 0 - (0 - 3) = 3$$

All net cost change values are positive therefore, current solution is optimal. Optimum cost of transportation is

$$= (28 \times 3) + (30 \times 2) + (20 \times 5) + (10 \times 0) + (2 \times 0) = \mathbf{244}$$

EXERCISE

4.1. What do you understand by transportation problem?

4.2 Define following terms:

(*i*) Feasible solution

(*ii*) Basic solution

(*iii*) Optimal solution

4.3. Explain degeneracy in transportation problem and method to resolve it.

4.4. Explain various methods of finding initial feasible solution of a transportation problem.

4.5. What is meant by an optimality test in a transportation problem?

4.6. What is unbalanced transportation problem? Explain how to convert the unbalanced transportation problem into a balanced problem.

4.7. Write the mathematical model of a transportation problem.

4.8. Explain the algorithm to solve a transportation problem.

4.9. Explain transshipment problem.

4.10. Explain how a transshipment problem can be solved as a transportation problem.

4.11. Explain how the profit maximization transportation problem can be converted to an equivalent cost minimization problem.

4.12. Explain in brief the procedure to solve a transportation problem.

4.13. Discuss methods to obtain the optimum solution of a transportation problem after finding initial feasible solution.

4.14. Show that a transportation problem can be considered as a Linear-programming problem.

4.15. Write practical applications of transportation problem.

4.16. Explain MODI method as applied to transportation problem giving example.

4.17. When does a transportation problem have a unique solution?

4.18. Determine an initial basic feasible solution to the following transportation problem using northwest-cell method.

		To					
		1	2	3	4	5	Available
	1	3	4	6	8	9	20
	2	2	10	1	5	8	30
From	3	7	11	20	40	3	15
	4	2	1	9	14	16	13
Demand		40	6	8	18	6	

[**Ans.** 878]

4.19. A company has three factories, say F_1, F_2 and F_3 that are feeding 5 zones – North, South, Eastern, Western and Central whose monthly demands are 20, 30, 25, 25 and 20 thousands units respectively. The cost of transporting one unit from each factory to each destination (*i.e.*, zone) is given below :

If the factories capacities are 40, 30 and 50 thousands units respectively, then

(*i*) Make a transportation model and find optimal solution

(*ii*) Make a LP model. [**Ans.** 745]

	Zones				
Factory	*North*	*South*	*Eastern*	*Western*	*Central*
F_1	5	4	7	10	12
F_2	13	10	7	5	6
F_3	12	9	8	7	10

4.20. Determine an initial basic feasible solution to the following transportation problem using :

(*a*) Northwest-corner cell method and

(*b*) Vogel's approximation method.

		A_1	B_1	C_1	D_1	E_1	Supply
		Destination					
	A	2	11	10	3	7	4
Origin	B	1	4	7	2	1	8
	C	3	9	4	8	12	9
Demand		3	3	4	5	6	

[**Ans.** *a* – 153, *b* – 68]

4.21. Find basic feasible (initial) solutions of the following transportation problem by using (*a*) North-west corner, (*b*) Least cost, and (*c*) Vogel's approximation methods:

(*i*)

	D_1	D_2	D_3	
O_1	30	50	45	40
O_2	40	45	38	30
O_3	37	35	39	30
	40	20	40	

(*ii*)

	R_1	R_2	R_3	R_4	
G_1	20	22	24	22	50
G_2	17	19	18	19	50
G_3	16	15	19	20	Any amount
	70	20	20	30	

4.22. Find the initial basic feasible solution to the following transportation problem by
(*a*) Northwest-corner cell method and
(*b*) Least cost cell method.

		To 1	2	3	Supply
	1	2	7	4	5
	2	3	3	1	8
From	3	5	4	7	7
	4	1	6	2	14
Demand		2	9	18	

[**Ans.** (*a*) 112, (*b*) 70]

State which of the methods is better.

4.23. A firm manufacturing a single product has three plants at locations *X*,*Y* and Z. The three plants have produced 60, 35 and 40 units respectively during this week. The firm has made commitments to sell 22, 45, 20, 18 and 30 units of the product to customers *A*, *B*,*C*, *D*, and *E* respectively. The net unit cost of transporting from the three plants to the five customers is given in the table below:

		Customer				
		A	B	C	D	E
	X	4	1	3	4	4
Plant	Y	2	3	2	2	3
location	Z	3	5	2	4	4

Use Vogel's approximation method to determine the cost of shifting the product from plant locations to the customers. Does your solution provide a least cost transportation schedule ? [**Ans.** 290]

4.24. A company has factories at four different places (1,2,3 and 4) which supply items to warehouses *A, B,C,D* and *E*. Monthly factory capacities are 200,175, 150 and 325 respectively. Monthly warehouse requirements are 110, 90, 120, 230 and 160 respectively. Unit shipping costs (in rupees) are given in the following table:

		To				
		A	*B*	*C*	*D*	*E*
	1	13	–	31	8	20
From	2	14	9	17	26	10
	3	25	11	12	17	15
	4	10	21	13	–	17

Shipments from 1 to *B* and from 4 to *D* are not possible. Determine the optimum distribution plan to minimize the shipping cost. [**Ans.** 7285]

4.25. The personnel department of a construction company recruits three categories of workers *A,B* and *C* of 100, 70 and 85 numbers respectively from centres *P, Q,R* and *S*. Their hourly rates (in Rs.) are given by the matrix below. Find the recruitment pattern at the lowest cost.

		Centre				
		P	*Q*	*R*	*S*	*Requirement*
	A	12	13	15	14	100
Category	*B*	15	14	13	12	70
	C	10	12	14	13	85
availability		35	70	100	50	255

4.26. Consider the following transshipment problem involving 4 sources and 2 destinations. The supply values of the sources S_1, S_2, S_3, and S_4 are 200 units, 250 units, 200 units and 450 units, respectively. The demand values of the destinations D_1 and D_2 are 550 units and 550 units, respectively. The transportation cost per unit between different sources and destinations are summarized in the following table. Solve the transshipment problem.

[**Ans.** 7950]

4.27. The *ABC* Tool Company has a sales force of 25 men who work out of three regional offices. The company produces four basic lines of hand tools. The sales manager feels that 6 salesmen are needed to distribute product line 1, 10 salesmen to distribute product line 2, 4 salesmen for product line 3 and 5 salesman for product line 4. The cost (in Rs.) per day of assigning salesman from each of the offices for selling each of the product lines are as follows :

		Product Lines			
		1	2	3	4
	A	Rs. 20	Rs. 21	Rs. 16	Rs. 18
Regional office	*B*	17	28	14	16
	C	29	23	19	20

At the present time, 10 salesmen are allocated to office A, 9 salesmen to office B, and 7 salesmen to office C. How many salesman should be assigned from each office to selling each product line in order to minimize cost ? Identify alternate optimal solutions, if any. [**Ans.** 472]

OBJECTIVE TYPE QUESTIONS

4.1. Which of the following conditions should be satisfied for the application of optimality test on an initial solution of transportation model?
(*a*) number of allocations should be less than $m + n - 1$
(*b*) number of allocations should be equal to $m + n - 1$
(*c*) number of allocations should be $m + n$
(*d*) number of allocations should be more than $m + n$

4.2. When there are m rows and n columns in a transportation problem, degeneracy is said to occur when the number of allocation is
(*a*) less than $m + n - 1$
(*b*) greater than $m + n - 1$
(*c*) equal to $m + n - 1$
(*d*) less than $m - n - 1$

4.3. In a 6 x 6 transportation problem, degeneracy would arise, if number of filled spots were
(*a*) equal to thirty six
(*b*) more than twelve
(*c*) equal to twelve
(*d*) less than eleven

4.4. The north-west corner rule is used to find an
(*a*) initial feasible solution
(*b*) optimal solution
(*c*) opportunity cost
(*d*) none of the above

4.5. In Vogel's approximation method, the penalty associated with the row is calculated by
(*a*) the difference between the smallest cost and the next smallest cost in that row
(*b*) the difference between the smallest unused cost and the next smallest unused cost in that row
(*c*) the difference between the smallest cost and the next smallest unused cost in that row
(*d*) none of the above

4.6. An unbalanced transportation problem is one in which
(*a*) the total supply is same as total requirement
(*b*) the number of jobs are not equal to number of facilities
(*c*) the total supply is not equal to total requirement
(*d*) none of the above

4.7. The maximum number of items that can be allocated to an unused route with the stepping - stone method is
(*a*) the maximum number in any cell
(*b*) the minimum number in any cell
(*c*) the minimum number in an increasing cell
(*d*) the minimum number in a decreasing cell on the closed path for that route

4.8. The dummy source or destination in a transportation problem is added to
   (*a*) to satisfy the rim conditions
   (*b*) prevent solution to become degenerate
   (*c*) ensure that total cost does not exceed a limit
   (*d*) none of the above

Answers

4.1. (*b*) 4.2. (*a*) 4.3. (*d*) 4.4. (*a*) 4.5 (*b*) 4.6 (*c*) 4.7 (*d*) 4.8. (*a*)

❐❐❐

Chapter 5

Assignment Problems

5.1. INTRODUCTION

As the name "assignment problem" implies that something is assigned to somebody and for maximum efficiency, (and thus maximum output or profit) it is required to assign that assignment to any single entity which can handle it best. The assignment allocation can be of any type *i.e.*, territories can be assigned to different salesman according to their capacity, loads to machines, subjects to teachers etc. Thus, one can define assignment problem as a special type of linear programming problem where the distributions or allocation is based on one to one basis *i.e.*, only single assignment be assigned to any person and a person can take only one job in his hand. In other words, it can be said that the number of demand and number of supply for each distribution is unity. Describing it in form of a cost function, it can be shown as,

Destinations

		D_1	D_2	------------	D_n	Supply (a)
	F_1	C_{11}	C_{12}	------------	C_{1n}	1
	F_2	C_{21}	C_{22}	------------	C_{2n}	1
Facilities	¦	¦	¦	------------	¦	¦
	F_m	C_{m1}	C_{m2}	------------	C_{mn}	1
	Demand (b)	1	1	------------	1	

An assignment problem can be formulated as a linear programming problem and can be solved by the simplex algorithm. In general,

let $\quad X_{ij} = 1$, if i^{th} person is assigned J^{th} job

$\quad\quad = 0$, if i^{th} person is not assigned J^{th} job

The mathematical formulation of the assignment problem is as given.

$$\text{Minimize } Z = \sum_{i=1}^{n} \sum_{j=1}^{m} C_{ij} X_{ij}$$

Subject to constraints

1. Each person should be assigned to one and only one job *i.e.*, $x_{i1} + x_{i2} + \ldots x_{in} = 1$ $(i = 1, 2, 3 \ldots n)$
2. Each job must be assigned to one and only one person, *i.e.*,
   $x_{1j} + x_{2j} + \ldots x_{nj} = 1, j = 1, 2, \ldots m$
3. $x_{ij} = 0$ or 1 (for all i, j values)

5.2. CLASSIFICATION OF THE SYSTEM OF ASSIGNMENT PROBLEMS

A. According to matrix type

1. Minimization type : The main aim is to minimise the cost or total time taken in completing the job.

2. Maximization type : The main aim is to maximize the profit, increase the efficiency, increase the rate of production etc.

B. According to the number of demand and supply

1. Balanced problem : Number of demand = Number of supply.
2. Unbalanced problem : Number of demand ≠ Number of supply.

C. Multiple Optimal Solution

D. Restricted assignment problem.

5.3. MINIMIZATION TYPE PROBLEMS

There are two methods of solving assignment problems.

5.3.1. Hungarian Method or Reduced Matrix Method or Flood's Technique

Steps

1. Prepare a suitable matrix having number of rows equal to number of columns *i.e.*, no. of demand is equal to number of supply.
2. Find the minimum number of cost matrix from all the elements in a row. Find it for all rows.
3. Prepare a new matrix by substracting the minimum number in a row from every element of that row. Do this for all rows.
4. Now find the minimum value of elements for each column and substract the corresponding minimum value from every element of the corresponding column. These steps are performed to reduce the matrix to simplex values.
5. Start with the first row. If there is any single zero in the row, mark it with some sign, traditionaly the zero is marked in a square (□) and cancel all other zero's of that column in which the zero is marked by putting (×) to each zero in the corresponding column.
6. Repeat step-5 for all rows. If the row contains more than one zero's, neglect that row and carry the job for next row.
7. Repeat the step-5 for all the columns of the matrix.
8. Count square marked zero's (*i.e.*, allocated cells). If the number of squared zero's is equal to the number of rows or columns and there is exactly one assignment in each row and in each column, it is the optimal solution. Allocate the demands to the respective allocated supplies.
   If number of zero (squared) ≠ Number of rows.

(i) Tick (✓) all rows that do not carry any marked zero.
(ii) Tick (✓) all columns that have zeros in the marked rows.
(iii) Tick (✓) all rows that have marked zero's in marked columns

9. Draw line through each unmarked row and each marked column.
10. Find out the minimum value of element out of undeleted (unmarked) cell values. Prepare a new matrix by copying the entries on lines as such, adding the minimum undeleted cell value to every number at the point of intersection of lines and substracting minimum undeleted cell value from all other undeleted cell values.
11. Repeat the procedure starting from step 5 till the number of marked zeros is equal to the number of rows.

Example 5.1. Solve the assignment problem for which the following cost matrix is given. Find its optimal solution.

	1	2	3	4
A	1	4	6	3
B	9	7	10	9
C	4	5	11	7
D	8	7	8	5

Solution : Step I : Check no. of rows = no. of columns. As it is 4 × 4 matrix, Finding the minimum row value for each row. For row 1, the elements are 1, 4, 6 and 3. Out of these, 1 is minimum. Substract the minimum value from each corresponding element.

	1	2	3	4	Min-Row Value
A	1	4	6	3	1
B	9	7	10	9	7
C	4	5	11	7	4
D	8	7	8	5	5

⟹

	1	2	3	4
A	1–1=0	4–1	6–1	3–1
B	9–7=0	7–7	10–7	9–7
C	4–4=0	5–4	11–4	7–4
D	8–5=0	7–5	8–5	5–5

Step 2 : Finding the minimum cell value for each column and substract the same from elements of each corresponding column.

	1	2	3	4
A	0	3	5	2
B	2	0	3	2
C	0	1	7	3
D	3	2	3	0
Minimum Column Value	0	0	3	0

⟹

	1	2	3	4
A	0	3	5–3	2
B	2	0	3–3	2
C	0	1	7–3	3
D	3	2	3–3	0
	0	0	3	0

⇓

	1	2	3	4
A	0	3	2	2
B	2	0	0	2
C	0	1	4	3
D	3	2	0	0

Step 3 : For row 1, there is only one zero. Mark it in square and cross the zero in the column (*i.e.*, zero in column 1, row 3). As there are two zero's in row 2, leave it. Only crossed zero is there, leave row 3. In row 4, again 2 zero's are there, therefore leave it also. Now for column 1, no free zero to mark, therefore leave it. For column 2, square the zero and put a cross on the zero in corresponding row (*i.e.*, row 2 and column 3). Similarly mark for other columns, then rows, again columns until no free zero's can be marked.

	1	2	3	4
A	[0]	3	2	2
B	2	[0]	~~0~~	2
C	~~0~~	1	4	3
D	3	2	[0]	~~0~~

⇒

	1 (②)	2	3	4	
A	[0]	3	2	2	✓
B	2	[0]	~~0~~	2	①
C	~~0~~	1	4	3	✓
D	3	2	[0]	~~0~~	③
	0 ✓	0	3	0	

Step 4 : As no. of allocations (3) ≠ No. of rows (4), tick row 3 as it has unmarked (only crossed) zero. Then tick (✓) column 1 and then row 1. Now no more unallocated (*i.e.*, square marked zero cell) cell are left. Therefore finding minimum undeleted value = 1.

Applying rule 10, we get the new matrix as

	1	2	3	4
A	0	3–1	2–1	2–1
B	2+1	0	0	2
C	0	1–1	4–1	3–1
D	3	2	0	0

⇒

	1	2	3	4
A	[0]	2	1	1
B	3	~~0~~	[0]	2
C	0	[0]	3	2
D	4	2	~~0~~	[0]

Step 5 : Repeat step 3. As number of rows = number of allocations *i.e.*, 4 = 4, this is the optimal solution. Adding allocated cells values from the original table we get Z_{min}.

$$\left.\begin{array}{l} A \to 1 \\ B \to 3 \\ C \to 2 \\ D \to 4 \end{array}\right\} Z_{min} = 1 + 10 + 5 + 5 = 21 \quad \textbf{Ans.}$$

Example 5.2. Four professors are each capable of teaching any one of the four different courses. Class preparation time in hours for different topics varies from professor to professor and is given in the table below. Each professor is assigned only one course so as to minimise the total course preparation time for all courses. Find the optimal solution.

Subject

Professor	M	N	P	Q
A	2	10	9	7
B	15	4	14	8
C	13	14	16	11
D	4	15	13	9

Solution :

	M	N	P	Q	Min - Value
A	2	10	9	7	2
B	15	4	14	8	4
C	13	14	16	11	11
D	4	15	13	9	4

$\Rightarrow$

	M	N	P	Q
A	0	8	7	5
B	11	0	10	4
C	2	3	5	0
D	0	11	9	5
Min - Value	0	0	5	0

New matrix,

	M	N	P	Q
A	[0]	8	2	5
B	11	[0]	5	4
C	2	3	[0]	~~0~~
D	~~0~~	11	4	5

After marking zero's and crossing zero's, we get the above matrix.

Number of allocations ≠ No. of rows

$$3 \neq 4$$

Therefore, it is not the optimal solution

As minimum undeleted cell value is 2, the new matrix will be formed as,

	M	N	P	Q	
A	[0]	8	2	5	③
B	11	[0]	5	4	
C	2	3	[0]	~~0~~	
D	~~0~~	11	4	5	①
	②				

$\Rightarrow$

	M	N	P	Q
A	~~0~~	6	[0]	3
B	13	[0]	5	4
C	4	3	~~0~~	[0]
D	[0]	9	2	3

Number of allocation = No. of rows

4 = 4. Therefore it is the optimal solution.

Professor		Courses	Min. Time
A	⟶	P	9
B	⟶	N	4
C	⟶	Q	11
D	⟶	M	4
		Z_{min}	28 hrs

Ans

5.3.2. Revised Hungarian Method

Steps 1 to 4 are same as previous method.

Step 5: Starting from row 1, put the zero in square if single and draw a wavy line through the whole column. It means this column values cannot be used further for allocation.

Step 6: To the same for columns cancelling rows here.

Step 7: Same as previous steps 10 and 11.

Example 5.3. Solve the Example 5.2 by new Hungarian method.

Solution :

	M	N	P	Q	Min - Value
A	2	10	9	7	2
B	15	4	14	8	4
C	13	14	16	11	11
D	4	15	13	9	4

$\Rightarrow$

	M	N	P	Q
A	0	8	7	5
B	11	0	10	4
C	2	3	5	0
D	0	11	9	5
Min - Value	0	0	5	0

No. of allocation ≠ No. of rows

$3 \neq 4$

Minimum undeleted value = 2

	M	N	P	Q
A	[0]	8	2	5
B	11	[0]	5	4
C	2	3	[0]	~~0~~
D	~~0~~	11	4	5

⇒

	M	N	P	Q
A	0	8	[0]	3
B	11	[0]	3	2
C	4	5	0	[0]
D	[0]	11	2	3

Number of allocations = Number of rows

4 = 4

Thus optimal solution is $A \to P$, $B \to N$, $C \to Q$, $D \to M$

$$Z_{min} = 9 + 4 + 11 + 4 = 28 \quad \textbf{Ans.}$$

5.4 MAXIMIZATION TYPE PROBLEMS

These problems are same as minimization type with the difference that here the aim is to maximize the objective function. The maximization problems are first reduced to minimization problems, then are solved by Hungarian method. They are reduced to minimization type by substracting all the elements of the matrix from the highest element in the matrix. But while calculating objective function, the element's values are taken from the original matrix.

Example 5.4. A department has four subordinates and 4 tasks to be performed. The subordinates differ in efficiency and tasks differ in their intrinsic difficulty. The estimates of the profit in rupees each man would earn is given in the effectiveness matrix. How should the tasks be allocated, one to each man, so as to maximize the total earnings?

Solution : It is a 4 × 4 matrix i.e. no of rows = no. of columns. Converting it to minimization type problem by substracting all elements from the maximum cell value of the matrix. Maximum (element) cell value 40. So substract each cell value from 40. It will give equivalent loss matrix.

Subordinate \ Tasks	5	6	7	8
1	5	40	20	5
2	25	35	30	25
3	15	25	20	10
4	15	5	30	15

	5	6	7	8	Min. Value
1	35	0	20	35	0
2	15	5	10	15	5
3	25	15	20	30	15
4	25	35	10	25	10

⇒

	5	6	7	8
1	35	0	20	35
2	10	0	5	10
3	10	5	5	10
4	15	0	0	15
Min. Value	10	0	0	10

New matrix is,

	5	6	7	8
1	25	[0]	20	25
2	0	0	5	[0]
3	[0]	5	5	5
4	0	25	[0]	5

No of allocations = No. of rows:

4 = 4

Therefore allocation is optimal.

Subordinate	Task Assigned	Earning
1	6	40
2	8	25
3	5	15
4	7	30
	Z_{max}	=Rs110

5.5. MULTIPLE OPTIMAL PROBLEMS

The assignment problem provides more than one solutions with the same cost or profit. It happens because in some cases there are more than one method to cross zero's which results in more than one solutions with the same value of objective function. Thus the any arrangement can be chosen out of the solution, according to the convenience.

Example 5.5. Find the optimal assignment for the given assignment problem with the following cost matrix.

	I	II	III	IV
A	5	3	1	8
B	7	9	2	6
C	6	4	5	7
D	5	7	7	6

Solution : 1. As it is a 4 × 4 matrix *i.e.*, number of columns = number of rows (4 = 4). It is a balanced matrix.

2. Apply Hungarian method

	I	II	III	IV	Min. Value
A	5	3	1	8	1
B	7	9	2	6	2
C	6	4	5	7	4
D	5	7	7	6	5

⇒

	I	II	III	IV
A	4	2	0	7
B	5	7	0	4
C	2	0	1	2
D	0	2	2	1
Min.Value	0	0	0	1

⇓

	I	II	III	IV
A	4	2	[0]	6
B	5	7	0	3
C	2	[0]	1	1
D	[0]	2	2	0

3. As the number of allocations (3) are not equal to the number of rows (4), thus solving further we get the following matrix.

Minimum undeleted cell value = 2, therefore,

	I	II	III	IV
A	3	2	[0]	5
B	4	7	0	2
C	1	[0]	1	0
D	[0]	3	3	0

↗

	I	II	III	IV
A	1	[0]	0	3
B	2	5	[0]	0
C	1	0	3	[0]
D	[0]	3	5	0

Solution

A ⟶ II

B ⟶ III

C ⟶ IV

D ⟶ I

$Z_{min} = 2+0+2+0 = 4$

↘

	I	II	III	IV
A	1	0	[0]	3
B	2	5	0	[0]
C	1	[0]	3	0
D	[0]	3	5	0

Solution

A ⟶ III

B ⟶ IV

C ⟶ II

D ⟶ I

Min Value = 0+4+0+0=4

4. As in the new matrix formed, there are two zero's in row, 1, 2, 3 and 4. Leaving them all, move to columns. In column 1, there is only one zero, mark it. Now column 2, 3 and 4 contains 2 zero's each. Similarly each row 1, 2 and 3 is also having 2 zero. So the condition of deadlock occurs.

Now choose any one zero in row 1 and allocate it by marking it with a square. Further solve as usual by Hungarian method. This will give one result as shown above. Now for the second solution, allocate the zero other than the previous one in row 1 and solve as usual. This will give a second solution.

Although the assignment pattern allocated is different for both solution, but their Z_{min} value remains the same. So any allocation can be used.

Example 5.6. A business executive must make the 4 round trips listed in table given, between the head office in Delhi and a branch office in Mumbai.

Departure date from Delhi	Return date to Delhi
Monday, June 3	Friday, June 7
Monday, June 10	Wednesday, June 12
Monday, June 17	Friday, June 21
Thursday, June 25	Friday, June 28

The price of round trip ticket from Delhi is Rs 400. A discount of 25% is granted if the date of the ticket span a weekend (Saturday and Sunday). If the stay in Mumbai lasts more than 21 days, the discount is increased to 30%. A one way ticket between Delhi and Mumbai (either direction) costs Rs 250. How should the executive purchase the tickets?

Solution: Use the code (city, date) to define rows and columns of matrix. (D, 3) (M, 7) means leaving Delhi on June 3 and returning from Mumbai on june 7. It will cost him Rs 400.

	M,7	M,12	M,21	M,28	Min. Value
D,3	400	300	300	280	280
D,10	300	400	300	300	300
D,17	300	300	400	300	300
D,25	300	300	300	400	300

4 × 4 matrix
(Balanced matrix)

New matrix is,

	M,7	M,12	M,21	M,28
D,3	120	20	20	[0]
D,10	0	100	0	0
D,17	0	0	100	0
D,25	0	0	0	100
Min value	0	0	0	0

As the minimum value for each column is zero, there will be no change in matrix.

Solution: 1 (D, 2) → (M, 28)
(D, 10) → (M, 7)
(D, 17) → (M, 12)
(D, 25) → (M, 21)

$$Z_{min} = 280 + 300 + 300 + 300 = \text{Rs } 1180$$

	M,7	M,12	M,21	M,28
D,3	120	20	20	[0]
D,10	0	100	0	0
D,17	0	0	100	0
D,25	0	0	0	100

(Dead look)

⇒

120	20	20	[0]
[0]	100	0	0
0	[0]	100	0
0	0	[0]	100

Solution 1

⇓

120	20	20	[0]
0	100	[0]	0
0	0	100	0
0	0	0	100

(Dead look)

⇗

120	20	20	[0]
0	100	[0]	0
[0]	0	100	0
0	[0]	0	100

⇘

120	20	20	[0]
0	100	[0]	0
0	[0]	100	0
[0]	0	0	100

Solution 2: (D, 3) → (M, 28)
(D, 10) → (M, 21)
(D, 17) → (M, 7)
(D, 25) → (M, 12)

$$Z_{min} = 280 + 300 + 300 + 300 = \text{Rs } 1180$$

Solution 3: (D, 3) → (M, 28)
(D, 10) → (M, 21)
(D, 17) → (M, 12)
(D, 25) → (M, 7)

$$Z_{min} = 280 + 300 + 300 + 300 = \text{Rs } 1180$$

All the three solutions have the same Z_{min} value. Therefore any alternative solution can be used.

5.6. UNBALANCED OR NON-SQUARE MATRIX PROBLEMS

The matrix is in form of $m \times n$ instead of $n \times n$. So the matrix is to be balanced by adding fictitious rows or columns (*i.e.*, demands and supplies) called "Dummies" such that the number of rows and columns become equal and thus balanced.

Example 5.7. Assign four trucks 1, 2, 3 and 4 to vacant spaces 5, 6, 7, 8, 9 so that the distance travelled is minimized. The matrix below shows the distance.

Solution :

Space \ Truck	1	2	3	4
5	50	40	60	20
6	40	30	40	30
7	60	20	30	20
8	30	30	20	30
9	10	20	10	30

1. No. of rows $\neq$ No. of columns

$5 \neq 4$

Therefore, adding an extra column having 'zero' distances between trucks and spaces.

	1	2	3	4	Dummy	Min. Value
5	50	40	60	20	0	0
6	40	30	40	30	0	0
7	60	20	30	20	0	0
8	30	30	20	30	0	0
9	10	20	10	30	0	0

Now, number of rows = number of columns

$5 = 5$

That means it has become balanced matrix. Therefore proceeding further, solve the problem as usual by Hungarian method.

	1	2	3	4	Dummy
5	50	40	60	20	0
6	40	30	40	30	0
7	60	20	30	20	0
8	30	30	20	30	0
9	10	20	10	30	0
Min. Value	10	20	10	20	0

$\Rightarrow$

	1	2	3	4	Dummy
5	40	20	50	[0]	0
6	30	10	30	10	[0]
7	50	[0]	20	0	0
8	20	10	10	10	0
9	[0]	0	0	10	0

	1	2	3	4	Dummy
5	40	20	50	$\boxed{0}$	10
6	20	0	20	0	$\boxed{0}$
7	50	$\boxed{0}$	20	0	10
8	10	0	$\boxed{0}$	0	0
9	$\boxed{0}$	0	0	10	10

Solution :

$$\left.\begin{array}{l} 1 \to 9 \\ 2 \to 7 \\ 3 \to 8 \\ 4 \to 5 \end{array}\right\}$$

$$Z_{min} = 10 + 20 + 20 + 20 = 70 \text{ Ans.}$$

Example 5.8. The marketing director of a multi-unit company is forced with a problem of assigning 5 senior managers to six zones. From the past experience he knows that the efficiency percentages judges by sales, operating costs etc. depends on manager - zone combination. The efficiency of different managers drop by the given percentage in different zones. The table shows the drop in officers efficiency.

Managers	I	II	III	IV	V	VI
A	73	91	87	82	78	80
B	81	85	69	76	74	85
C	75	72	83	84	78	91
D	93	96	86	91	83	82
E	90	91	79	89	69	76

Find out which zone will be managed by a junior manager due to non availability of senior manager.

Solution : It is not a balanced matrix as it is a 5 × 6 matrix. So adding a row of zeros to the matrix to convert it into balanced one.

Managers	I	II	III	IV	V	VI	Min. Value
A	73	91	87	82	78	80	73
B	81	85	69	76	74	85	69
C	75	72	83	84	78	91	72
D	93	96	86	91	83	82	82
E	90	91	79	89	69	76	69
Dummy	0	0	0	0	0	0	0

As it is a efficiency drop matrix, the objective function is to minimize the drop, *i.e.*, it is a minimization type of problem.

Managers	I	II	III	IV	V	VI
A	[0]	18	14	9	5	7
B	12	16	[0]	7	5	16
C	3	[0]	11	12	6	19
D	11	14	4	9	1	[0]
E	21	22	10	20	[0]	7
Dummy	0	0	0	[0]	0	0

Solution :

Manager	Zone
A	I
B	III
C	II
D	VI
E	V
Dummy	IV

Senior manager A will be assigned I zone
Senior manager B will be assigned III zone
Senior manager C will be assigned II zone
Senior manager D will be assigned VI zone
Senior manager E will be assigned V zone
Junior manager (Dummy) will be assigned IV zone **Ans.**

5.7. RESTRICTION TYPE ASSIGNMENT PROBLEM

Sometimes there are some restrictions on assigning a particular facility to a particular job. The restrictions may be due to high toll taxes (in case of transportation problem), inefficiency of a worker. To overcome this difficulty a very high cost (infinite cost) in case of minimization type or very poor (approx. zero) performance in case of maximization type of problems is assigned to that particular cell which automatically excludes this activity from the optimal solution.

Example 5.9. The secretary of a school is taking bids on the city's four school bus routes. Four companies have made the bids. A detailed version is given in the following table

Company	Route 1	Route 2	Route 3	Route 4
1	Rs 4000	Rs 5000	–	–
2	–	Rs 4000	–	Rs 4000
3	Rs 3000	–	Rs 2000	–
4	–	–	Rs 4000	Rs 5000

Suppose each bidder can be assigned only one route. Use the assignment model to minimize the school's cost of running the four bus routes.

Solution : It is a balanced matrix having some, number of rows and columns. Further there are some restrictions for companies to allocate certain routes as shown by dash (–) in the problem. So assign 'α' cost to these cells then solve as usual.

Company	Route 1	Route 2	Route 3	Route 4	Min. Value
1	4000	5000	α	α	4000
2	α	4000	α	4000	4000
3	3000	α	2000	α	2000
4	α	α	4000	5000	4000

Company	Route 1	Route 2	Route 3	Route 4
1	0	1000	α	α
2	α	0	α	0
3	1000	α	0	α
4	α	α	0	1000
Min. Value	0	0	0	0

As the α is too high a value, substracting any amount from them wouldn't cause any effect on cost. It will remain α (infinity).

Company	Route 1	Route 2	Route 3	Route 4
1	[0]	1000	α	α
2	α	[0]	α	~~0~~
3	1000	α	[0]	α
4	α	α	~~0~~	1000

No. of allocated cells ≠ No. of rows

$$3 \neq 4$$

Minimum undeleted cell value = 1000

Company \ Route	1	2	3	4
1	[0]	0	α	α
2	α	[0]	α	0
3	1000	α	[0]	α
4	α	α	0	[0]

Solution :

Company	Route
1	1
2	2
3	3
4	4

Minimum cost (Z_{min}) = 4000 + 4000 + 2000 + 5000 = Rs 15000 **Ans.**

Example 5.10. Four engineers are available to design four projects. Engineer Z is not competent enough to design project B. Given the following time estimates to each engineer to design a given project, find how should the engineers be assigned to projects so as to minimize the total design time of four projects.

Solution : Balanced matrix. Assign α to cell (2, B)

Engineer \ Project	A	B	C	D
1	16	14	14	12
2	16	–	17	13
3	11	15	21	9
4	8	10	9	7

⇒

	A	B	C	D	Min Value
1	16	14	14	12	12
2	16	α	17	13	13
3	11	15	21	9	9
4	8	10	9	7	7

⇓

	A	B	C	D
1	4	2	2	0
2	3	α	4	0
3	2	6	12	0
4	1	3	2	0
Min Value	1	2	2	0

	A	B	C	D
1	3	[0]	0	0
2	2	α	2	[0]
3	1	4	10	0
4	[0]	1	0	0

As number of allocations are not equal to number of rows

$3 \neq 4$

therefore, solving further, we get, minimum undeleted cell value = 1

	A	B	C	D
1	3	[0]	0	1
2	1	α	1	[0]
3	[0]	3	9	0
4	0	1	[0]	1

Solution:

Engineer	Project
1 →	B
2 →	D
3 →	A
4 →	C

minimum design time (Z_{min}) = 14 + 13 + 11 + 9

= 47 hours

Example 5.11. A car rental company has one car at each of its five depots D_1, D_2, D_3, D_4, and D_5. A customer in each of the five towns A, B, C, D and E requires a car. The distances between the depots and the towns are indicated below:

Find the optimal assignment of cars to the customers and find the total distance travelled by the five cars. **(UPTU: 2003-2004)**

Towns/Depots	D_1	D_2	D_3	D_4	D_5
A	160	130	176	190	200
B	135	120	130	160	175
C	140	110	155	170	185
D	50	50	80	80	110
E	55	35	70	80	105

Solution: (1) Substracting minmum row value from each element of that row.

Towns/Depots	D_1	D_2	D_3	D_4	D_5
A	30	0	46	60	70
B	15	0	10	40	55
C	30	0	45	50	75
D	0	0	30	30	60
E	20	0	35	45	70

(2) Substracting minimum column value from each element of that column

Towns/Depots	D_1	D_2	D_3	D_4	D_5
A	30	[0]	36	30	15
B	15	0	[0]	10	0
C	30	0	35	20	20
D	[0]	0	20	0	5
E	20	0	25	15	15

No. Of rows 5 ≠ (No. Of allocated cells)
Minimum undeleted value = 15
(3)

Towns/Depots	D_1	D_2	D_3	D_4	D_5
A	15	0	21	15	[0]
B	15	15	[0]	10	0
C	15	[0]	20	5	5
D	[0]	15	20	0	5
E	5	0	10	[0]	0

On substracting minimum row value from each element of corresponding row and minimum column value from each element of corresponding column we will get the same table. Thus proceeding further, we find that, No. of rows = No. of allocations *i.e.*, 5 = 5.

Therefore, this is the optimal solution.

	Total distance travelled
A → D_5	200
B → D_3	130
C → D_2	110
D → D_1	50
E → D_4	80
	Σ = 570

Total distance travelled = 570 units **Ans.**

Example 5.12. The table shown below is a profit matrix, entries of which indicate the profit generated by sales person working in various zones.

Sales prrsons

Sales zones		S_1	S_2	S_3
	1	10	17	14
	2	16	8	11
	3	14	15	16

Find assigned zones of sales persons, so that profit is maximum.

(UPTU. 2004-2005)

Solution: (1) Convert above maximization problem into minimization type by subtracting all elements from highest profit of matrix (17)

Sales persons

Sales zones		S_1	S_2	S_3
	1	7	0	3
	2	1	9	6
	3	3	2	1

cost matrix

(2) Subtracting smallest number in each row from every number in that row.

Sales zones		S_1	S_2	S_3
	1	7	0	3
	2	0	8	5
	3	2	1	0

(3) Subtracting smallest number in each column from every number in that column.

Sales persons

Sales zones		S_1	S_2	S_3
	1	7	[0]	3
	2	[0]	8	5
	3	2	1	[0]

(4) Make assignments in above matrix. Since there is one assignment in each row and in each column. Optimal solution is reached.

Assignments

Salesperson	Sales zone	Profit
S_1	2	16
S_2	1	17
S_3	3	16
	Total profit =	49

Example 5.13. Suggest optimal assignment of the four salesmen to four sale-territories given the estimated sales (in lakhs of Rs) as follows:

Sale Territories

		Allahabad	Bombay	Calcutta	Delhi
Salesman	Abhishek	10	11	15	12
	Benjamin	18	17	15	16
	Chaitanya	20	15	18	19
	Devendra	12	14	14	13

Solution:

	Allahabad	Bombay	Calcutta	Delhi
Abhishek	10	11	15	12
Benjamin	18	17	15	16
Chaitanya	20	15	18	19
Devendar	12	14	14	13

	Allahabad	Bombay	Calcutta	Delhi
Abhishek	10	9	5	8
Benjamin	2	3	5	4
Chaitanya	0	5	2	1
Devendar	8	6	6	7

	Allahabad	Bombay	Calcutta	Delhi
Abhishek	5	4	0	3
Benjamin	0	1	3	2
Chaitanya	0	5	2	1
Devendar	2	0	0	1

	Allahabad	Bombay	Calcutta	Delhi
Abhishek	5	4	[0]	2
Benjamin	[0]	1	3	1
Chaitanya	~~0~~	5	2	[0]
Devendar	2	[0]	~~0~~	~~0~~

Optimal solution $A \rightarrow$ calcutta, B $\rightarrow$ Allabhabad, C $\rightarrow$ Delhi, D $\rightarrow$ Bombay
Maximum sales 15 + 18 + 19 + 14 = 66 Lakhs

Example 5.14. Assignement Problem
Problem is 5 × 5 balanced

	1	2	3	4	5
1	[–2]	–4	–8	–6	–1
2	0	[–9]	–5	–5	–4
3	–3	–8	[–9]	–2	–6
4	–4	–3	–1	[0]	–3
5	–9	–5	–8	–9	[–5]

2. Basic feasible solution by N.W.

$$Z = -2 - 9 - 9 + 0 - 5 = -25 \text{ Units}$$

3. Hungarian method to get optimal solution. Now each row is having one zero but there is no zero in 5th column.

	1	2	3	4	5
1	6	4	0	2	7
2	9	0	4	4	5
3	6	1	0	7	3
4	0	1	3	4	1
5	0	4	1	0	4

New table for column. Now each column is also having one zero.

	1	2	3	4	5
1	6	4	0	2	6
2	9	0	4	4	4
3	6	1	0	7	2
4	0	(1)	(3)	4	0
5	0	(4)	(1)	0	3

Check for optimal solution; no. of lines = 4 < 5 (n)

∴ Solution is not optimal

	1	2	3	4	5
1	4	4	[0]	~~0~~	4
2	7	[0]	4	2	2
3	4	1	~~0~~	5	[0]
4	[0]	(1+2)	(3+2)	4	~~0~~
5	~~0~~	(4+2)	(1+2)	[0]	3

Number of lines = 5 = $n(5)$

∴ Solution is optimal.

∴ allocations are

1	⟶	3
2	⟶	2
3	⟶	5
4	⟶	1
5	⟶	4

$$Z = -8 - 9 - 6 - 4 - 9$$
$$Z = -36 \text{ Units}$$
$$\therefore \text{ Improvement} = -36 - (-25)$$
$$= -36 + 25 = -11$$

EXERCISE

5.1. Describe the assignment problem with a suitable example.

5.2. Write the mathematical formulation of an assignment problem.

5.3. Explain the similarities and differences between a transportation problem and an assignment problem.

5.4. Describe the procedure to solve an assignment problem.

5.5. How can an objective function in the assignment problem be maximized?

5.6. Discuss the practical applications of an assignment problem.

5.7. Describe steps of the Hungarian method to solve an assignment problem.

5.8. What is an unbalanced assignment problem?

5.9. How can an assignment problem be solved where a particular assignment is prohibited?

5.10. Can there be multiple optimal solutions to an assignment problem? How to identify such situations?

5.11. Show that in an assignment problem, if we multiply each element of the cost matrix by some fixed number, then the optimal solution remains same.

5.12. A director in a management institute has the problem of assigning courses to teachers with a view to maximizing educational quality in his institute. He has available to him one Professor, two Associate professors and one teaching assistant (TA). Four courses must be offered and after appropriate evaluation, has arrived at the following relative ratings (100 = best rating) regarding the ability of each instructor to teach each of the four courses.

	Course 1	Course 2	Course 3	Course 4
Prof. 1	60	40	60	70
Prof. 2	20	60	50	70
Prof. 3	20	30	40	60
TA	30	10	20	40

How should he assign his staff to the courses to realize his objective?

[Ans. 210]

5.13. XYZ transport company has a surplus of one truck in each of the cities 1, 2, 3, 4, 5 and a deficit of one truck in each of the cities A, B, C, D, E and F. The distance (in km) between the cities with a surplus and cities with deficit are given below:

		To					
		A	B	C	D	E	F
	1	80	140	80	100	56	98
	2	48	64	94	126	170	100
From	3	56	80	120	100	70	64
	4	99	100	100	104	80	90
	5	64	80	90	60	60	70

How should the trucks be dispatched so as to minimize the total distance traveled? Which city will not receive a truck?

5.14. Four new machines M_1, M_2, M_3 and M_4 are to be installed in a machine shop. There are five vacant places A, B, C, D and E available. Because of limited space machine M_2 cannot be placed at **C** and machine M_3 cannot be placed at A. Find the optimal assignment schedule. **[Ans. 12]**

	A	B	C	D	E
M_1	4	6	10	5	6
M_2	7	4	–	5	4
M_3	–	6	9	6	2
M_4	9	3	7	2	3

5.15. Solve the following assignment problem using Hungerian method. The matrix entries are processing times in hours.

		Operator				
		1	2	3	4	5
	1	20	22	35	22	18
	2	4	26	24	24	7
Job	3	23	14	17	19	19
	4	17	15	16	18	15
	5	16	19	21	19	25

5.16. A college is having an undergruduate programme for which the effective semester time available is very less and the degree course requires field work. Hence, the savings in the total number of class hours handled can be utilized for such field work. Based on past experience, the college has established the number of hours required by each faculty to teach each subject. The course in its present semester has 4 subjects and the college has considered 6 existing faculty members to teach these courses. The objective is to assign the best 4 teachers, out of these 6 faculty to teach 4 different subjects such that the total number of class hours required is minimized. The data for this problem is summarized below. Solve and optimize the assignment problem.

		Subject 1	2	3	4
	1	25	44	33	35
	2	33	40	40	43
	3	40	35	33	30
Faculty	4	44	45	28	35
	5	45	35	38	40
	6	40	49	40	46

5.17. Consider the problem of assigning four sales persons to four different sales regions as shown below such that the total sales is maximized.

		Sales region 1	2	3	4
	1	5	11	8	9
	2	5	7	9	7
Salesman	3	7	8	9	9
	4	6	8	11	12

The cell entries represents annual sales figures in crores of rupees. Find the optimal allocation of the sales persons to different regions. [**Ans.** 39]

5.18. In a textile sales emporium, four salesmen *A*,*B*,*C*, and *D* are available to four counters *W*, *X*,*Y*, and *Z*. Each salesman can handle any counter. The service (in hour) of each counter when manned by each salesman is given below:

		Salesman A	B	C	D
	W	41	72	39	52
Counter	X	22	29	49	65
	Y	27	39	60	51
	Z	45	50	48	52

How should the salesmen be allocated appropriate counters so as to minimize service time ?(Each salesman must handle only one counter) [**Ans.** 147]

5.19. Suggest optimal assignment of the four salesmen to the four sale-territories, given the estimated sales (in lakhs of rupees) as follows:

		SALE TERRITORIES Allahabad	Bombay	Calcutta	Delhi
Salesman	Abdullah	10	11	15	12
	Benjamin	18	17	15	16
	Chetan	20	15	18	19
	Devinder	12	14	14	13

5.20. A captain of a circket team has to allot five middle batting positions to five batsmen. The average runs scored by each batsmen at these positions are as follows:

		Batting positions				
		III	IV	V	VI	VII
Batsmen	A	40	40	35	25	50
	B	42	30	16	25	27
	C	50	48	40	60	50
	D	20	19	20	18	25
	E	58	60	59	56	53

Make the assignment so that the expected total averagea runs scored by these batsmen are maximum. **[Ans. 232]**

5.21. In a job shop operation, five jobs may be performed on any of four machines. The hours required for each job on each machine are presented in the following table.

		Machine			
		1	2	3	4
Job	A	13	14	16	10
	B	12	13	15	12
	C	11	12	12	9
	D	16	16	18	14
	E	10	12	13	12

The plant foreman would like to assign the jobs so that the total time is minimized. Find the optimal solution. Which job will be left unassigned ?

5.22. An automobile dealer wishes to put four repairmen to four different jobs. The repairmen have somewhat different kinds of skills and they exhibit different levels of efficiency from one job to another. The dealer has estimated the number of man-hours that would be required for each job-man combination. This is given in the matrix form in the following table :

Man	Job			
	A	B	C	D
1	5	3	2	8
2	7	9	2	6
3	6	4	5	7
4	5	7	7	8

Find the optimum assignment that will result in minimum man-hours needed.

5.23. Suggest optimal assignment of the workers to jobs if the completion time (in hours) of different jobs by different workers is as given below :

		Workers			
		W	*X*	*Y*	*Z*
Jobs	A	8	7	9	10
	B	7	9	9	8
	C	10	8	7	11
	D	10	6	8	7

(*ii*) What are the similarities between transportation and assignment techniques?

5.24. Solve the following assignment problems:

	M_1	M_2	M_3	M_4
J_1	5	8	3	2
J_2	10	7	5	8
J_3	4	10	12	10
J_4	8	6	9	4

5.25. There are 4 jobs *A, B, C* and *D* and these are to be performed on four machine centres *I, II, III, IV*. One job is to be allocated to a machine centre, though each machine is capable of doing any job, at different costs given by the matrix below:

	I	*II*	*III*	*IV*
A	15	14	12	16
B	23	22	25	24
C	31	34	32	33
D	21	32	44	53

(*i*) Find an allocation of jobs to the machine centres so that the total cost of processing is minimum. [**Ans.** 88]

(*ii*) If the machine centre IV goes out of order, which job will then be allocated to the rest of the centres at a minimum cost.

5.26. Given the following data, determine the least cost allocation of the available machines to five jobs.

		Job				
		A	*B*	*C*	*D*	*E*
Machine	1	25	29	31	42	37
	2	22	19	35	18	26
	3	39	38	26	20	33
	4	34	27	28	40	32
	5	24	42	36	23	45

OBJECTIVE TYPE QUESTIONS

5.1. The assignment problem

(*a*) Is a special case of the transportation problem.

(*b*) Can be solved with the simplex algorithm.

(*c*) When treated as an *LP*, always has an optimal integer solution.

(*d*) All of the above.

5.2. The assignment problem consists of the following elements—

(*a*) A set of *n* jobs (*b*) A set of *n* Facilities

(*c*) A set of cost, one for each pair of job facility

(*d*) All of the above (*e*) *a*, *b* but not *c*

(*f*) None of the above.

5.3. Which of the following is true for the assignment problem?

(*a*) The number of rows equals the number of columns.

(*b*) All circled values in a stone are 1. (*c*) All rim requirements are 1.
(*d*) All of the above.

5.4. The method used to solve an assignment problem is called
(*a*) reduced matrix method (*b*) MODI method
(*c*) Hungarian method (*d*) All of the above.

5.5. If there were n workers and n jobs, there would be
(*a*) $n!$ solutions (*b*) $(n-1)!$ Solutions
(*c*) $(n!)n$ solutions (*d*) n solutions

5.6. For a salesman who has to visit n cities, following are the ways of his tour plan
(*a*) n ! (*b*) $(n + 1)!$
(*c*) $(n - 1)$ (*d*) n

5.7. The purpose of a dummy row or column in an assignment problem is to
(*a*) obtain balance between total activities and total resources
(*b*) prevent a solution from becoming degenerate
(*c*) provide the means of representing a dummy problem
(*d*) none of the above

5.8. An optimal assignment requires that the maximum number of lines which can be drawn through squares with zero opportunity cost be equal to the number of
(*a*) rows or columns (*b*) rows and columns
(*c*) rows + column - 1 (*d*) none of the above

Answers

5.1. (*d*) 5.2. (*d*) 5.3. (*d*) 5.4. (*c*) 5.5. (*a*) 5.6. (*c*) 5.7. (*a*)
5.8. (*a*)

❑❑❑

Chapter 6

Dynamic Programming

6.1. INTRODUCTION

Dynamic programming is a mathematical technique dealing with the optimization of multi-stage decision process. Dynamic - decisions are taken at several distinct stages (everyday). Programming - used in mathematical sense of selecting an optimal allocation of resources. Richard Bellman developed this technique in 1950. It is also known as Recursive optimization. In Dynamic programming, a large problem, which has 'n' decision variables, is split into 'n' sub problems (stages) each in one variable. An optimum solution is obtained in an orderly manner starting from one stage to next.

6.2. TERMINOLOGY OF DYNAMIC PROGRAMMING

(1) **Stage:** Each point in the problem where a decision must be made is known as stage. The problem is broken down into sub problems and each sub problem is referred to as stage. At each stage there are number of alternatives and the decision making process involves the selection of one feasible alternative which is called as stage decision. Associated with each stage is the return function of the decision variables, which evaluates the worth of each alternative.

(2) **State:** Information describing the problem at each stage - values of state variables. These values are referred to as states. State variables link up two stages.

(3) **Alternative:** In a given stage, there may be more than one choice of carrying out task. Each choice is known as alternative.

(4) **Policy:** A decision making rule which, at any stage, allows a feasible sequence of decisions. A policy transforms the state at a given stage into a state associated with next stage.

(5) **Optimal policy:** A policy, which optimizes the value of a criterion, objective or return function is known as optimal policy. Optimal decision at any stage is in no way dependent on the previous history of the system.

6.3. APPLICATIONS OF DYNAMIC PROGRAMMING

The Dynamic programming can be applied to many real-life problems. Some of the applications of dynamic programming are given below:

1. To allocate salesmen to different sales zones.
2. Reliability improvement problem
3. Linear Programming problem
4. Inventory control problem

5. Capital budgeting problem
6. Shortest path problem
7. Selection of an advertising media.
8. To determine at which stage equipment is to be replaced for optimal return.
9. For production, scheduling and employment smoothening.

6.4. DECISION TREE AND BELLMAN'S PRINCIPLE OF OPTIMALITY

A multi stage decision system, in which each decision and state variable can take only finite number of values, can be represented graphically by a decision tree.

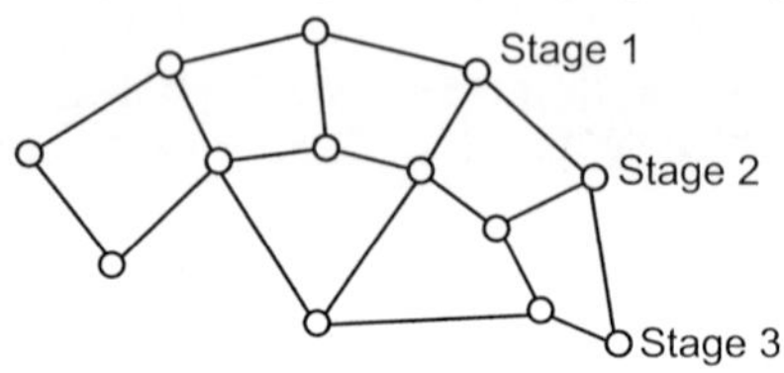

Fig. 6.1. Decision tree

Circles represent nodes corresponding to stages and lines between circles denotes arcs corresponding to decisions.

The concept of dynamic programming is mainly based on Principle of optimality given by Bellman.

"An optimal policy has the property that, what ever the initial state and initial decisions are, the remaining decisions must constitute an optimal policy with regard to the state resulting from the first decision process".

6.5. CHARACTERISTICS OF DYNAMIC PROGRAMMING

Basic characteristics of dynamic programming problem are as follows:

1. The problem can be sub-divided into stages with a policy decision required at each stage.
2. Every stage consists of a number of states associated with it.
3. Decision at each stage converts the current state into state associated with the next stage.
4. The state of the system at a stage is described by a set of variables, known as state variables.
5. Given the current stage, an optimal policy for remaining stages is independent of the policy adopted in previous stages.
6. A recursive equation with n stages remaining is formed to identify the optimum policy for each state of the system, given the optimal policy for each state with n -1 stages left.
7. Using recursive equation approach, each time the solution procedure moves backward stage by stage for obtaining the optimum policy of each state for that particular stage, till it attains the optimum policy beginning at the initial stage.

6.6. DYNAMIC PROGRAMMING ALGORITHM

The solution of a multi-stage problem by dynamic programming involves the following steps:

1. Identify the decision variables and specify the objective function to be optimized under certain limitation, if any.
2. Decompose the given problem into a number of smaller sub-problems (stages). Identify the state variable at each stage.
3. Develop the general recursive relationship for calculating the optimal policy. Decide whether forward or backward method is to be followed to solve the problem.
4. Construct appropriate stages to show the required values of the return function at each stage.
5. Determine the overall optimal policy or decisions and its value at each stage. There may be more than one such optimal policy.

Example 6.1. A firm has divided its marketing area into three zones. Amount of sales depends upon the number of salesman in each zone. The firm has been collecting the data regarding sales and salesman in each area over a number of past years. The information is summarised in table given below. For the next year the firm has only 9 salesman. Allocate salesman to different zones to maximize sales.

Solution : The problem is to allocate these salesman to 3 different zones so that the total sales are maximized.

Profits in thousands of Rs.

Table

No. of salesman	Zone 1	Zone 2	Zone 3
0	30	35	42
1	45	45	54
2	60	52	60
3	70	64	70
4	79	72	82
5	90	84	95
6	98	93	102
7	105	98	110
8	100	100	110
9	90	100	110

Objective is to maximize total sales objective fun $Z_{min} = F_1(x_1) + F_2(x_2) + F_3(x_3)$ subject to condition $x_1 + x_2 + x_3 \leq 9$ and no. of salesman = 9.

Stage (1) Zone (1)

no. of salesman	0	1	2	3	4	5	6	7	8	9
Profit	30	45	60	70	79	90	98	105	100	90

We consider first two zones (1) and (2), 9 salesman can be divided into 2 zones in 10 different ways each combination will have associated with certain return. This is given in table (2).

Zone (1) and (2)

No. of Salesman	Zone 1	X_1	9	8	7	6	5	4	3	2	1	0
	Zone 2	X_2	0	1	2	3	4	5	6	7	8	9
Profit in 1000 Rs.	Zone 1	$F_1(X_1)$	90	100	105	98	90	79	70	60	45	30
	Zone 2	$F_2(X_2)$	35	45	52	64	72	82	93	98	100	100
$S = F_1(X_1) + F_2(X_2)$			125	145	157	162	162	161	163	158	145	130

where x_1 and x_2 denotes no of salesman in zone (1) and (2). $f(x_1), f_2(x_2)$ denotes the profit or return in zone (1) and (2) respectively. If S denotes the total profit from each combination then the various alternatives are

$$S = f_1(9) + f_2(0)$$
$$S = f_1(8) + f_2(1)$$
$$S = f_1(0) + f_2(9)$$

In general form, $S = f_1(x) + f_2(9 - x)$

or $\boxed{S = f_1(x) + f_2(A - x)}$

A = total number of salesman

Objective functions $S = f(A)\,[f_1(x) + f_2(A - x)$ ...(1)

Stage II

Table

Zone 2	Zone 1 →	0	1	2	3	4	5	6	7	8	9
	Profit →	30	45	60	70	79	90	98	105	100	90
0	35	65	80	95	105	114	125	133	140	135	125
1	45	75	90	105	115	124	135	143	150	143	
2	52	82	97	112	122	131	142	150	157		
3	64	94	109	124	134	143	154	162			
4	72	102	117	132	142	151	162				
5	82	112	127	142	151	162					
6	93	123	138	153	163						
7	98	128	143	158							
8	100	130	145								
9	100	130									

Optimum return for various combinations are tabulated below :

	Maximum return from optimum allocation of sales man									
No. of sales man	0	1	2	3	4	5	6	7	8	9
Total Profit $F_1(X_1) + F_2(X_2)$	65	80	95	105	115	125	135	143	154	163
$X_1 + X_2$	0 + 0	1 + 0	2 + 0	$\begin{bmatrix} 3+0 \\ 2+1 \end{bmatrix}$	3 + 1	5 + 0	5 + 1	$\begin{bmatrix} 6+1 \\ 4+3 \end{bmatrix}$	5 + 3	3 + 6

Computation of selecting the optimum selection of zone (1), (2) and (3) can be carried as follows.

Stage III

No. of Sales man in Zone 3		No. of sales man in Zone 1 & 2									
		0	1	2	3	4	5	6	7	8	9
		65	80	95	105	115	125	135	143	154	168
0	42	107	122	137	147	157	167	177	185	196	205
1	54	119	134	149	159	169	179	189	199	208	
2	60	125	140	155	165	175	185	195	203		
3	70	135	150	165	175	185	195	205			
4	82	147	162	177	187	197	207				
5	93	160	175	190	200	210					
6	102	167	182	197	207						
7	110	175	190	205							
8	110	175	190								
9	110	175									

Salesman Problem

Example 6.2. A company has 8 salesmen, who have to be allocated to 4 mkt. zones. The return of profit from each zone depends upon the no. of salesman working in that zone. The expected return for different no. salesman in different zones, as estimated from past records are given below. Determine the optimum allocation policy.

No. of Sales man	Mkt. Zones 1	2	3	4
0	45	30	35	42
1	58	45	45	54
2	70	60	52	60
3	82	70	64	70
4	93	79	72	82
5	101	90	82	95
6	108	98	93	102
7	113	105	98	110
8	118	110	100	110

$A = 8,$ Stages $= 4$, $A \leq 8$

Solution: Zone (1) and (2)

(1)

No. of Sales Man	Zone 1	X_1	8	7	6	5	4	3	2	1	0
	Zone 2	X_2	0	1	2	3	4	5	6	7	8
Profit in 1000 Rs.	Zone 1	$F(X_1)$	118	113	108	101	93	82	70	58	45
	Zone 2	$F(X_2)$	30	45	60	70	79	90	98	105	110
$S = F_1(X_1) + F_2(X_2)$			148	158	168	171	172	172	168	163	155

(2) Table for optimal distribution for any no of salesman.

Zone 1		0	1	2	3	4	5	6	7	8
Zone 2	Profit	45	58	70	82	93	101	108	113	118
0	30	75	88	100	112	123	131	138	143	148
1	45	90	103	115	127	138	146	153	158	
2	60	105	118	130	142	153	161	168		
3	70	115	128	140	152	163	171			
4	79	124	137	149	161	172				
5	90	145	148	160	172					
6	98	143	156	168						
7	105	150	163							
8	110	155								

(3) Maximum return from optimum allocation of salesman in zone (1) and (2)

No. of sales man	0	1	2	3	4	5	6	7	8
Total Profit $F_1(X_1) + F_2(X_2)$	75	90	105	118	130	145	153	163	172
$X_1 + X_2$	(0+0)	(0+1)	(0+2)	(1+2)	(2+2)	(0+5)	(4+2)	(4+3)	(3+5) or (4 + 5)

		No. of Salesman in Zone 1 + 2								
No. of Salesman in Zone 3		0	1	2	3	4	5	6	7	8
		75	90	105	118	130	145	153	163	172
0	35	110	125	140	153	165	180	188	198	207
1	45	120	135	150	163	175	190	198	208	
2	52	127	142	152	170	182	197	205		
3	64	139	154	169	182	194	209			
4	72	147	162	177	190	202				
5	82	157	172	187	200					
6	93	168	183	198						
7	98	173	188							
8	100	175								

Maximin return from optimum allocation of salesman in zone 1 + 2 + 3

No. of sales man(A)	0	1	2	3	4	5	6	7	8
Profit $F_1(X_1) + F_2(X_2) + F_3(X_3)$	110	125	140	153	165	180	190	198	209
$X_1 + X_2 + X_3$	0+0+0	0+1+0	0+2+0	1+2+0	2+2+0	0+5+0	0+5+1	[4+3+0 or 4+2+1]	0+5+3

		No. of Salesman in Zone 1 + 2 + 3								
No. of Salesman in Zone 4		0 110	1 125	2 140	3 153	4 165	5 180	6 190	7 198	8 209
0	42	(152)	(167)	(182)	(195)	(207)	(222)	232	240	251
1	54	164	179	194	(207)	(219)	(234)	244	(252)	
2	60	170	185	200	213	225	240	250		
3	70	180	195	210	223	235	250			
4	82	192	207	222	235	247				
5	95	205	220	235	248					
6	102	212	227	242						
7	110	220	235							
8	110	220								

Optimum return is Rs. 252/-

$$x_1 + x_2 + x_3 = 7, x_4 = 1$$

$$x_1 = 4, x_2 = 3 \quad \text{or} \quad 2, x_3 = 0 \text{ or } 1. \text{ (From previous table)}$$

Example 6.3. A company has 6 salesmen and 3 market areas *A*, *B*, *C*. It is desired the number of salesmen to allocate to each market area to maximize profit. The following table gives the profits from each market area as a function of the number of salesmen allocated.

Area	Sales man						
	0	1	2	3	4	5	6
A	38	41	48	58	66	72	33
B	40	42	50	60	66	75	82
C	60	64	68	78	90	102	109

Use dynamic programming technique to solve the above problem.

Solution : Let the three areas be considered as three-stage problem. The decision variables x_j denote the number of salesmen and let $f_j(x_j)$ be the expected profit from the allocation of x_j salesmen to area $j (j = A, B, C)$.

Start with area *A*. The returns corresponding to different number of salesmen allocated to area A are:

No. of salesmen (x_1)	:	0	1	2	3	4	5	6
Profit $f_1(x_1)$	:	38	41	48	58	66	72	83

Now, consider areas *A* and *B* combined. Six salesmen can be divided into two areas in seven different ways as shown below:

Area A :	x_1	0	1	2	3	4	5	6
Area B :	x_2	6	5	4	3	2	1	0
Profits :	$f_1(x_1)$	38	41	48	58	66	72	83
	$f_2x_2)$	82	75	66	60	50	42	40

If p denotes the total profits from area *A* and *B* considered, various alternatives are :

$$p = f_1(0) + f_2(6), \qquad p = f_1(1) + f_2(5), \text{ etc.}$$

Thus, in general the problem is to

Maximize $\quad p = f_1(x_j) + f_2(s - x_j)$

where s stands for total number of salesmen.

The above equation can be applied to determine the optimum distribution of any number of salesmen. This can be done in the tabular form:

x_2	$f_2(x_2)$	x_1 → / $f_1(x_1)$ →	0 / 38	1 / 41	2 / 48	3 / 58	4 / 66	5 / 72	6 / 83
0	40		78*	81*	88*	98*	106*	112*	123*
1	42		80	83	90	100	108	114	
2	50		88*	91	98	108	116		
3	60		98*	101	108	118			
4	66		104	107	114				
5	75		113*	116					
6	82		120						

* Indicates maximum number along each diagonal

Expected profits are computed for all the possible combinations. For a particular number of salesmen, the profits for different combinations can be read along the diagonal. For example, if two salesmen are to be distributed between two areas, the values of sales for the combination 2 + 0, 1 + 1, 0 + 2 can be read along the diagonal 2 – 2. The maximum profit of 88 for two salesmen results from the combination of 2 salesmen in *A* and none in *B* or 2 salesmen in *B* and none in *A*. The optimum returns from various combinations are summarized below:

s	0	1	2	3	4	5	6
$F(x) = f_1(x_1) + f_2(x_2)$	78	81	88	98	66	113	123
$x = x_1 + x_2$	0 + 0	1 + 0	2 + 0	3 + 0	4 + 0	0 + 5	6 + 0
			0 + 2	0 + 3			

Now, we move to the third stage and consider the distribution of 6 salesmen into three areas, namely *A*, *B* and *C*. So, we maximize

$$p = F(x_j) + f_3(s - x_j); \qquad 0 \le s_j \le x.$$

This is given in the following table:

x_3	$f_1(x_3)$	x / F(x)→	0 / 78	1 / 81	2 / 88	3 / 98	4 / 66	5 / 113	6 / 123
0	60		138*	141	148*	158*	126	173	183*
1	64		142*	145	152	162	130	177	
2	68		146	149	156	166	134		
3	78		156	159	166	176			
4	90		168*	171	178				
5	102		180*	183*					
6	109		187						

The number marked* are the maximum along each diagonal.

As we see along the diagonal 6 – 6, the optimum combination is 6 salesmen to area *A* and *B* combined and none in area *C*, or one salesmen in area *A* and *B* combined and 5 salesmen in area *C*. This gives a maximum profit of 183.

Now proceeding backwards, from the previous table we find that the optimum allocation of

(*i*) 6 salesmen in *A* and *B* implies 6 salesmen in *A* and none in *B*.

(*ii*) One salesmen in *A* and *B* implies one salesmen in *A* and none in *B*.

Thus, the best possible allocation among all the three areas is as follows:

Area	:	*A*	*B*	*C*
No. of salesmen	:	6	0	0
to be allocated	:	1	0	5

The maximum profit on the basis of this allocation is 183.

Example 6.4. A cosmetic mfg co. is interested in selecting the advertising media for its product and the frequency of advertisement in each in the data collected over past two years regarding the frequency of advertisement in 3 media of Newspaper, radio and T.V. and real sales of the product gives the following result.

	Expected sales in thousand of Rs.		
Frequency/week	Television-2000	Radio-1000	Newspaper-500
1	220	150	100
2	275	250	175
3	325	300	225
4	350	320	250

The cost of advertisement in Newspaper is Rs. 500/- per appearance while in Radio and is T.V. it is Rs. 1000/- and Rs. 2000 resp. the budget provides Rs. 4500/ per week for advertisement. The problem is of determining the optimal combination of advertising media and use the advertisement frequency.

Solution: (1) Problem can be decomposed into 3 stages corresponding to 3 medias of advertisement. In each media alternatives (frequencies) are possible. Each alternative is associated with certain cost and return (expected sales). Here the capital marked to allocation for different medias is the state variables, state $x = 4500$/-.

A combination of media and frequency is be to selected in such a way to maximise the total sales with the expenditure not exceeding the sp. limit of Rs. 4500/.

Let us consider the advt. media of TV at the first stage. The table for various levels of expenditure for various alternatives are evaluated and the one giving largest expected sale is selected. The selected frequency and optimal return for different values of x are given in table

Stage 1: (Television)

State x_1	Cost per appearance Rs 2000 Return (in 1000'of Rs)	Frequency
0	—	—
500	—	—
1000	—	—
1500	—	—
2000	220	1
2500	220	1
3000	220	1
3500	220	1
4000	275	2
4500	275	2

For $x_1 = 0, 500, 1000, 1500$ it is not possible to advt. in this media since cost of appearance per week is Rs. 2000/- for $x_1 = 2000, \ldots. 3500$ the product can be advt. once in a week giving a return of Rs. 220,000/; with $x_1 = 4000, 4500$/- two appearance can occur giving a return of Rs. 2,75,000/-

Stage 2: (Radio) – cost of appearance = Rs. 1000/

State x_3	Return (1000's of Rs) 0	1	2	3	4	Return	Freq.
500	0	0	—	—	—	0	1
1000	0	150	—	—	—	150	1
1500	0	150+0	—	—	—	150	1
2000	220	150+0	250	—	—	250	2
2500	220	150+0	250+0	—	—	250	2
3000	220	150+220	250+0	300	—	370	1
3500	220	150+220	250+0	300+0	—	370	1
4000	275	150+220	250+220	300+0	320	470	2
4500	275	150+220	250+220	300+0	320+0	470	2

Here state x_2 shows expd. incurred in the first stage and the current stage.

Stage 3: News paper (cost of appearance = Rs. 500/-)

State x_3	Return (1000's of Rs) 0	1	2	3	4	Return	Freq.
500	0	100	—	—	—	100	1
1000	150	100+0	175	—	—	175	2
1500	150	100+150	175+0	225	—	250	1
2000	250	100+150	175+150	225+0	250	325	2
2500	250	100+250	(175+150)	225+150	250+0	375	3
3000	370	100+250	(175+250)	225+150	250+150	425	2
3500	370	100+370	(175+250)	225+250	250+150	475	3
4000	470	100+370	175+370	225+250	250+250	545	2
4500	470	100+470	175+370	225+370	250+250	595	3

Optimal solution is

News paper — 3 papers
Radio — 1
T.V. — 1

CAPITAL BUDGETING PROBLEM

Example 6.5. An organization is planning to diversify its business with a maximum outlay of Rs. 5 crores. It has identified three different locations to instal plants. The organization can invest in one or more of these plants subject to the availability of the fund. The different possible alternatives and their investment (in crores of rupees) and present worth of returns during the useful life (in crores of rupees) of each of these plants are summarized in table. The first row of table has zero cost and zero return for all the plants. Hence, it is known as *do-nothing* alternative. Find the optimal allocation of the capital to different plants which will maximize the corresponding sum of the present worth of returns.

	Plant 1		Plant 2		Plant 3	
Alternative	Cost	Return	Cost	Return	Cost	Return
1	0	0	0	0	0	0
2	1	15	2	14	1	3
3	2	18	3	18	2	7
4	4	28	4	21	–	–

Solution: Maximum capital amount, C = Rs. 5 crores. Each plant is treated as a stage. So, the number of stages is equal to 3. The plants 1, 2 and 3 are defined as Stage 1, Stage 2 and Stage 3, respectively. So, the forward recursive function is used for this problem.

Stage 1: The recursive function for a given combination of the state variable, x_1 and alternative, m_1 in the first stage is presented below. The corresponding returns are summarized in table. For each value of the state variable, the best return and the corresponding alternative are presented in the last two columns, respectively.

$$f_1(x_1) = R(m_1)$$

Table: Calculations for Stage 1 (Plant 1)

	Alternative m_j					
Stable variable x_1	1	2	3	4		
	C R	C R	C R	C R		
	0 0	1 15	2 18	4 28	$f_1{}^*(x_1)$	$m_1{}^*$
0	0	–	–	–	0	1
1	0	15	–	–	15	2
2	0	15	18	–	18	3
3	0	15	18	–	18	3
4	0	15	18	28	28	4
5	0	15	18	28	28	4

C = cost, R = return.

Stage 2: The recursive function $f_2(x_2)$ for a given combination of the state variable, x_2 and alternative, m_2 in the second stage is given by

$$f_2(x_2) = R(m_2) + f_1[x_2 - C(m_2)]$$

The corresponding returns are summarized in table. For each value of the state variable, the best return and the corresponding alternative (s) are presented in the last two columns, respectively.

Table: Calculations for Stage 2 (Plant 2)

	Alternative m_j					
Stable variable x_1	1	2	3	4		
	C R	C R	C R	C R		
	0 0	1 14	3 18	4 21	$f_2{}^*(x_2)$	$m_2{}^*$
0	0	–	–	–	0	1
1	0 + 15 = 15	–	–	–	15	1
2	0 + 18 = 18	14 +0 = 14	–	–	18	1
3	0 + 18 = 18	14 +15 = 29	18 + 0 = 18	–	29	2
4	0 + 28 = 28	14 + 18 = 32	18 + 15 = 33	21 + 0 = 21	33	3
5	0 + 28 = 28	14 + 18 32	18 + 18 = 36	21 + 15 = 36	36	3 and 4

Stage 3. The recursive function $f_2(x_3)$ for different combinations of the state variable, x_3 and alternative, m_3 in the third stage is:

$$f_3(x_3) = R(m_3) + f_2[x_3 - C(m_3)]$$

The corresponding returns are summarized in table . For each value of the state variable, the best return and the corresponding alternative (s) are presented in the last two columns, respectively.

Table: Calculations for Stage 3 (Plant 3)

	Alternative m_j				
State variable x_3	1	2	3		
	C R	C R	C R		
	0 0	1 3	2 7	$f_3(x_3)^*$	$m_3{}^*$
5	0 + 36 = 36	3 + 33 = 36	7 + 29 = 36	36	1, 2 and 3

The final results of the original problem is traced in table below. Form this table, one can visualize the fact that the original problem has four alternate optimal solutions.

Table: Final Results

Stage 3		Stage 2		Stage 1		Optimal alternatives stage		
C^*	m_3^*	C^*	m_2^*	C^*	m_1^*	1	2	3
5	1	5 – 0 = 5	3	5 – 3 = 2	3	3 –	3 –	1
			4	5 – 4 = 1	2	2 –	4 –	1
5	2	5 –1 = 4	3	4 – 3 = 1	2	2 –	3 –	2
5	3	5 – 2 = 3	2	3 – 2 = 1	2	2 –	2 –	3

Linear Programming Problem

Example 6.6. Use dynamic programming to solve

Maximum $Z = Y_1 \cdot Y_2 \cdot Y_3$

Subject to constraints

$$Y_1 + Y_2 + Y_3 = 5$$

and $Y_1, Y_2, Y_3 \geq 0$

Solution: Since there are 3 decision variables, the given problem is a 3 stage problem, with the state variables S_1, S_2, S_3 defined as

$$S_3 = Y_1 + Y_2 + Y_3 = 5$$
$$S_2 = Y_1 + Y_2 = S_3 - Y_3$$
$$S_1 = Y_1 = S_2 - Y_2$$

The recursive equations are

$$f_1(S_1) = \max (Y_1) = S_1 = (S_2 - Y_2)$$
$$0 \leq Y_1 \leq X_1$$
$$f_2(S_2) = \max (Y_1 \cdot Y_2) = \max (Y_2 \cdot f_1(S_1))$$
$$0 \leq Y_2 \leq X_2$$
$$f_3(S_3) = \text{Max} (Y_1 \cdot Y_2 \cdot Y_3)$$
$$0 \leq Y_3 \leq X_3$$
$$= \text{Max} (Y_3 \cdot f_2(S_2))$$
$$0 \leq Y_3 \leq X_3$$

To solve the recursive equation

$$f_1(S_1) = S_2 - Y_2 = S_1$$
$$f_2(S_2) = \text{Max } Y_2 f_1(S_1)$$
$$0 \leq Y_2 \leq X_2$$
$$= \text{Max} (Y_2 \cdot f_1(S_2 - Y_2)) \; (\because S_1 = S_2 - Y_2)$$
$$= \text{Max} (Y_2 (S_2 - Y_2))$$

The function $Y_2(S_2 - Y_2)$ will attain its maximum at $Y_2 = \dfrac{S_2}{2}$

$$\therefore \quad f_2(Y_2) = \text{Max} (Y_2(S_2 - Y_2))$$
$$0 \leq Y_2 \leq X_2$$

$$= \frac{S_2}{2}\left(S_2 - \frac{S_2}{2}\right) = \left(\frac{S_2}{2}\right)^2$$

$$f_3(S_3) = \underset{0 \le Y_3 \le X_3}{\text{Max}} (Y_3 \cdot f_2(S_2))$$

$$= \underset{0 \le Y_3 \le X_3}{\text{Max}} (Y_3 \cdot f_2(S_3 - Y_3)) \;(\because S_2 = S_3 - Y_3)$$

$$= \underset{0 \le Y_3 \le x_3}{\text{Max}} \left(Y_3 \left(\frac{S_3 - Y_3}{2}\right)^2\right)$$

The function $Y_3\left(\frac{S_3 - Y_3}{2}\right)^2$ will attain its maximum when $Y_3 = \frac{S_3}{3}$

$$\therefore \quad f_3(S_3) = \frac{S_3}{3}\left(\frac{S_3 - \frac{S_3}{3}}{2}\right)^2 = \left(\frac{S_3}{3}\right)^2$$

But $\quad S_3 = 5 \Rightarrow Y_3 = \frac{S_3}{3} = \frac{5}{3}$

$$S_2 = S_3 - Y_3 = 5 - \frac{5}{3} = \frac{10}{3} \Rightarrow Y_2 = \frac{S_2}{2}$$

$$= \frac{10}{3 \times 2} = \frac{5}{3}$$

$$S_1 = S_2 - Y_2 = \frac{10}{3} - \frac{5}{3} = \frac{5}{3}$$

$$Y_1 = S_1 = \frac{5}{3}$$

and $\quad f_3(S_3) = f_3(5) = \left(\frac{5}{3}\right)^3 = \frac{125}{27} = \text{Max } Z.$

$\therefore$ The optimal policy is $\left(\frac{5}{3}, \frac{5}{3}, \frac{5}{3}\right)$ and Max $Z = \frac{125}{27}$.

Example 6.7. A student has to take examination in three courses X, Y and Z. He has three days available for study. He feels it would be best to devote a whole day to study the same course, so that he may study a course for one day, two days or three days or not at all. His estimates of grades he may get by studying are as follows.

Study days / Course	X	Y	Z
0	1	2	1
1	2	2	2
2	2	4	4
3	4	5	4

How should he plan to study so that he maximizes the sum of his grades?

Solution: Let n_1, n_2 and n_3 be the number of days he should study the course x, y and z respectively. If $f_1(n_1), f_2(n_2), f_3(n_3)$ be the grades earned by such a study then the problem becomes

$$\text{Max } Z = f_1(n_1) + f_2(n_2) + f_3(n_3)$$

Subject to $n_1 + n_2 + n_3 \leq 3$ and integers.

Here n_j are the decision variables and $t_j(n_j)$ are the corresponding return functions for $j = 1, 2, 3$. Now, introducing state variables S_j as follows

$$S_3 = n_1 + n_2 + n_3 \leq 3$$
$$S_2 = n_1 + n_2 = S_3 - n_3$$
$$S_1 = n_1 = S_2 - n_2$$

Thus state transformation function are defined as

$$S_{j-1} = T_j(S_j, n_j) \qquad \text{where,} \quad j = 2, 3$$

Recursive equations applicable here are

$$f_1(S_j) = \text{Max}[f_j(n_j) + F_{j-1}(S_{j-1})] \quad \text{where,} \quad j = 2, 3$$

or $\quad F_1(S_1) = F_1(n_1)$

where $\quad F_3(S_3) = \text{Max } [f_1(n_1) + f_2(n_2) + f_3(n_3)]$

for any feasible value of S_3. Then the required solution would become

$$\text{Max } f_3(S_3)$$

Recursive operations leading to the answer are tabulated as follows.

	$f_2(n_2)$				$f_1(s_1) = f_1(n_1)$				$f_2(n_2) + f_1(s)$				$F_2(s_2)$
s_2 \ n_2	0	1	2	3	0	1	2	3	0*	1	2	3	
0	2	–	–	–	1	–	–	–	3	–	–	–	3
1	2	2	–	–	2	1	–	–	4	3	–	–	4
2	2	2	4	–	2	2	1	–	4	4	5	–	5
3	2	2	4	5	4	2	2	1	6	4	6	6	6
	$f_3(n_3)$				$f_2(s_2) = f_2(n_2)$				$f_3(n_3) + F_2(S_2)$				$F_3(S_3)$
S_2 \ n_2	0	1	2	3	0	1	2	3	0	1	2*	3	
0	1	–	–	–	3	–	–	–	4	–	–	–	4
1	1	2	0	–	4	3	–	–	5	5	–	–	5
2	1	2	4	–	5	4	3	–	6	6	7	–	7
3	1	2	4	4	6	5	4	3	7	7	8	7	8

Stage returns $f_j(n_j)$

j \ nj	0	1*	2	3
1	1	2	2	4
2	2	2	4	5
3	1	2	4	4

Stage transformation S_{j-1}
$j = 2, 3$

s_j \ n_j	0	1*	2	3
0	0	–	–	–
1	1	0	–	–
2	2	1	0	–
3	3	2	1	0

Proceeding backwards through enclosed type numbers, the optimal policy is obtained as $n_3 = 2, n_2 = 0, n_1 = 1$, keeping in view $n_1 + n_2 + n_3 \leq 3$.

The required maximum return is 8.

Example 6.8. A distance network consists of eleven nodes which are distributed as shown in figure 6.2. Find the shortest path from Node 1 to Node 11 and also the corresponding distances.

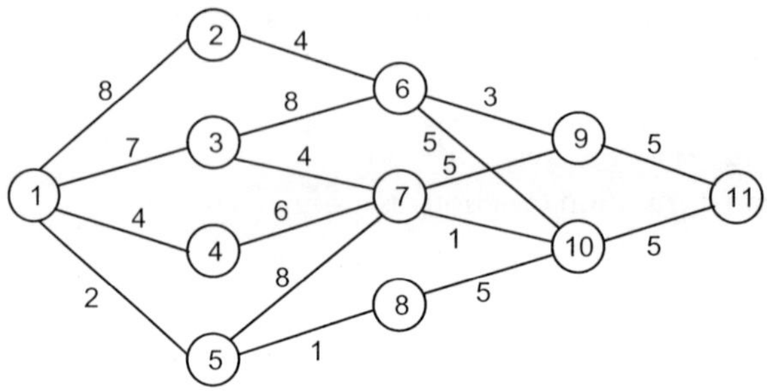

Fig. 6.2. Distance network.

Solution: Each pair of adjacent vertical columns of nodes is treated as a stage. As shown in figure 6.3, there are four stages in this problem. Since the stages are defined from right to left, backward recursive function is to be used.

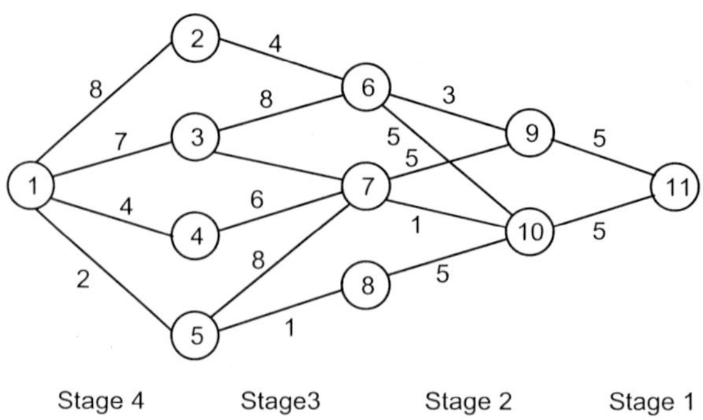

Fig. 6.3. Distance network with stages.

In Stage 1, the possible alternative is only one, *i.e.*, Node 11. In the same stage (stage 1), the possible state variables are Nodes 9 and 10.

The recursive function $f_1(x_1)$ for a given combination of the state variable, x_1 and alternative, m_1 in the first stage is:

$$f_i(x_1) = d(x_1, m_1)$$

where $d(x_1, m_1)$ is the distance between Node x_1 and Node m_1.

The *recursive function* for a given combination of the state variable, x_i and alternative, m_i in the stage i for i more than 1 is presented below.

$$f_i(x_i) = d(x_i, m_i) + f_{i-1}(x_{i-1} = m_i)$$

where d(x_i, m_i) be the distance between the nodes x_i and m_i, $f_2(x_i)$ is the shortest distance from Node x_i in the current Stage i to the destination node in Stage 1 (node 11).

Stage 1: The recursive function $f_1(x_1)$ for Stage 1 is:

$$f_1(x_j) = d(x_1, m_1)$$

The corresponding distances are shown in table. For each value of the state variable, the best distance and the corresponding alternative are presented in the last but one and the last columns, respectively of table.

Table: Calculations for Stage 1

State variable x_1	Alternative m_1 11	$f_1(x_1)^*$	m_1^*
9	5	5	11
10	8	8	11

Stage 2: The recursive function $f_2(x_2)$ for a given combination of the state variable, x_2 and alternative, m_2 in the second stage is:

$$f_2(x_2) = d(x_2, m_2) + f_1((x_1 = m_2)$$

The corresponding distances are summarized in table. For each value of the state variable, the best distance and the corresponding alternative are presented in the last two columns of table, respectively.

Table: Calculations for Stage 2

State variable	Alternative m_2			
x_2	9	10	$f_2(x_2)^*$	m_2^*
6	3 + 5 = 8	5 + 8 = 13	8	9
7	5 + 5 =10	1 + 8 = 9	9	10
8	–	5 + 8 = 13	13	10

Stage 3: The recursive function for $f_3(x_3)$ a given combination of the state variable, x_3 and alternative, m_3 is:

$$f_3(x_3) = d(x_3, m_3) + f_2(x_2 = m_3)$$

The corresponding distances are summarized in table. For each value of the state variable, the best distance and the corresponding alternative are presented in the last two columns of table, respectively.

Table: Calculations for Stage 3

State variable	Alternative m_3				
x_3	6	7	8	$f_3(x_3)^*$	m_3^*
2	4 + 8 = 12	–	–	12	6
3	8 + 8 = 16	4 + 9 = 13	–	13	7
4	–	6 + 9 = 15	–	15	7
5	–	8 + 9 = 17	1 + 13 = 14	14	8

Stage 4: The recursive function $f_4(x_4)$ for a given combination of the state variable, x_4 and alternative, m_4 in the forth stage is:

$$f_4(x_4) = d(x_4, m_4) + f_3(x_3 = m_4)$$

The corresponding distances are summarized in table. For each value of the state variable, the best distance and the corresponding alternative are presented in the last two columns of table, respectively.

Table: Calculations for Stage 4

State variable x_4	Alternative m_4 2	3	4	5	$f_4(x_4)^*$	m_4^*
1	8 + 12 = 20	7 + 13 = 20	4 + 15 = 19	2 + 14 = 16	16	5

The final results of the original problem are traced in tables towards backwards. Therefore, the shortest path is 1 – 5 – 8 – 10 – 11. Hence, the corresponding shortest distance = 16 units.

An Optimal Routing Problem

Example 6.9. Figure 6.4 shows a network of cities spread over a state. *A* co. has to transport some goods from city *A* to *J*. Cost of transportation between different cities is given along lines connecting nodes. *A* node represents city, it is required to determine the optimal route connecting *A* and *J*.

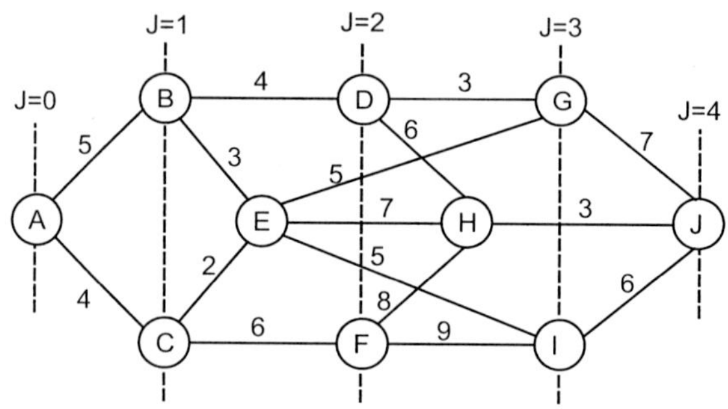

Fig. 6.4.

Stages $J_1, J_2, J_3,$

State value x_j $J = 0$

$$x_0 = A, \quad J = 1, \quad x_1 = B \,\&\, C$$

$$J = 2, \quad x_2 = D, E, \& F$$

Decision variables are let u_j

State 1 $\rightarrow u_1 = 5 \,\&\, 4$

2 $\rightarrow u_2 = 2, 3, 4 \,\&\, 6$

(2) F_2 (x_2) — Transportation cost $\rightarrow$ optimum (minimum)

$\downarrow$ $\downarrow$

Stage State $(D, E \,\&\, F)$

$$= F_1(x_1) + u_2 \text{ (previous + current)}$$

$J-1$ to J, x_{J-1} to x_J

Above problem can be divided into 4 stages. At each stage decision will comprise of selecting a path out of different possible alternatives. At stage 1 one of the two paths *AB* or *AC* is to be selected and at stage 3, one of alternatives *DH, EG, DG, EH,* DE *EI, FH, FI*. The optimal policy consists of a set of paths connecting *A* with *J*. The node reached represents the stage of the system. If x_j represents the stage then at $J = 0$, $x_0 = A$, at $J = 1$, $x_1 = B \,\&\, C$ in given problem. It means x_1 will have two values *B* & *C*. Possible alternative paths from one stage to next will be referred to as decision variables.

Let these be represented by U_j (decision variables) which takes us from stage $J-1$ to J and with this state changes from x_{J-1} to x_j and the minimum transportation cost from x_0 to x_j is $f_j(x_j)$ using backward computation: Let us start by selecting node J, stage 4, $x_4 = J$, $u_4 = 3, 6, 7$ to $\rightarrow x_3 = G, H$ & I.

Least of values of $F_4(x_4)$ is to be selected

$$F_4(x_4) = F_4(J) = \min \begin{cases} 7 + F_3(G) & \text{previous} \\ 3 + F_3(H) \\ 6 + F_3(I) \end{cases}$$

Stage 3, $J = 3$, $x_3 = G, H, I$

$$F_3(G) = \min \begin{cases} 3 + F_2(D) \\ 5 + F_2(E) \end{cases} \quad F_3(H) = \min \begin{cases} 6 + F_2(D) \\ 7 + F_2(E) \\ 8 + F_2(F) \end{cases}$$

$$F_3(I) = \min \begin{cases} 5 + F_2(E) \\ 9 + F_2(F) \end{cases}$$

Stage 2 $\quad J = 2, \quad x_2 = D, E$ & F, $u_2 = 2, 3, 4, 6$

$$F_2(D) = 4 + F_1(B) \qquad F_2(F) = 6 + F_1(C)$$

$$F_2(E) = \min \begin{cases} 3 + F_1(B) \\ 2 + F_1(C) \end{cases}$$

Stage 1, $J = 1$, $x_1 = B$ & C, $u_1 = 5$ & 4

$$F_1(B) = 5, F_1(C) = 4$$

$$u_4 = u_4 + F_3(x_3)$$

$$u_3 = u_3 + F_2(x_2), u_2 = u_2 + F_1(x_1)\ u_1 = u_1$$

General form $F_j(x_j) = \min U_J = U_J + F_{J-1}(x_{J-1})$

recursion equation

Forward Pass

Stage (1) $x_1 = B$ & C, $\qquad u_1 = 4, 5$

Table (1)

State x_1	Decision variable u_1	$f_1(x_1)$
B	5	5
C	4	4

Stage (2) $x_2 = D, E, F$ $\qquad u_2 = 4, 3, 2, 6$

Table (2)

u_2	$f_1(x_1)$				$u_2 + f_1(x_1)$				Optimal
x_2	2	3	4	6	2	3	4	6	$F_2(x_2)$
D	–	–	5	–	–	–	9	–	9
E	4	5	–	–	6	8	–	–	6
F	–	–	–	4	–	–	–	10	10

Stage (3) $J = 3, x_3 = G, H \& I$ $u_3 = 3, 6, 5, 7, 8, 9$

Table (3)

u_3	$f_2(x_2)$						$f_2(x_2) + u_3$						Optimal
x_3	3	6	5	7	8	9	3	6	5	7	8	9	$F_3(x_3)$
G	9	–	6	–	–	–	12	–	11	–	–	–	1
H	–	9	–	6	10	–	–	15	–	13	18	–	13
I	–	–	6	–	–	10	–	–	11	–	–	19	11

Stage (4) $J = 4, x_4 = J$

Table (4)

u_4	$F_3(x_3)$			$u_4 + f_3(x_3)$			Optimal
x_4	3	6	7	3	6	7	$f_4(x_4)$
J	13	11	11	16	17	18	16

Least cost of connecting A to J is **16**

$$F_4(J) = 16 = u_4 + F_3(x_3)$$
$$= 3 + 13 \qquad \text{Path } H - J$$
$$F_3(x_3) = u_3 + F_2(x_2) = 7 + 6 \qquad \text{Path } E - H$$
$$F_2(x_2) = u_2 + F_1(x_1) = 2 + 4 \qquad \text{Path } C - E$$
$$F_1(x_1) = u_1, \text{Path } AC$$

$A \to C \to E \to H \to J$ **is optimal route.**

EXERCISE

6.1. Define dynamic programming problem. What are applications of dynamic programming?

6.2. State and explain Bellman's principle of optimality in dynamic programming.

6.3. Write mathematical formulation of a dynamic programming problem.

6.4. Define dynamic programming problem. List and explain the terminologies of dynamic programming problem, also applications of this technique.
(UPTU: 2004-2005)

6.5. Discuss the use of Bellman's principle of optimality to solve multistage decision problems.

6.6. What do you understand from multistage decision process? Give example. State the "principle of optimality" and explain it giving examples. What do you understand from the "Concept of Sub-optimization"? **(UPTU: 2003-2004)**

6.7. Define following terms in dynamic programming:
(*i*) Stage
(*ii*) State
(*iii*) State variable
(*iv*) Decision variable
(*v*) Optimal return

6.8 What are characteristics of dynamic programming problem?

6.9. What are advantages and disadvantages of dynamic programming?

6.10. What are similarities and differences between dynamic programming and linear programming?

6.11. An investor has Rs. 6000 to invest. This amount can be invested any of three ventures available to him. But, he must invest in units of Rs. 1000. The potential return from investment in any one venture depends upon the amount invested according to the following table (all figure in thousands).

The investor wishes to invest Rs. 6000 so that the return from investment is maximum. Formulate the above problem as a dynamic programming problem and find the optimum investment policy.

[**Ans.** Rs. 3000 in A, Rs. 1000 in B and Rs. 2000 in C; Max. return Rs. 6900].

Amount Invested	Return from Venture		
	A	*B*	C
0	0	0	0
1	0.5	1.5	1.2
2	1.0	2.0	2.4
3	3.0	2.2	2.5
4	3.1	2.3	2.6
5	3.2	2.4	2.7
6	3.3	2.5	2.8

6.12. An oil company has 8 units of money available for exploration of 3 sites. If oil is present at site, the probability of finding it (depends upon the amount allocated for exploiting the site) is given below.

Units of money allocated

Site	0	1	2	3	4	5	6	7	8
1	0.0	0.0	0.1	0.2	0.3	0.5	0.7	0.9	1
2	0.0	0.1	0.2	0.3	0.4	0.6	0.7	0.8	1
3	0.0	0.1	0.1	0.2	0.3	0.5	0.8	0.9	1

The problem that oil exists at the site 1,2 and 3 is 0.4, 0.3 and 0.2 respectively. Find the optimum allocation of money. [**Ans.** 40%]

6.13. An electronic item has three components in series. So, the reliability of the system is equal to the product of the reliabilities of the three components ($R = r_1 r_2 r_3$). It is a known fact that the reliability of the system can be improved by providing standby units at extra cost. The details of costs and reliability for different number of standby units for each of the components of the system are summarized in the following table.

No. of standby units	Component 1		Component 2		Component 3	
	Cost (Rs)	Reliability	Cost (Rs.)	Reliability	Cost (Rs.)	Reliability
1	1	0.70	3	0.85	2	0.85
2	2	0.85	4	0.95	3	0.92
3	3	0.95	6	0.98	5	0.97

The total capital budgeted for this purpose is Rs. 8. Determine the optimal number of standby units for each of the components of the system such that the total reliability of the system is maximized. [**Ans.** Reliability = 0.6864]

6.14. The owner of a chain of 4 grocery stores has purchased 6 crates of fresh strawberries. The estimated profits at each store when it is allocated various no. of boxes is given below. Find allocation of 6 crates so as to maximize the profit.

Stores

No. of boxes	1	2	3	4
0	0	0	0	0
1	4	2	6	2
2	6	4	8	3
3	7	6	8	4
4	7	8	8	4
5	7	9	8	4
6	7	10	8	4

6.15. A distance network consists of eleven nodes which are distributed as shown in the following table. Find the shortest path from Node 1 to Node 11 and the corresponding distance. [**Ans.** 1 – 4 – 7 – 9 – 11; 19]

Arc	Distance	Arc	Distance
1–2	8	5 – 8	12
1–3	7	5 – 9	7
1–4	1	6 – 9	9
2–5	5	7 – 9	6
3–5	9	7 – 10	13
3–6	2	8 – 11	4
3–7	8	9 –11	2
4–7	10	10 –11	15

6.16. Solve the following linear programming problem using dynamic programming technique.

Maximize $Z = 30\,x_1 + 15x_2$

subject to $6x_1 + 8x_2 \leq 180$

$15x_2 \leq 210$

x_1 and $x_2 \geq 0$ [**Ans.** $x_1 = 30, x_2 = 0, z = 900$]

6.17. A Company has 3 medias available for advertising its product. The data collected over the past year about the relationship between sales frequency of advertising in different media is as follows.

	Estimate sales (unit) per month		
Freq/week	A	B	C
1	125	180	300
2	225	290	350
3	260	340	450
4	300	370	500

The cost of advt. is Rs. 5000/- in media A, Rs. 10,000 in B and Rs. 20,000/- in C. Total budget allocated for advt. the product is Rs. 40,000/-. Determine the optimal combination of advt. frequency. [**Ans.** A – 2, B – 1, C – 1]

❑❑❑

Unit III

Chapter 7

Decision Theory

7.1. INTRODUCTION

A decision is a selection from two or more courses of actions. It can be described as a process, which results in the selection from a set of alternative courses of action; considered to be best according to some pre-designed standard. Decision theory is the study of techniques for making decisions. Now a days, engineers, management students, businessmen, persons from industries and government, economists etc. are giving much emphasis over decision making under conditions of uncertainty as much of life involves making choices under uncertainty. In such situations, we are uncertain about the actual consequences that will occur for each course of action being considered. The study of making decisions to choose the best among a number of alternative courses of action after logical and quantitative analysis is known as **decision theory** or **statistical decision theory** or **Bayesian decision theory.**

So, in short, it can be said that decision analysis is concerned with the making of rational, consistent decisions, especially under condition of uncertainty. It helps the decision maker to choose the best alternative with the information available (which normally is incomplete and unreliable). It also enables the decision maker to analyze a complex situation with many different alternatives, states and consequences.

7.2. APPLICATIONS

1. Design and development of a new product. The development from invention to commercialization is an expensive process having uncertainty regarding both technical and commercial success.
2. Selecting the best from among several job offers.
3. An agriculture firm selecting the mixture of crops for upcoming season with uncertain weather conditions.
4. Determining whether or not to expend a manufacturing facility.
5. Determining the best process for manufacturing a particular product.
6. Selecting the most profitable investment portfolio.
7. Determine the number of computers required in an office.
8. In Research & Development, decision maker might be faced with problem of choosing whether to pursue two or more designs simultaneously or developing the most promising design first, and if it fails, going for other designs.

7.3. STEPS IN DECISION THEORY

A problem of making decision can be solved by using following steps:

1. Clearly identify and define the given problem.
2. Specify objectives and decision criteria.
3. Identify and evaluate the possible alternative courses of actions.
4. Select one decision theory model.
5. Use the selected model and find the best course of action.
6. Carry out the sensitivity analysis.
7. Implement the decision.
8. Take feedback of the decision taken.

7.4. ELEMENTS OF A DECISION PROBLEM

1. **The decision maker:** The decision maker is the one who makes decisions. The decision maker can be an individual, any company, industrial body, any country etc.
2. **Alternative courses of action:** There can be different ways in which a problem can be solved or a task can be performed. These different courses/ways are termed as Alternative courses of action. Each course of action has a specified status. The decision maker has options of different courses of action and his job is to determine the action, which will give the maximum benefit or the minimum loss.
3. **Event:** Also known as state or states of nature or the states of the world. Events are mutually exclusive and collectively exhaustive i.e. only one of all the possible events will occur. These are the states of environment, which are not under the control of decision maker. Under condition of uncertainty decision maker is completely unknown or ignorant of which event will occur. Uncertainty is measured in terms of probabilities assigned to the events. *e.g.* The demand of a product during a specified time period.
4. **Pay-off or Consequences:** It is a quantitative measure of the value, to the decision maker, of the consequences or the outcomes. It is the net benefit received by the decision maker from a given combination of course of action and event. Also known as outcome, benefits or losses. They can be easily summarized in a payoff matrix or decision matrix, which displays the consequences of all course of action-event combinations. There is a consequence associated with each course of action-event pair.

   The pay-off usually represents the net monetary gain (profit), but some other measures can also be used, as cost is negative profit. If the consequences of the outcome are not completely certain then the pay-off becomes an 'expected value' of the measure of the consequence.
5. **Pay-off table:** It is commonly used to provide the pay-off for each combination of a course of action and a state of nature.

Events	*Pay-off (Rs.) Courses of actions/strategies*				
	A_1	A_2	A_3	$A_{4...}$	A_n
S_1	p_{11}	p_{12}	p_{13}	p_{14}	p_{1n}
S_2	p_{21}	p_{22}	p_{23}	p_{24}	p_{2n}
$\vdots$	$\vdots$	$\vdots$	$\vdots$	$\vdots$	$\vdots$
S_m	p_{m1}	p_{m2}	...	...	p_{mn}

6. **Regret or opportunity loss:** It is the difference between the highest possible profit for a state of nature and the actual profit obtained for a particular action taken i.e. the opportunity loss is the loss incurred due to failure of not adopting the best possible course of action.

   Opportunity loss of a given course of action = pay-off of best course of action – pay-off of that course of action

7. **Regret table or opportunity loss table:** It gives the regret or opportunity loss values for each combination of a course of action and a state of nature.

States of nature	*Opportunity loss (Rs.) courses of actions / strategies*			
	A_1	A_2	A_3 ...	A_n
S_1	$M\text{-}p_{11}$	$M\text{-}p_{12}$	$M\text{-}p_{13}$ ...	$M-p_{1n}$
S_2	$M\text{-}p_{21}$	$M\text{-}p_{22}$	$M\text{-}p_{23}$ ...	$M-p_{2n}$
$\vdots$	$\vdots$	$\vdots$	$\vdots$	
S_m	$M\text{-}p_{m1}$	$M\text{-}p_{m2}$	$M\text{-}p_{m3}$ ...	$M-p_{mn}$

Where M = maximum profit, p_{ij} = actual profit

7.5. CLASSIFICATION OF DECISION MAKING

1. **Decision making without experimentation:** In this case, the decision maker decides the decision immediately without performing any type of experiment.

2. **Decision making with experimentation:** In this case, an experimentation or testing is done before reaching on the decision to reduce the level of uncertainty about the net outcome of decision e.g. if a new product is planned to be launched in the market and if the customer reaction, his expectation and demands are studied before going for full production of the product, it is said to be making an experimentation. And the decision that is made after this experimentation, comes under the category of decision making with experimentation.

   e.g. An oil company has a piece of land that may contain oil. Based on the report of a geologist, it appears to have some chances of containing oil. Relying on the geologist report, another company wants to buy this land. Now the oil company, which has land, is in dilemma whether to dig wells for oil or to sell the land. Any decision taken at this stage will be the - decision without experimentation. The second option is to conduct a detailed seismic survey of the land to have a better estimate of finding oil in the land. The decision made after this additional data comes under the - decision making with experimentation.

Depending upon the level of uncertainty of the outcome of each alternative courses of action, the decision making environment can be further classified in following three categories:

(*A*) Decision making under condition of certainty (deterministic decision)

(*B*) Decision making under condition of risk

(*C*) Decision making under condition of uncertainty (probabilistic decision)

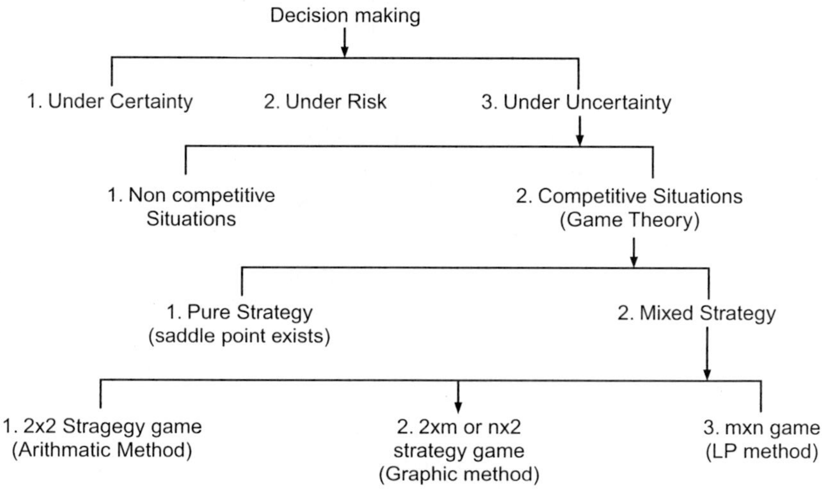

Fig. 7.1.

7.6. DECISION MAKING UNDER CONDITION OF CERTAINTY

If there is complete information about the outcome of each alternative known to decision maker, then the decision taken is known as deterministic decision. There will be only one outcome for each alternative. Various techniques used to solve such decision problems are Linear programming, Dynamic programming, Integer programming, Queuing models, inventory models, Break-even analysis etc.

The example of such type of problems are-

1. In case of maintenance, if the annual maintenance cost and the annual operating costs of an equipment is known in advance and if these values are not going to vary in future, then the decision taken on the basis of this information is called decision under certainty.
2. For the economic evaluation of alternatives, if the annual revenues of different alternatives are known and that are not subjected to change in future, then the decision taken under such condition is a deterministic decision. Like, an investor wish to invest Rs. 70,000 for 6 years period. One alternative is to open a saving bank account in a bank giving 3.5 % interest and the other alternative is to invest in PPF paying 8.5 % interest. If both investments are secure then PPF will be the better option.

7.7. DECISION MAKING UNDER CONDITION OF RISK

In this case, the availability of information is partial and it can be described in form of a probability distribution. From the past experience, past data of decision environment, the parameters of probability distribution can be estimated for likely possible occurrence of each state of nature. The best decision is to select that course of action, which gives the largest expected payoff value. e.g.

1. The arrival pattern and service pattern in queuing theory cannot be ascertained with 100 percent surety so some approximation is made about these patterns using the various distribution models as Poisson distribution, exponential distribution etc. These decisions are the decision under condition of risk.
2. In the economic evaluation of alternatives, if the annual revenue of different alternatives is described in form of probability distribution, then the decisions come under the category of decisions under risk

Following decision criterion are used to evaluate the alternative courses of action:

1. Expected Monetary value (EMV) criterion

In this decision criterion, the objective is to optimize the expected payoff.

(*i*) Construct the payoff table listing all possible courses of action and states of nature.

(*ii*) Enter the pay-off values (conditional profit values) for each decision-event (alternative course of action - state of nature) combination and their associated probabilities.

(*iii*) Calculate expected pay-off values for each course of action (*i.e.*, EMV) by multiplying the pay-off values by their assigned corresponding probabilities and add these weighted values for each course of action.

(*iv*) Select the alternative having highest EMV.

$$EMV(A_i) = p_{i1}P(S_1) + p_{i2}\,P(S_2) + p_{i3}\,P(S_3) + p_{i4}\,P(S_4) + \dots + p_{im}\,P(S_m)$$

Where $P(S_j)$ = probability of occurrence of jth state of nature

Example 7.1. Determine EMV for each alternative

Solution: (*i*) Conditional pay off table

States of nature S_j	*Probability* $P(S_j)$	*Conditional pay off (Rs) Courses of action* $A_1(0)$	$A_2(1)$	$A_3(2)$	$A_4(3)$
$S_1(0)$	0.1	0	– 30	– 60	– 90
$S_2(1)$	0.2	0	10	– 20	– 50
$S_3(2)$	0.3	0	10	20	– 10
$S_4(3)$	0.4	0	10	20	30

(*ii*) Expected pay off table

States of nature S_j	*Probability* $P(S_j)$	*Expected pay off (Rs.) courses of action* A_1 $(P(S_j) \times 0)$	A_2 $(P(S_j) \times 1)$	A_3 $(P(S_j) \times 2)$	A_4 $(P(S_j) \times 3)$
$S_1(0)$	0.1	0	– 3	– 6	– 9
$S_2(1)$	0.2	0	2	– 4	– 10
S_3 (2)	0.3	0	3	6	– 3
S_4 (3)	0.4	0	4	8	12
Expected monetary value (EMV) →		0	6	4	– 10

The maximum value of EMV is 6 corresponding to course of action A_2.

2. Expected Opportunity Loss (EOL) criterion:

In this decision criterion, approach is to minimize expected opportunity loss (EOL). EOL is the loss, which the decision maker is suffering due to lack of perfect information. EMV and EOL, both criterion lead to the choice of the same course of action.

(*i*) Construct the opportunity loss table listing all possible courses of action and states of nature.

(*ii*) Enter the conditional opportunity loss value and their probabilities for each decision - event (alternative course of action - state of nature).

(*iii*) Calculate conditional opportunity loss (COL) by subtracting the pay-off from the maximum pay-off for that state of nature.

(*iv*) Calculate EOL by multiplying COL by associated probabilities.

(*v*) Select the alternative having lowest EOL.

$$EOL(A_i) = (M_1 - p_{1j})\,P(S_1) + (M_2 - p_{2j})\,P(S_2) + \dots \dots + (M_m - p_{mj})\,P(S_m)$$

Where M_j = maximum payoff of jth state of nature.

Example 7.2. Find EOL for each alternative using data of example 7.1. Which is optimal strategy?

Solution: (*i*) Conditional opportunity loss table

States of nature	*Probability*	*Conditional opportunity loss (Rs.) courses of action*			
S_j	$P(S_j)$	A_1 (0)	A_2 (1)	A_3 (2)	A_4 (3)
S_1 (0)	0.1	0	30	60	90
S_2(1)	0.2	10	0	30	60
S_3 (2)	0.3	20	10	0	30
S_4(3)	0.4	30	20	10	0

COL = $(M_j - p_{ij})$

(*ii*) Expected opportunity loss table

States of nature	*Probability*	*Expected opportunity loss (Rs.) courses of action*			
S_j	$P(S_j)$	A_1 $P(S_j) \times (0)$	A_2 $P(S_j) \times 1$	A_3 $P(S_j) \times 2$	A_4 $P(S_j) \times 3$
S_1(0)	0.1	0	3	6	9
S_2(1)	0.2	2	0	6	12
S_3(2)	0.3	6	3	0	9
S_4(3)	0.4	12	8	4	0
Expected opportunity loss (EOL) ⟶		24	14	16	30

Minimum value of EOL is 14 corresponding to A_2. Therefore optimal strategy is A_2 course of action.

3. Expected value of perfect information (EVPI):

Perfect information is complete and accurate information about the future event. Some amount is to be paid to receive this extra information. So EVPI gives that value (maximum) which the decision maker spends to get additional information.

EPPI can be calculated by finding sum of product of payoff of best outcome of each state of nature and its probability of occurrence.

EPPI = Σ Best outcome of jth state of nature × probability of jth state of nature

As this value cannot go beyond the minimum EOL value so,

Min. EOL = EVPI

or EVPI = EPPI – max. EMV

Where EPPI = Expected profit with perfect information.

maximum EMV = Expected profit without perfect information.

Example 7.3. Calculate EVPI for pay off table of example 7.1.

Solution: Table for EPPI

States of nature S_j	*Probability* $P(S_j)$	*Conditional pay off (Rs.) courses of action* A_1 (0)	A_2 (1)	A_3 (2)	A_3 (3)	$P(S_j)$ × *best outcome*
$S_1(0)$	0.1	0	– 30	– 60	– 90	0.1 × 0 = 0
$S_2(1)$	0.2	0	10	–20	– 50	0.2 × 10 = 2
$S_3(2)$	0.3	0	10	20	– 10	0.3 × 20 = 6
$S_4(3)$	0.4	0	10	20	30	0.4 × 30 = 12
Expected profit with perfect information (EPPI)→		EPPI = 0 + 2 + 6 + 12 = 20				

EVPI = EPPI – max EMV
= 20 – 6
= 14
= Minimum EOL (Example 7.2)

4. Incremental or Marginal analysis:

For large demand and stock conditions, a lot of calculations are to be performed in finding pay-off values, which results in wastage of time and also the possibilities of error making increases. Therefore to cut short lengthy calculations, marginal analysis is performed. This analysis gives the information about the condition in which the additional units are beneficial to be stored or stocked.

Let P = Probability of selling one additional unit

$(1 - P)$ = Probability of not selling this additional unit

There will be two conditions, if any additional unit is purchased

1. This additional unit is sold. Thus the total profit will increase because of the addition of the profit earned by the selling of this unit. This increase in profit is called incremental (marginal) profit (IP).
2. This additional unit is not sold. The profit will reduce and this reduction is called incremental (marginal) loss (IL).

The decision of storing the additional units is reasonable only when the incremental profit from stocking each of them is greater then or equal to the incremental loss from stocking each *i.e.*, expected incremental profit > expected incremental loss.

$$P(IP) > (1 - P)(IL)$$

or at least, $$P(IP) = (1 - P)(IL)$$

$$P(IP) = IL - (P)IL$$

or $$P(IP + IL) = IL$$

or $$P = IL/(IP + IL)$$

Thus the additional units should be stored so long as the probability of selling at least one additional unit is greater than *P*.

7.8. DECISION MAKING UNDER CONDITION OF UNCERTAINTY

If the information about the outcomes is incomplete and the information available cannot be described in terms of probability density, then the decision taken is called the decision under conditions of uncertainty. Multiple outcomes are there for each alternative and there is no knowledge of how to assign probability attached to each. *e.g.* (1) when a new product is introduced in the market or a new plant is set up, there is no historical data available. (2) in game theory there are two competitors. For each combination of alternatives of players, there will be an associated outcome. Both players will play the game simultaneously for their own benefits. Hence it is not possible to describe any probability distribution to define this decision environment. There are mainly five criteria for decision making under uncertainty.

1. **Maximax (optimistic) criterion:** It was suggested by Leonid Hurwitz. It maximizes the maximum payoff. Steps involved are:
   (*i*) Determine the maximum possible pay-off for each alternative course of action.
   (*ii*) Choose the alternative course having maximum pay-off from the previously selected maximum pay-off group.

Example 7.4. Determine maximax strategy in pay off matrix of example 7.1

Solution:

States of nature or demand	*Courses of action (supply)*			
	A_1 *(0)*	A_2*(1)*	A_3*(2)*	A_4*(3)*
$S_1(0)$	0	– 30	– 60	– 90
$S_2(1)$	0	10	– 20	– 50
$S_3(2)$	0	10	20	– 10
$S_4(3)$	0	10	20	30
Maximum in columns (Rs.) ⟶	0	10	20	30 (Maximax)

Therefore, Maximax strategy is A_4 of having 3 items with pay off 30.

2. **Maximin (Pessimistic) criterion:** It was suggested by Arabian Wald, hence also called as Waldian criterion. It maximizes the minimum possible. Steps involved are:
   (*i*) Find the minimum possible pay-off for each alternative course of action.

(*ii*) Then select the maximum pay-off alternative out of these selected minimum pay-offs.

Example 7.5. Find maximin strategy in example 7.4.

Solution:

States of nature or Demand (S_j)	*Courses of action (Supply)-Pay off* $A_1(0)$	$A_2(1)$	$A_3(2)$	$A_4(3)$
$S_1(0)$	0	– 30	– 60	– 90
$S_2(1)$	0	10	– 20	– 50
$S_3(2)$	0	10	20	– 10
$S_4(3)$	0	10	20	30
Minimum in columns (Rs.) →	0 ↑ Maximin	– 30	– 60	– 90

3. **Minimax regret criterion (Savage criterion):** It was given by Leonard Savage. Various steps involved are:
   (*a*) Construct the regret table.
   Regret (opportunity loss) payoff = maximum payoff for a state of nature – payoff
   (*b*) Find the maximum regret for each alternative course of action.
   (*c*) Select the course of action with minimum regret out of these maximum regrets.

Example 7.6. Find minimax regret strategy in example 7.2.

Solution: **Regret Table**

States of nature or Demand (S_j)	*Courses of action (supply) Opportunity loss (regret)* $A_1(0)$	$A_2(1)$	$A_3(2)$	$A_4(3)$
$S_1(0)$	0	30	60	90
$S_2(1)$	10	0	30	60
$S_3(2)$	20	10	0	30
$S_4(3)$	30	20	10	0
Maximum in column (Rs.) →	30 ↑ Minimax	30 ↑ Minimax	60	90

4. **Hurwicz criterion/criterion of realism/weighted average criterion:** It is a compromise between maximax criterion and maximin criterion. According to degree of optimism (or Pessimism), weightage is given to different pay-offs. Maximum sum of these weighted pay-off is selected.

Decision index $(D) = \alpha M_1 + (1 - \alpha)M_2$

Where α = degree of optimism,

M_1 = maximum pay-off in column

M_2 = minimum pay-off in column

$\alpha = 0$ (for maximin criterion)

$\alpha = 1$ (for maximax criterion)

Example 7.7. Find optimal strategy in example 7.1 using Hurwicz criterion with $\alpha = 0.6$.

Solution: Pay off table

States of nature (S_j)	*Conditional pay off (Rs) of alternatives* $A_1(0)$	$A_2(1)$	$A_3(2)$	$A_4(3)$
$S_1(0)$	0	– 30	– 60	– 90
$S_2(1)$	0	10	– 20	– 50
$S_3(2)$	0	10	20	– 10
$S_4(3)$	0	10	20	30

$$D = \alpha\, M_1 + (1 - \alpha)\, M_2$$
$$D_{(A1)} = 0.6 \times 0 + (1 - 0.6) \times 0 = 0$$
$$D_{(A2)} = 0.6 \times 10 + (1 - 0.6) \times (-30) = -6$$
$$D_{(A3)} = 0.6 \times 20 + (1 - 0.6) \times (-60) = -12$$
$$D_{(A4)} = 0.6 \times 30 + (1 - 0.6) \times (-90) = -18$$

$D_{(A1)} = 0$ is maximum, therefore A_1 is optimal strategy.

5. **Laplace or Bayes criterion (criterion of rationality):** It was developed by Thomas Bayes and Laplace. Equal probabilities to all events of each alternative course are assigned, as there is no information about probabilities of occurrence. Steps involved are:
   (*a*) Assign equal probabilities to each state of nature.
   (*b*) Calculate expected value for each alternative by multiplying each payoff by its probability and then adding.
   (*c*) Select the alternative course having maximum expected pay-off value.

Example 7.8. Find optimal strategy using Laplace criterion in example 7.1.

Solution: Since, there are 4 states of nature.

Equal probability of each $= \frac{1}{4}$

States of nature (S_j)	*Probability* $P(S_j)$	*Expected pay off (Rs.) of alternatives* A_1 $P(S_j) \times 0$	A_2 $P(S_j) \times 1$	A_3 $P(S_j) \times 2$	A_4 $P(S_j) \times 3$
$S_1(0)$	0.25	0	– 7.5	– 15	– 22.5
$S_2(1)$	0.25	0	2.5	– 5	– 12.5
$S_3(2)$	0.25	0	2.5	5	– 2.5
$S_4(3)$	0.25	0	2.5	5	7.5
Σ(expected pay off) ⟶		0	0	– 10	– 30

Optimal strategy (A_1 and A_2)

Example 7.9. A news paper boy has following probabilities of selling a magazine:

No. of copies	Probability
10	0.10
11	0.15
12	0.20
13	0.25
14	0.30
	1.00

Cost of a copy is Rs. 30 and sales price is Rs. 50.

He can not return unsold copies. How many copies should he order?

(UPTU: 2004-2005)

Solution: (*i*) The profit resulting from any possible combination of supply and demand of magazine is shown in conditional profit table below:

Possible demand (no. of copies)	*Probabilities*	*Conditional profit (Rs.) Possible stock action*				
		10	*11*	*12*	*13*	*14*
10	0.10	200*	170	140	110	80
11	0.15	200	220	190	160	130
12	0.20	200	220	240	210	180
13	0.25	200	220	240	260	230
14	0.30	200	220	240	260	280

* Pay off = (SP – CP) × no. of copies sold – CP × unsold copies

= (50 – 30) × 10 – 30 × 0 = 200 – 0 = 200

(*ii*) Expected profit is obtained by multiplying conditional profit by associated probabilities.

Expected profit table

Possible demand (no. of copies)	*Probabilities*	*Expected Profit (Rs.)*				
		10	*11*	*12*	*13*	*14*
10	0.10	20	17	14	11	8
11	0.15	30	33	28.5	24	19.5
12	0.20	40	44	48	42	36
13	0.25	50	55	60	65	57.5
14	0.30	60	66	72	78	84
Expected value	Monetary (EMV) ⟶	200	215	222.5	220	205

News paper boy should order 12 copies to maximise his net profit (Rs. 222.5) over a period of time.

Example 7.10. A certain output is manufactured at Rs. 80 and sold at Rs. 140 per unit. The product is such that if it is produced but not sold during a day's time it becomes worthless. The daily sales records in the past are as follows :

Sales per day :	30	40	50	60	70
No. of days :	24	24	36	24	12

(*i*) Prepare a payoff and a regret table. (*ii*) Find the expected payoffs and regret. (*iii*) Find the optimal act and the *EVPI.*

Solution: Since number of items to be stocked is under control of decision-maker, number of items to be stocked per day is considered as an 'act' or 'course of action' (choice) and the daily demand of the items is uncertain and only known with probability, it is considered as an 'event' or state of nature'. Also each item sold within a day yields a profit of Rs. (140 – 80) = Rs. 60 and otherwise it is a loss of Rs. 80.

The various conditional payoff (or profit) values for each combination of act—event are given by:

Conditional profit =MP (items sold) –ML (items not sold)

$$= \begin{cases} (140-80)\,D = 60\,D & \text{if } D \geq S \\ (140-80)\,D - 80\,(S-D) = 140\,D - 80S, & \text{if } D < S \end{cases}$$

where D denotes the number of items sold within a day and S the number of items stocked.

The resulting conditional payoff values and corresponding expected payoffs are calculated in following table.

Table 7.1. Computation of Expected Pay-offs

State of Nature (Demand) S_j	*Prob $P(S_j)$*	*Conditional payoff (Rs) Courses of Action (Stock)* A_1 30	A_2 40	A_3 50	A_4 60	A_5 70	*Expected payoff (Rs.) Courses of Action (Stock)* A_1 30	A_2 40	A_3 50	A_4 60	A_5 70
	(1)	*(2)*	*(3)*	*(4)*	*(5)*	*(6)*	*(1)×(2)*	*(1)×3)*	*(1)×(4)*	*(1)×(5)*	*(1)×(6)*
S_1:30	0.2	1800	1000	200	–600	–1400	360	200	40	–120	–280
S_2:40	0.2	1800	2400	1600	800	0	360	480	320	160	0
S_3:50	0.3	1800	2400	3000	2200	1400	540	720	900	660	420
S_4:60	0.2	1800	2400	3000	3600	2800	360	480	600	720	560
S_5:70	0.1	1800	2400	3000	3600	4200	180	240	300	360	420
Expected Monetary Value (EMV):							1800	2120	2160	1780	1120

Since the course of action 'Stock 50 items' yields the highest EMV of Rs 2,160, the optimal course of a action would be to produce or stock 50 items every day.

Table 7.2. Expected Regret (Opportunity Loss Values) Table

State of Nature (Demand) S_j	*Prob $P(S_j)$*	*Conditional opportunity loss (Rs) Courses of Action (Stock)* A_1 30	A_2 40	A_3 50	A_4 60	A_5 70	*Expected opportunity loss (Rs.) Courses of Action (Stock)* A_1 30	A_2 40	A_3 50	A_4 60	A_5 70
	(1)	*(2)*	*(3)*	*(4)*	*(5)*	*(6)*	*(1)×(2)*	*(1)×(3)*	*(1)×(4)*	*(1)×(5)*	*(1)×(6)*
S_1:30	0.2	0	800	1600	2400	3200	0	160	320	480	640
S_2:40	0.2	600	0	800	1600	2400	120	0	160	320	480
S_3:50	0.3	1200	600	0	800	1600	360	180	0	240	480
S_4:60	0.2	1800	1200	600	0	800	360	240	120	0	160
S_5:70	0.1	2400	1800	1200	600	0	240	180	120	60	0
Expected Opportunity Loss(EOL):							1080	760	720	1100	1760

Minimum EOL is 720. It gives same optimal course of action i.e., stock 50 items.

Example 7.11. The sales manager of Star Ltd. is highly experienced in the food market. He is sure that the sales of JUMBO (during the period it has special appeal) will not be less than 25,000 units. Plant capacity limits total production to a maximum of 80,000 units during JMBO's brief life. According to the sales manager, there are 2 chances in 5 for a sales volume of 50,000 units. The probability that it will be more than 50,000 is four times the probability that it will be less than 50,000. If sales exceed 50,000 units, volumes of 60,000 and 80,000 are equally likely. A 70,000 unit volume is 4 times as likely as either. It costs Rs. 30 to produce a unit of JUMBO whereas its selling price is estimated at Rs. 50 per unit. Initial investment is estimated at Rs. 8.00,000 (*i*) Should venture of production be undertaken?(*ii*) What plant capacity would you suggest?

Solution: Let probability of sale of less than 50,000 units = x so that prob. of sale of more than 50,000 units = $4x$. Also prob. of sale of 50,000 units $= \frac{2}{5}$

Thus $\quad x + \frac{2}{5}x + 4x = 1 \Rightarrow x = 0.12$

Let prob. of sale of 60,000 units = Prob. of sale of 80,000 units = y, so that prob. of sale of 70,000 units = $4y$

$\therefore \quad y + 4y + y = 4x = 0.48 \Rightarrow y = 0.08$

Now, we have complete probability distribution of sales except that for sales less than 50,000, we have a summarised probability of 0.12.

Table 7.3.

State of Nature (Demand)	*Prob.*	*Conditional Payoff (Rs.) Courses of Action*		*Expected Payoff (Rs.) Courses of Action*	
		Produce	*Not produce*	*Produce*	*Not produce*
25,000	0.12	– 3,00,000	—	– 36.000	—
50,000	0.40	2,00,000	—	80,000	—
60,000	0.08	4,00,000	—	32,000	—
70,000	0.32	6,00,000	—	1,92,000	—
80,000	0.80	8,00,000	—	65,000	—
Expected Monetary Value (EMV):				3,32,000	Nil

Since EMV of the act (to produce) is positive, Jumboo should be produced.

Table 7.4. Payoff Table and calculation of EMV Values

State of Nature (Demand)	*Prob.*	*Conditional Payoff (Rs.) Courses of Action (Strategies)*				
		A_1 *25,000*	A_2 *50,000*	A_3 *60,000*	A_4 *70,000*	A_5 *80,000*
S_1 : 25,000	0.12	5,00,000	–25,000	55,000	–85,000	–115,000
S_2 : 50,000	0.40	5,00,000	1,00,000	70,000	40,000	10,000
S_3 : 60,000	0.08	5,00,000	10,00,000	1,20,000	90,000	60,000
S_4 : 70,000	0.32	5,00,000	10,00,000	1,20,000	1,40,000	1,10,000
S_5 : 80,000	0.08	5,00,000	10,00,000	1,20,000	1,40,000	1,60,000
Expected Monetary Value (EMV):		5,00,000	8,50,000	79,000	69,000	43,000

Since EMV of the course of action A_2 is greatest, the plant capacity should be 50,000 units.

Example 7.12. Professional investment is considering which of two methods it should use to market its investment services to the public : one is direct mailing and the other is newspaper advertisement. It regards these forms of marketing as mutually exclusive. It has a budget for expenditure on marketing of Rs. 2,00,000. Cost of direct mailing is 25 paise for each 'shot'. Previous experience leads the company to expect a response rate of between 6% and 12% with an average of 8%. The chances of the lower, higher and average response rate actually occurring are estimated to be 15%, 20% and 65% respectively. Newspaper advertisements have also been used in the past and these also produce varying response rates. Company's budget would also allow it to run a campaign of weekly insertions, (*i.e.*, 52 in total) in certain suitable Sunday newspapers. Again, based on past experience, it can expect response rates varying between 700 and 2,000 per insertion, with an average response rates actually occurring are 20%, 25% and 55% respectively. In either case only 40% of the responses can be expected to produce a sale. Each sale generates a net income of Rs.10. Which method would you recommend?

Solution: Direct mailing :

$$\text{Number of shots} = \frac{2,00,000}{0.25} = 8,00,000$$

Number of responses:

Low : $0.06 \times 8,00,000 = 48,000$

High : $0.12 \times 8,00,000 = 96,000$

Average : $0.08 \times 8,00,000 = 64,000$

Each shot produces a net income of 0.4×10 = Rs. 4 so that the expected net annual incomes are:

Low : $48,000 \times 4$ = Rs. 1,92,000

High : $96,000 \times 4$ = Rs. 3,84,000

Average : $64,000 \times 4$ = Rs. 2,56,000

Newspaper advertisement:

Using newspaper advertisements, the net incomes for each level of response are:

Low : $700 \times 52 \times 4$ = Rs. 1,45,600

High : $2,000 \times 52 \times 4$ = Rs. 4,16,000

Average : $1,400 \times 52 \times 4$ =Rs. 2,91,200

Table 7.5. Calculation of EMV Values

States of Nature (Response)	*Direct Mail (A_1)* *Net income (In Rs.'000)*	*prob.*	*Expected payoff*	*Newspaper Advertisement (A_2)* *Net income (in Rs.000)*	*Prob.*	*Expected payoff*
	(1)	(2)	(1)× (2)	(3)	(4)	(3) ×(4)
Low	192	0.15	28.8	145.6	0.20	29.12
High	384	0.20	76.8	416.0	0.25	104.00
Average	256	0.65	166.4	291.2	0.55	150.6
Expected MonetaryValue (EMV):			Rs. 272.0			Rs. 293.28

Since EMV is highest for courses of action A_2, newspaper advertisement is the beast option.

Example 7.13. A wholesaler of sports goods has an opportunity to buy 5,000 pairs of gloves that have been declared surplus by the government. The wholesaler will pay Rs. 50 per pair and can obtain Rs. 100 a pair by selling gloves to retailers. The price is well-established, but the wholesaler is in doubt as to just how many pairs he will be able to sell. Any gloves left over, he can sell to discount outlets at Rs. 20 a pair. After a careful consideration of the data, the wholesaler assigns probabilities to the demand as follows :

Retailer's demand	*Probability*
1,000 pairs	0.6
3,000 pairs	0.3
5,000 pairs	0.1

(*a*) Calculate the conditional monetary and expected monetary values.

(*b*) Calculate the expected profit with a perfect predicting device.

(*c*) Calculate the EVPI.

Solution : We are given:

$$\begin{aligned} \text{Cost per pair} &= \text{Rs. } 50 \\ \text{Selling price per pair} &= \text{Rs. } 100 \\ \text{Profit per pair} &= \text{Rs. } 50 \text{ (if sold)} \\ \text{Disposal selling price} &= \text{Rs. } 20 \text{ (if unsold)} \\ \text{Loss on each unsold pair} &= (50\text{–}20) = \text{Rs. } 30 \end{aligned}$$

Conditional profit values are therefore computed by

$$CP = \begin{cases} 50S & \text{when } D > S \\ 50D - 30\,(S - D) & \text{when } D < S \end{cases}$$

where CP = conditional profit, D = pairs demanded and S= pairs stocked.

(*a*) The resulting conditional payoffs and corresponding expected payoffs are calculated in the table below:

		Conditional payoffs (Rs.) (Stock per week)			*Expected payoffs (Rs.) (stock per week)*		
Retailer's demand	*Probability*	*1,000 pairs*	*3,000 pairs*	*5,000 pairs*	*1,000 pairs*	*3,000 pairs*	*5,000 pairs*
1,000 pairs	0.6	50	– 10	– 70	30	– 6	– 42
3,000 pairs	0.3	50	150	90	15	45	27
5,000 pairs	0.1	50	150	250	5	15	25
				EMV	50	54	10

(*b*) The expected profit under perfect information (*EPPI)* is computed below:

		Conditional payoffs			*Under perfect information*	
Retailer's demand	*Probability* (*i*)	*1,000 pairs* (*ii*)	*3,000 pairs* (*iii*)	*5,000 pairs* (*iv*)	*Maximum payoff* (*v*)	*Expected payoff* (*i*)×(*v*)
1,000 pairs	0.6	50	–10	–70	50	30
3,000 pairs	0.3	50	150	90	150	45
5,000 pairs	0.1	50	150	250	250	25
						EPPI 100

(*c*) $EVPI = EPPI - EMV = 100 - 54 = 46$

Thus, $EVPI$ = Rs. 46, 000.

Example 7.14. A man has the choice of running either a hot-snack stall or an ice-cream stall at a seaside resort during the summer season. If it is a fairly cool summer, he should make Rs. 5,000 by running the hot-snack stall, but if the summer is quite hot he can only expect to make Rs. 1,000. On the other hand, if he operates the ice-cream stall, his profit is estimated or Rs. 6,500 if the summer is hot, but only Rs. 1000 if it is cool. There is a 40% chance of the summer being hot. Should he opt for running the hot-snack stall or the ice-cream stall ? Give mathematical argument.

Solution. The following payoff table shows the conditional pay-off resulting from the given act-event combinations:

Event E_i (Weather)	Probability $P(E_i)$	*Conditional payoff (Rs.) (Action)* Hot-snack stall	Ice-cream stall
	(i)	*(ii)*	*(ii)*
Cool summer	0.60	5,000	1,000
Hot summer	0.40	1,000	6,500

The expected conditional payoffs are computed as below:

Event Ei (Weather)	*Probability(Ei)* (i)	*Conditional payoff (Rs.) (Action)* Hot-snack stall (i)×(ii)	Ice-cream stall (i)×(iii)
Cool summer	0.6	3,000	600
Hot summer	0.4	400	2,600
	EMV	Rs 3,400	Rs 3,200

Since the expected monetary value of running a hot-snack stall is higher, the man should opt for running a hot-snack stall.

Example 7.15. The Goodwill company owns a land that may contain oil. A according to geologist report there is 1 chance in 4 of land having oil. The company is considering to drill the land for oil. The cost of drilling is Rs. 1,000,00. If oil is found, the resulting expected revenue will be Rs. 8,000,00. So the company's expected profit (after deducting the cost of drilling) will be Rs. 7,00,000. A loss of Rs. 1,00,000 (the drilling cost, will be incurred if the land is dry (*i.e.*, having no oil). The company has a second option as well to sell the land to another company which is interested to buy it for a amount of Rs. 90,000. Prepare the payoff table and determine best action using all five approaches. Take Hurwicz criterion with degree of optimization (α) = 0.7.

Solution: The payoff table for the decision analysis is given as,

	Alternative course of action	*State of nature* oil	dry
(1)	Drill for oil	700	–100
(2)	Sell the land	90	90
	Prior probability	0.25	0.75

Probability of oil = 1 chance in 4 *i.e.*, $\frac{1}{4} = 0.25$

Probability of being dry = 1 – (prob. of oil)
= 1 – 0.25
= 0.75

(–) ve sign indicates "loss".

1. Maximax Criterion

Alternative course	*States of nature*		*Maximum value in row*
	Oil	*Dry*	
Drill	700	– 100	(700) → Maximax
Sell	90	90	90

Its maximum payoff value out of max. Payoff group is 700 therefore the alternative to be selected is to drill the land.

2. Maximin Criterion

Alternative course of action	*States of nature*		*Minimum pay off value for each row*
	Oil	*Dry*	
Drill	700	– 100	–100
Sell	90	90	(90) → Maximin value

The maximum value for minimum payoff group is 90, therefore the best option is to sell the land. **Ans.**

This criterion is normally of interest only to a very cautious decision maker.

3. Minimax Regret Criterion

(*a*) If payoffs represent profit, then i^{th} regret = max. payoff –ith payment for the jth event.

(*b*) If payoffs represent cost, then i^{th} regret = i^{th} profit – maximum payoff for the jth event.

For this question, it is a profit matrix.

∴ Regret = (Max. payoff – i^{th} payoff) (for all j events)

Max. payoff value for oil state = 700

Max. payoff value for dry state = 90

Alternative course of action	*State of nature*		*Max. Regret (Rs.)*	*Max. Regret (Rs.)*	*Decision / Optimal solution*
	Oil	*Dry*			
Drill	700 – 700 = 0	90–(–100)=190	190	(190)	→ Drill
Sell	700–90=610	90–90 = 0	610	610	

Min. Regret is to be considered for each nature for different course of action. Therefore min regret is 190 for the option drill well.

4. Hurwiz Criterion

Taking $\alpha = 0.7$ and $(1 - \alpha) = 0.3$

Alternative	*Max. profit*	*Minimum profit*	*Payoff = α(max) + (1 – α) (Min)*	*Decision*
Drill	700	– 100	0.7 (700) + 0.3 (–100) = (360) →	As it is the maximum pay off decsion, ∴ decision is to drill
Sell	90	90	0.7(90) – 0.3 (90) = 36	

As the maximum payoff is 360, therefore the option of drill is to be chosen.

5. Laplace Criterion

As these are only two states of nature, therefore the equal probability assigned to both of them will be $\frac{1}{2}$.

Alternative	*State of nature*		*Equal probability value* $= \frac{1}{2}(P_1 + P_2)$	*Decision*
	Oil	*Dry*		
Drill	700	–100	$\frac{1}{2}$ (700–100) = (300) →	Drill
Sell	90	90	$\frac{1}{2}$ (90 + 90) = 90	

As the maximum payoff comes out to be 300, therefore it is beneficial to drill the wells for oil.

Example 7.16. The cost of making an item is Rs. 25, the SP is Rs. 30, if it is sold within a week, and it could be disposed off at Rs. 20 per piece at the end of week if unsold. Frequency of weekly sales is given as,

Weekly sales :	(<3)	4	5	6	7	(> 8)
No. of week :	0	10	20	40	30	0

Find the optimum number of items per week the industry should make. Solve by all the four criterion of decision under risk condition.

Solution: Cost of producing an item (CP) = Rs. 25

Selling price of an item (SP) = Rs. 30

If the piece is sold, profit = 30 – 25 = Rs. 5

If the piece is unsold, disposal price = Rs. 20

Disposal price = Rs. 20

Therefore loss (if an item is unsold) = – Rs 25 – Rs 20 = Rs 5

Also, the probability of weekly sales of less than three items or greater than 8 items is zero.

$$\text{Prob. of selling 4 items per week} = \frac{10}{10+20+40+30} = \frac{10}{100} = 0.1$$

$$\text{Prob. of selling 5 items per week} = \frac{20}{10+20+40+30} = \frac{20}{100} = 0.2$$

Similarly we can find out the prob. for different weekly sales.

1. EMV Criterion

Table 7.6. Conditional Profit table (Payoff values (in the table) are in Rs.)

Sales (Rs.)	*Probab-ility*	*Production of items (in numbers)*			
		4	5	6	7
4	0.1	Rs. 20	15	10	15
5	0.2	20	25	20	15
6	0.4	20	25	30	25
7	0.3	20	25	30	35

[Production = 5 items
sale = 4 items
loss = 5–4 = 1 item
i.e., less = 1 × 5 = 5
Profit = 20 – 5 = 15]

[Production = 5 items
sales = 5 items
Profit = 5 × 5 = 25]

Production = 4 items
SP = 5 Rs
∴ Profit = 4 × 5 = 20 Rs.

Production = 4 items
sale needed = 5 items
But sales can't exceed the number of produced items. Therefore maximum profit = 4 × 5 = Rs. 20

Table 7.7. Expected Profit Table

Possible demand	*Prob.*	*Expected profit (Rs.)*			
		4	5	6	7
4	0.1	20 × 0.1 = 2	1.5 × 0.1 = 1.5	1	0.5
5	0.2	20 × 0.2 = 4	25 × 0.2 = 5	4	3
6	0.4	20 × 0.4 = 8	25 × 0.4 = 10	12	10
7	0.3	20 × 0.3 = 6	25 × 0.3 = 7.5	9	10.5
Total Expected profit		Σ = 20	Σ = 24	Σ = 26 (max)	Σ = 24

[Expected profit value = (Payoff value) × respective prob. value)]

As the highest EMV value is 26, therefore the optimum number of items to be produced = 6. **Ans.**

2. EOL Criterion

From table 7.7 *i.e.*, from conditional profit table, prepare conditional loss table.

Table 7.8. Conditional Loss table

Sales	*Prob.*	*Production (Alternative)*			
		4	5	6	7
4	0.1	20 – 20 = 0	20 – 15 = 5	20 – 10 = 10	20 – 6 = 15
5	0.2	25 – 20 = 5	25 – 25 = 0	25 – 20 = 5	25 – 15 = 10
6	0.4	30 – 20 = 10	30 – 25 = 5	30 – 30 = 0	30 – 25 = 5
7	0.3	35 – 20 = 15	35 – 25 = 10	35 – 30 = 5	35 – 35 = 0

These values come out by substracting each row element from respective row maximum payoff value. *i.e.*, by substracting all payoff values for an event/alternative course of state from the maximum payoff value of that respective event. For example for the event, when sale is of 4 items from table 7.6 we can find out the maximum payoff value is 20. Thus substracting every payoff value for sale of 4 items from 20, we get the loss values.

Expected loss values are found by multiplying the conditional loss values (from table 7.8) with corresponding prob. values.

Table 7.9. Expected loss table

Sales	*Prob.*	*Production*			
		4	5	6	7
4	0.1	0 × 0.1 = 0	5 × 0.1 = 0.5	10 × 0.1 = 1	1.5 × 0.1 = 1.5
5	0.2	5 × 0.2 = 1	0 × 0.2 = 0	5 × 0.2 = 1	10 × 0.2 = 2
6	0.4	10 × 0.4 = 4	5 × 0.4 = 2	0 × 0.4 = 0	5 × 0.4 = 2
7	0.3	15 × 0.3 = 4.5	10 × 0.3 = 3	5 × 0.3 = 1.5	0 × 0.3 = 0
EOL	→	Σ = 9.5	Σ = 5.5	Σ = 3.5 (minimum)	Σ = 5.5

Optimum stock of 6 items be produced as the minimum loss or the minimum value of EOL is 3.5 for it.

3. EVPI Criterion

Table 7.10. Conditional profit table under certainty (in Rs.)

Possible demand	*Possible* 4	*stock* 5	6	7
4	20	–	–	–
5	–	25	–	–
6	–	–	30	–
7	–	–	–	35

Table 7.11. Expected profit table with perfect information (in Rs.)

Market size (No. of items)	*Prob. of market size*	*Conditional profit under certainty*	*Expected Profit with perfect information*
4	0.1	20	20 × 0.1 = 2
5	0.2	25	25 × 0.2 = 5
6	0.4	30	30 × 0.4 = 12
7	0.3	35	35 × 0.3 = 10.5
			EPPI = 29.5

Already calculated maximum expected profit value (from table 7.7) = Rs. 26

EPPI = Rs. 29.5

Difference = EPPI – EMV (man.)

= 29.5 = 26

= Rs. 3.50

Thus EVPI = EPPI – EMV (maximum)

EVPI = 3.50 Rs.

Also EVPI = Min. EOL

∴ Min. EOL = Rs. 3.50 (can be verified from table 7.8)

4. Incremental or marginal analysis

Incremental profit (IP) = Rs. 30 – Rs 25 = Rs 5

Incremental loss (IL) = Rs. 25 – Rs. 20 = Rs. 5

The condition for storing an additional item is,

$$P = \frac{IL}{IP + IL} = \frac{5}{5+5} = \frac{5}{10} = 0.5$$

i.e., to justify the storing of an item, these must be at least 0.5 cumulative probability of selling that unit.

Sales	*Probability of the corresponding sale level*	*Cumulatively prob. that sales will be at the level or higher*
4	0.1	0.9 + 0.1 = 1
5	0.2	0.7 + 0.2 = 0.9
6	0.4	0.3 + 0.4 = 0.7
7	0.3	0.3

As the cumulatively prob. for sale of 6 items (0.7) is nearer to the value of $P = 0.5$, therefore there should be 6 pieces stored.

Example 7.17. A retail store desires to determine the optional daily order size for a perishable item. The store buys at the rate of Rs. 80/kg and sells at the rate of Rs. 100/kg. If the order size is more than demand, the excess of quality can be sold at Rs. 70/kg in a secondary market, otherwise the opportunity cost for the store is Rs. 15/kg for the unsatisfied portion of demand. The demand varies from 50 kg to 250 kg in steps of 50 kg. The possible values of order size are from 75 kg to 300 kg in steps of 75 kg. Determine the optimal order size which will maximize the daily profit of the store. Solve by all five criterion. Take $\alpha = 0.5$.

Solution: Cost price of perishable item = Rs. 80 per kg

Selling price in primary market = Rs. 100 per kg

Selling price in secondary market = Rs. 70 per kg

Profit in primary market = Rs. 100 – Rs. 80 = Rs. 20/kg

Profit in secondary market = Rs. 70 – Rs. 80 = Rs. – 10/kg [i.*e*., loss]

Opportunity cost of not meeting the demand = Rs. 15/kg

Daily net profit = (demand) × 20 – (Order size – Demand) 10 → if order size ≥ demand

= (demand) 20 – (demand – order size) 15 → if demand > order size

The net outcomes are summarized in the given table.

Table 7.12. Daily net profit

(Demand)

Order size	*Order size*	*50*		*100*	*150*	*200*	*250*
a_1	75	50 × 20	– 25 × 10 = 750	1125	375	– 375	– 1125
a_2	150	50 × 20	– 100 × 10 = 0	1500	3000	2250	1500
a_3	225	50 × 20	– 175 × 10 = – 750	750	2250	3750	4125
a_4	300	50 × 20	– 250 × 10 = – 1500	0	1500	3000	4500

Maximax (optimistic) criterion; maximin criterion; Laplace criterion

State of nature = Demand (as retail store has no control over it). Alternative course of action/strategy = order size (retail store's possible decision).

	Optimistic criterion (maximum value)	*Pessimistic criterion (minmum value)*	*Equal prob. value* $= \frac{1}{n}(P_1 + P_2 + \dots + P_n)$ *(Laplace criterion)*
A_1	+ 1125	– 1125	$\frac{1}{5}$ [750 + 1125 + 375 – 375 – 1125] = 150
a_2	3000	(0) — Maxi-min value	$\frac{1}{5}$ [0 + 1500 + 3000 + 2250 + 1500] = 1650
a_3	4125	– 725	$\frac{1}{5}$ [– 750 + 750 + 2250 + 3750 + 4125] = (2025)
a_4	+ (4500)	– 1500	$\frac{1}{5}$ [– 1500 + 0 + 1500 + 3000 + 4500] = 1500
	Optimal decision a_4	a_2	a_3

1. Maximax (optimistic) criterion: The max. payoff value is Rs. 4500 out of maximum values for each event. Thus the optimal daily order size should be a_4 *i.e.,* 300 kg.

2. Maximin (Pessimistic) Criterion : The maximin value is zero for a_2 option. Thus the optimal daily order size should be 150 kg.

3. Laplace criterion : As here are 5 demands (state of nature) therefore the prob. $= \frac{1}{5}$. The maximum value comes out to be 2025 for strategy a_3. Thus the optimal daily order size should be 225 kg.

4. Minimax regret criterion

Table 7.13. Regret Payoff Table

Demand

Order size	50	100	150	200	250	maximum value
75	0	375	2625	4125	5625	5625
150	750	0	0	1500	3000	3000
225	1500	750	750	0	375	(1500) → Mimimax value
300	2250	1500	1500	750	0	2250

[Regret payoff = Maximum payoff from course of action] – (payoff of that cell)

The minimax value is 1500 corresponding to row 3. Thus daily order size should be 225 kg.

5. Hurwicz Criterion

$$\alpha = \frac{1}{2}$$

$$(1 - \alpha) = 1 - \frac{1}{2} = \frac{1}{2}$$

	50	*100*	*150*	*200*	*250*	*Maximum value*	*Minimum value*	*Payoff* = α *(max)* + $(1-\alpha)$ *(min)*
75	750	1125	375	– 375	– 1125	1125	– 1125	$\frac{1}{2}(1125-1125)=0$
150	0	1500	3000	2250	1500	3000	0	$\frac{1}{2}(3000+0)=1500$
225	– 750	750	2250	3750	4125	4125	– 750	$\frac{1}{2}(4125-750)=1687.5$
300	– 1500	0	1500	3000	4500	4500	– 1500	$\frac{1}{2}(1500-1500)=1500$

The maximum payoff of 1687.50 Rs. is selected for which daily size (optimal) is 225 kg. **Ans.**

7.9. DECISION TREE

It is the graphical representation of the various stages that occured during making a final decision. It consists of different alternatives of decision, states of nature, their probabilities, conditional benefits or losses. Decision tree is a simple method for making a decision where all the options are open in front of the decision maker in concise form and he can easily visualize them. It is equally beneficial for a simple everyday problem to complex decision-making situations in any field.

Let us take an easy problem. A person, on a cloudy day, is leaving his house for office. He has two options whether to take an umbrella with him or to move without it. If it rains on the way to office he will be satisfied and relieved that he is carrying an umbrella to save himself from rain. If there is no rain he will be annoyed that uselessly he carried an extra weight with him. If it doesn't rain, but sun shines brightly, he can still use the umbrella although the purpose will be totally different from the one for which the umbrella was carried. This problem can be represented by a decision tree as shown in Figure 7.2.

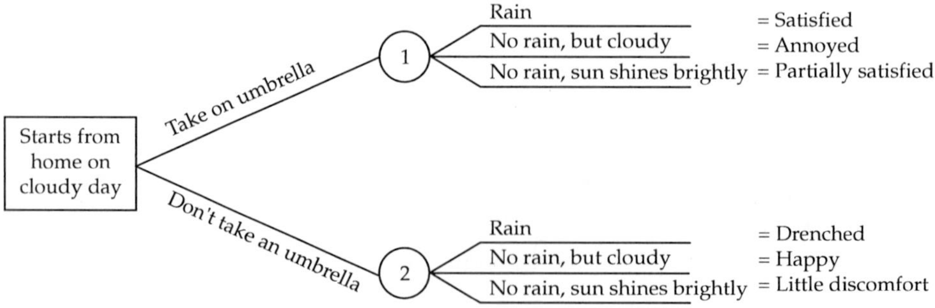

Fig. 7.2.

Basically decision tree is drawn for those problems where more than one decisions are to be taken and the decision taken at one stage affects the subsequent decision. For example a not so established company is doing well and wants to increase its profit. It has 2 options whether to invest money in starting new auxiliary industry or to expand the established one. Further decisions will be based over this decision. The problem can be shown in term of decision tree.

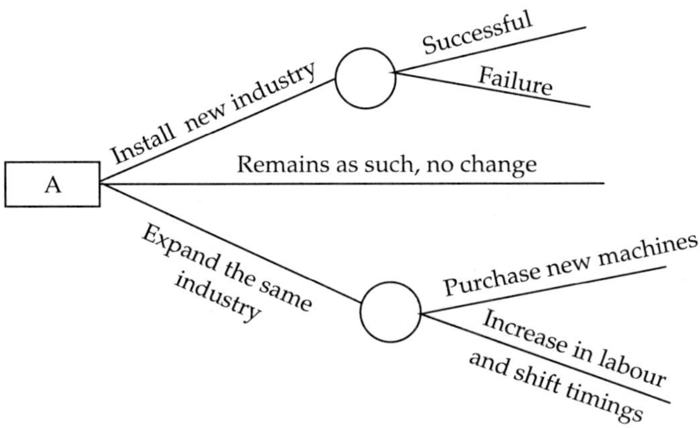

Fig. 7.3.

Representation of a decision tree

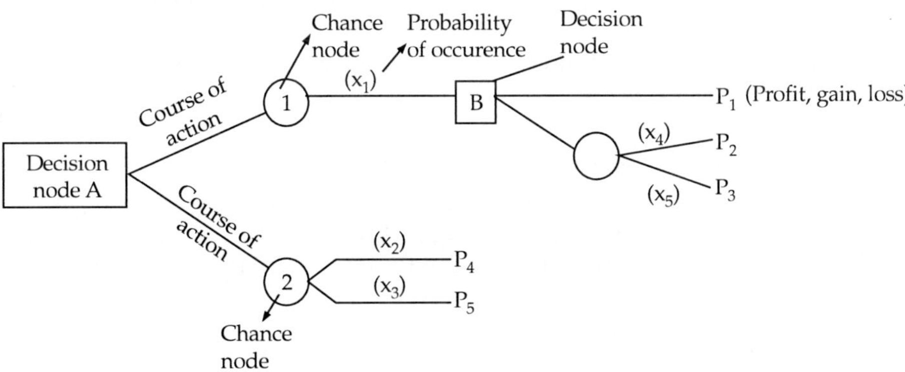

Fig. 7.4.

Comparing a decision tree with a natural tree, the decision node can be described as the trunk of the tree. Various strategies acts as branches, states of nature are the sub-branches, probability values are the leaves and conditional profits or losses shift places with fruits. Decision tree consists of 2 types of nodes.

Decision node = Represented by a square (□) = shows an action or a decision.

Chance node = Represented by a circle (O) = shows point of uncertainty. Some of the important points to be noted are,

1. Sum of the probabilities of occurence of events at a node is always one. For example in figure 7.4, $x_2 + x_3 = 1$, $x_4 + x_5 = 1$
2. At any of the chance node, the EMF's of different alternatives at that particular tree are always added to obtain single EMV value.

At decision node (□), any one decision is chosen out of different options present on the basis of their EMV values.

Roll back Technique : A decision tree is analysed using the roll-back technique. The calculations are started from the last decision in the sequence and works back to the first for each of the possible decisions.

Problems in Decision Tree

1. It becomes complex when there are three or more decision alternatives.
2. It provides the average valued solutions.
3. There can be inconsistency in assigning probabilities for different events.

Although some problems are there, still a decision tree provides the solution of the decision better than any other technique. It helps the person to enumerate all feasible alternatives and outcomes (even the alternatives not worth considering) into a workable pattern and to analyze the problem in an orderly manner.

Steps for drawing a decision tree

1. First work is to identify the decision nodes, chance nodes and alternative courses of action.
2. Place them systematically in an appropriate order/sequence. Keep the decision node first followed by alternative actions and chance nodes.
3. Assign probabilities to each course of action and find out the Pay-off associated with them.
4. Starting from extreme right end, calculate the expected payoff (EMV) for each course of action.
5. Choose the course of action having maximum value of EMV. Then move backwards to next stage.
6. Repeat the same method till the first decision point is reached.
7. Mark the sequence of action of courses adopted from beginning to end.

Example 7.18. The management of a company has to make a decision regarding company's upliftment. It has 3 alternatives. A_1, A_2 and A_3. The decision nodes leads to the events such that there occurs 3 results. Results are R_1, R_2 and R_3. Probability of occurrence of these results are as follows.

$$R_1 = 0.2$$
$$R_2 = 0.3$$
$$R_3 = 0.5$$

Matrix payoffs between decisions and results are given as

	R_1	R_2	R_3
A_1	10	14	8
A_2	6	10	7
A_3	8	5	6

Draw a decision tree of the problem and find the most preferred decision and its corresponding expected value.

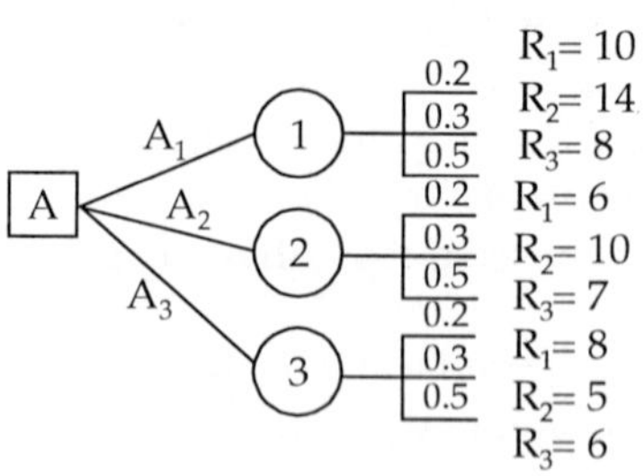

Expected payoff

Expected payoff = 10×0.2 = 2
Expected payoff = 0.3 × 14 = 4.2
Expected payoff = 0.5 × 8 = 4.0
Expected payoff = 0.2 × 6 = 1.2
Expected payoff = 0.3 × 10 = 3.0
Expected payoff = 0.5 × 7 = 3.5
Expected payoff = 0.2 × 8 = 1.6
Expected payoff = 0.3 × 5 = 1.5
Expected payoff = 0.5 × 6 = 3.0

$$\text{EMV at chance node } 1 = \Sigma\,(\text{EMV})_1 = 2 + 4.2 + 4 = 10.2$$
$$\text{EMV at chance node } 2 = \Sigma\,(\text{EMV})_2 = 1.2 + 3 + 3.5 = 7.7$$
$$\text{EMV at chance node } 3 = \Sigma\,(\text{EMV})_3 = 1.6 + 1.5 + 3 = 6.1$$

Since node 1 has highest EMV, so the management will opt for decision A_1.

Example 7.19. A manager has a choice between (i) A risky contract promising Rs 7 lakh with probability 0.6 and Rs. 4 lakhs with prob. 0.4 (*ii*) A diversified portfolio consisting of two contracts with independent outcomes each Promising Rs. 3.5 lakhs with probability of 0.6 and Rs. 2 lakhs with probability of 0.4. Construct a decision tree using EMV criterion.

Solution:

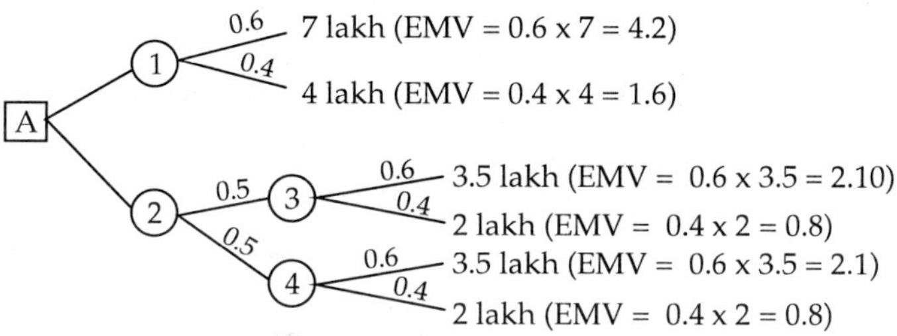

EMV at node 1 = 4.2 + 1.6 = 5.8

EMV at node 4 = 2.1 + 0.8 = 2.9

EMV at node 4 = 2.1 + 0.8 = 2.9

EMV at node 2 = (0.5) (2.9) + (0.5) (2.9) = 1.45 + 1.45 = 2.9

Since the EMV for node 1 is higher than others, manager will opt for choice 1.

Example 7.20. A company is currently working with a process, which after paying for materials labour etc. brings a profit of Rs. 12000. The company has the following alternatives.

1. The company can conduct the research (R_1) which is expected to cost Rs. 10,000 and having 90% probability of success. If successful, the gross income will be Rs. 26,000.
2. The company can conduct the research (R_2) expected to cost Rs. 6000 and having a probability of 60% success. If successful, the gross income will be Rs 24,000.
3. The company can pay Rs. 5,000 as royalty of a new process which will bring a gross income of Rs. 20,000.
4. The company may continue the current process.

Because of the limited sources, only one of the two types of research can be carried out at a time. Draw the decision tree and find the optimal decision of the company.

Solution: Point 1 is the decision node. Company has 4 alternatives. Upon failure of a particular research say R_1, there are again 3 alternatives, conduct research R_2, pay royalty or continue the existing process. If R_2 fails after the failure of R_1, the company is left with only two choices; *i.e.*, either to pay royalty or continue the existing process.

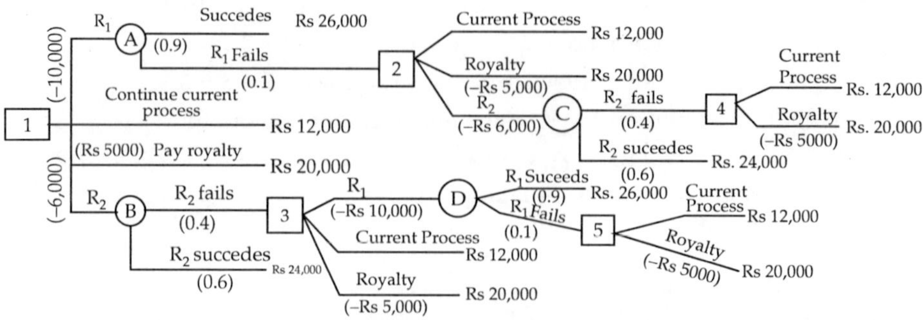

EMV of decision node 4 = max Rs. [12,000, (20,000 – 5000)] = Rs. 15,000

EMV of decision node 5 = max Rs. [12,000, (20,000 – 5000)] = Rs. 15,000

EMV of chance node C = Rs. [0.4 × 15000 + 0.6 × 24000] = Rs. 20,400

EMV of chance node D = Rs. (0.9 × 26,000 + 0.1 × 15,000) = Rs. 24,900

EMV of decision node 2 = Max Rs. [(20,000 – 5000), (20,400 – 6000)] = Rs. 15,000

EMV of decision node 3 = Max Rs. [12,000, (20,000 – 5000), (24,900 – 10,000)]
= Rs. 15,000

EMV of chance node A = Rs. [0.9 × 26,000 + 0.1 × 15000] = Rs. 24,900

EMV of chance node B = Rs. [0.6 × 24,000 + 0.4 × 15000] = Rs. 20,400

EMV of decision node 1 = max Rs. (24,900 – 10,000), 12,000, (20,000 – 5000), (20,400 – 6000)] = Rs. 15000

The company should pay Rs. 5,000 as royalty to the new process to earn maximum expected profit of Rs. 15000.

Example 7.21. Diamond Company is currently working with a process which after paying for materials, labour, etc., brings a profit of Rs. 12,000. The following alternatives are made available to the company:

(*i*) The company can conduct research (R_1) which is expected to cost Rs. 10,000 having 90% chances of success. If it proves a success, the company gets a gross income of Rs. 25,000.

(*ii*) The company can conduct research (R_2) which is expected to cost Rs. 8,000 having a probability of 60% success, the gross income will be Rs. 25000.

(iii) The company can pay Rs. 6,000 as royalty for a new process which will bring a gross income of Rs. 20,000.

(*iv*) The company continues the current process.

Because of limited resources, it is assumed that only one of the two types of research can be carried out at a time.

Use decision-tree analysis to locate the optimal strategy for the company.

Solution: The various act-event combinations and resulting payoffs of the problem are introduced in the following decision-tree diagram.

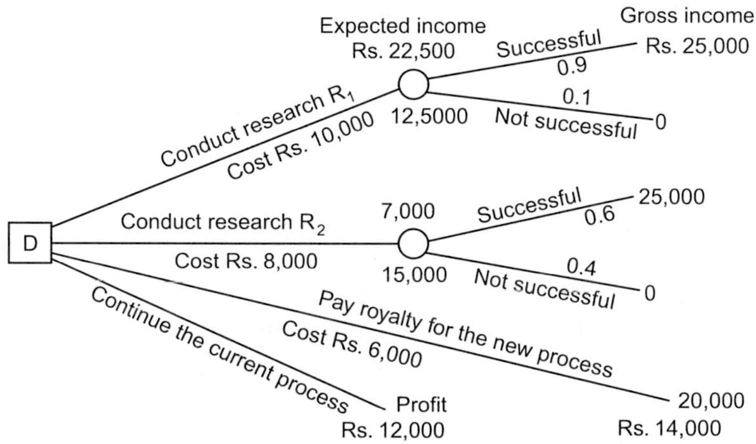

Fig. 7.5. Diamond company's decision-tree.

Decision analysis at point D

Decision	*Even*	*Probability*	*Gross income*	*Expected income*
1. Conduct research R_1	Successful	0.9	(Rs.) 25,000	(Rs.) 22,500
	Not successful	0.1	0	0
			Total expected income = Rs. 22,500	
				Less cost Rs. 10,000
				Net EMV Rs. **12,500**
2. Conduct research R_2	Successful	0.6	25,000	15,000
	Not successful	0.4	0	0
			Total expected income = Rs. 15,000	
				Less cost Rs. 8,000
				Net EMV Rs. **7,000**
3. Pay royalty for the new process	Certain	1	20,000	Rs. 20,000
				Less cost Rs. 6,000
				Net EMV Rs. **14,000**
4. Continue the current process	Certain	1	12,000	EMV Rs. **12,000**

As the net EMV is highest for the alternative 'pay royalty for the new process', the optimal decision would be to procure new process on royalty basis.

EXERCISE

7.1. What is decision theory? Write its applications.

7.2. Explain various ingredients of a decision problem.

7.3. What are types of decisions? Explain them in brief with suitable example.

7.4. Explain different approaches for decision under uncertainty.

7.5. What are differences between decision under risk and decision under uncertainty?

7.6. Explain the terms:
(*i*) Expected Monetary Value (EMV)
(*ii*) Expected value of perfect information (EVPI)
(*iii*) Expected Opportunity Loss (EOL)
(*iv*) Payoff table
(*v*) Regret table

7.7. What is a decision tree? What is node in a decision tree? Illustrate with an example.

7.8. Write notes on 'Bayesian decision theory'.

7.9. What are basic steps of a decision making process?

7.10. Write a note on Hurwiz criterion. Define the index of optimism related to it.

7.11. Write short note on following:
(*i*) Maximax (optimistic) criterion
(*ii*) Maximin (Pessimistic) criterion
(*iii*) Minimax regret criterion (Savage criterion)
(*iv*) Hurwicz criterion/criterion of realism/weighted average criterion
(*v*) Laplace or Bayes criterion (criterion of rationality)

7.12. Distinguish between decision under certainty and decision under uncertainty.

7.13. Describe various techniques of problem solving for decision under certainty.

7.14. Consider the details of two competing alternatives as shown in the following table. The initial outlay of each of the alternatives is Rs. 50,00,000. The life of each alternative is 15 years. Find the best alternative when the interest rate is 0% using the expected value criterion. (**Ans.** Alternative 2)

Alternative 1		*Alternative 2*	
Annual revenue (Rs.)	*Probability*	*Annual revenue (Rs)*	*Probability*
12,00,000	0.25	22,00,000	0.20
19,00,000	0.45	24,00,000	0.45
28,00,000	0.30	32,00,000	0.35

7.15. A retail store desires to determine the optimal daily order size for a perishable item. The store buys the perishable item at the rate of Rs. 60 per kg and sells at the rate of Rs.90 per kg. If the order size is more than demand, the excess quantity can be sold at Rs. 75 per kg in a secondary market; otherwise, the opportunity cost for the store is Rs. 10 per kg for the unsatisfied portion of the demand. Based on the past experience, it is found that the demand varies from 50 kg to 200 kg in steps of 50 kg. The possible values of the order size are from 100 kg to 300 kg in steps of 100 kg. Determine the optimal order size which will maximize the daily profit of the store using: (*a*) Laplace criterion (*b*) maximin criterion. [**Ans.** *a* – 300 kg, *b* – 300 kg]

7.16. Consider the following cost matrix and determine the best order size using the minimax criterion.

		Demand (D_j)				
		50	100	150	200	250
	75	50	125	375	375	125
Order size	150	40	500	100	250	500
(Qi)	225	750	550	250	750	125
	300	500	40	500	400	540

[**Ans.** 75 units]

7.17. Consider the following data on daily net profit. Find the best order size based on the savage minimax regret criterion.

		Demand (D_j)				
		50	100	150	200	250
	75	950	1200	575	–675	–1425
order size	150	50	1700	2000	2250	1600
(Q_i)	225	–850	850	2550	3550	4525
	300	–1800	600	1800	2000	5000

Also, obtain the best order size based on the Hurwicz criterion.

[**Ans.** 225 Units]

7.18. A company owns a lease on a certain property. It may sell the lease for Rs. 75,000 or may drill the said property for oil. Various possible drilling results are as under along with the probabilities of happening and rupee consequences.

Possible results	*Probability*	*Rupee consequences*
Dry well	0.10	–1,00,000
Gas well only	0.40	45,000
Oil and gas combination	0.30	98,000
Oil well	0.20	1,99,000

Draw a decision tree for the above problem and determine whether the company should drill or sell. [**Ans.** should dril; 77,200]

7.19. An oil company may bid for only one of the two contracts for oil drilling in two different areas. It is estimated that a profit of Rs. 30,000 would be realized from the first field and Rs. 40,000 from the second field. These profit amounts have been determined ignoring the costs of bidding which amount to Rs. 2,500 for the first field and Rs. 5,000 for the second field. Which oil field the Co. should bid for if the probability of getting contract for first field is 0.7 and that of second field is 0.6? [**Ans.** second oil field]

7.20. A company wants to know whether or not a new shaving cream should be marketed. The present value of all future profits for the success of the cream is Rs. 10,00,000 and its failure would result in a net loss of Rs. 5,00,000. Not marketing it would not change the profits. The chances of the success of the new cream are 50%:

(*i*) Construct the payoff table.

(*ii*) Which act will maximize the payoff ?

(*iii*) What is expected payoff under uncertainty ?

(*iv*) What is EVPI ? What is the minimum expected opportunity loss ?

[**Ans.** (*ii*) Market cream; (*iii*) 2,50,000 (*iv*) 2, 50, 000]

7.21. A newspaper vendor has to decide how many copies of a particular magazine he should buy for the coming month. Each magazine cost him Rs. 5 and sells for Rs. 10. At the end of the month the sold magazines are thrown away. The demand distribution of the magazines is as follows :

No. of copies demanded :	10	11	12
Probability :	$\frac{1}{3}$	$\frac{1}{3}$	$\frac{1}{3}$

Construct a payoff table. According to maximin criteria, how many copies should he stock ? Which number of copies will maximize the expected payoff?

[**Ans.** (*i*) 45 units (*ii*) Rs. 160]

7.22. A certain output is manufactured at Rs. 8 and sold at Rs. 14 per unit. The product is such that if it is produced but not sold during a day's time it becomes worthless. The daily sales record in the past are:

Demand per day :	30	40	50	60	70
No. of days each sales Level was recorded :	24	24	36	24	12

(*i*) Calculate the expected sales per day

(*ii*) Prepare a payoff and a regret table.

(*iii*) Find the expected payoffs and regret.

(*iv*) Find the optimal act and the EVPI.

7.23. A farmer wants to decide which of the three crops he should plant on his 100acre farm. The profit from each is dependent on the rainfall during the growing season. The farmer has categorized the amount of rainfall as high, medium and low. His estimated profit for each is shown in the table below

Rainfall	*Estimated conditional profit (Rs.)*		
	Crop A	*Crop B*	*Crop C*
High	8,000	3,500	5,000
Medium	4,500	4,500	5,000
Low	2,000	5,000	4,000

If the farmer wishes to plant only one crop, decide which should be his 'best crop' using:

(*a*) Maximax criterion.

(*b*) Maximin criterion.

(*c*) Hurwicz criterion (farmer's degree of optimism being 0.6)

(*d*) Laplace criterion.

(*e*) Minimax regret criterion. [**Ans.** (*a*) *A* (*b*) C (*c*) *A* (*d*) *A* (e) *A* or C]

7.24. The *ABC* Company is faced with four decision alternatives relating to investments in a capital expansion programme. Since these investments are made in future, the company foresees different market conditions as expressed in the form of states-of-nature. The following table summarizes the decision alternatives, the various states-of-nature and the rate of return associated with each state-of-nature

Decision	*States-of-nature* θ_1	θ_2	θ_3
D_1	17%	15%	8%
D_2	18%	16%	9%
D_3	21%	14%	9%
D_4	19%	12%	10%

If the company has no information regarding the probability of the occurrence of the three states-of-nature, give the recommended decision for the decision criteria listed below:

(*i*) Maximax criterion.
(*ii*) Maximin criterion.
(*iii*) Minimax regret criterion.
(*iv*) Laplace rational criterion.
(*v*) Hurwicz criterion ($\alpha = 0.75$) [**Ans.** (*i*) D_3 (*ii*) D_4 (*iii*) D_3 (*iv*) D_3]

7.25. A producer of boats has estimated the following distribution of demand for a particular kind of boat:

No. demanded :	0	1	2	3	4	5	6
Probability :	0.14	0.27	0.27	0.18	0.09	0.04	0.01

Each boat cost him Rs 7,000 and he sells them for Rs, 10,000 each. Any boats that are left unsold at the end of the season must be disposed of for Rs. 6,000 each. How many boats should be in stock so as to maximize his expected profit? [**Ans.** 3 boats per season, 4080]

7.26. A modern home appliances dealer finds that the cost of holding a mini-cooking range in stock for a month is Rs. 200 (insurance, minor deterioration, interest on borrowed capital, etc.). Customer who cannot obtain a cooking range immediately tends to go to other dealers and he estimates that for every customer who cannot get immediate delivery, he loses an average of Rs. 500. The probabilities of a demand of 0, 1, 2, 3, 4, and 5 cooking ranges in a month are 0.05, 0.10, 0.20, 0,30 0.20 and 0.15 respectively. Determine the optimum stock level of cooking range. Also find EVPI.

[**Ans.** 4 cooking ranges, EVPI = 315]

7.27. A Finance Manager is considering drilling a well. In the past, only 70% of wells drilled were successful at 20 meters depth in that area. Moreover on finding no water at 20 meters, some persons in that area drilled it further upto 25 metres but 20% struck water at that level. The prevailing cost of drilling is Rs. 500 per meter. The Finance Manager estimated that in case he does not get water from the well he will have to pay Rs. 15,000 to buy water from outside. The following decisions are considered:

(*i*) Do not drill any well.
(*ii*) Drill upto 20 metres, and
(*iii*) If no water is found at 20 metres, drill further upto 25 metres.

Draw an appropriate decision-tree and determine the Finance Manager's optimal strategy.

7.28. A businessman has two independent investments *A* and *B* available to him, but he lacks the capital to undertake both of them simultaneously. He can choose to take *A* first and then stop, or if *A* is successful then take *B*, or viceversa. The probability of success on *A* is 0.7 while for *B* it is 0.4. Both investments require an initial capital outlay of Rs. 2,000, and both return nothing if the venture is unsuccessful. Successful completion of *A* will return Rs. 3,000 (over cost), successful completion of *B* will return Rs. 5,000 (over cost).

Draw the decision-tree and determine the best strategy.

[**Ans.** Accept A, if succesful, B]

7.29. A client asks an estate agent to sell three properties (*A*, *B* and *C*) for him and agrees to pay him a 5% commission on each sale. He specifies certain conditions. The estate agent must sell *A* first, and this he must do within 50 days. If and when *A* is sold, the agent receives his 5% commission on that sale. He can then either back out at this stage or nominate and try to sell one of the two remaining properties within 60 days. If he does not succeed in selling the nominated property in that period, he is not given the opportunity to sell the other. If he does sell it in the period, he is given the opportunity to sell the third property on the same conditions. The following table summarizes the prices, selling costs (incurred whenever a sale is attempted) and the estate agent's estimated probability of making a sale.

Property	*Price of property*	*Selling costs*	*Probability of sale*
A	Rs. 12,000	Rs. 400	0.7
B	25,000	225	0.6
C	50,000	450	0.5

(*i*) Draw up an appropriate decision-tree for the estate agent.
(*ii*) What is the estate agent's best strategy under EMV approach?

[**Ans.** Accept offer sell A, then B, then c]

7.30. Growfast Company is evaluating four alternative single-period investment opportunities whose returns are based on the state of the economy. The possible states of the economy and the associated probability distribution is as follows:

State :	*Fair*	*Good*	*Great*
Probability :	0.2	0.5	0.3

The returns for each investment opportunity and each state of the economy are as follows:

Alternative	*State of economy*		
	Fair	*Good*	*Great*
W	Rs. 1,000	Rs. 3,000	Rs. 6,000
X	Rs. 500	Rs. 4,500	Rs. 6,800
Y	Rs. 0	Rs. 5,000	Rs. 8,000
Z	Rs. 4,000	Rs, 6,000	Rs. 8,500

Using the decision-tree approach, determine the expected return for each alternative. Which alternative investment proposal would you recommend if the expected monetary value criterion is to be employed?

[**Ans.** Alternative *Y*, 4900]

OBJECTIVE TYPE QUESTIONS

7.1. Decision theory is concerned with
(*a*) the amount of information that is available
(*b*) criteria for measuring the "goodness" of a decision
(*c*) selecting optimal decisions in sequential problems
(*d*) all of the above

7.2. Concerning decision-making under risk, which of the following is not true?
(*a*) We assume that the decision-maker knows the probability with which each state of nature will occur.
(*b*) We use the criterion of maximizing return.
(*c*) We use the criterion of maximizing expected return.
(*d*) We use the criterion of minimizing expected regret.

7.3. Which of the following criteria does not apply to decision-making under uncertainty?
(*a*) maximin return (*b*) maximax return
(*c*) minimax regret (*d*) maximize expected return

7.4. Maximin return, maximax return, and minimax regret are criteria that
(*a*) lead to the same optimal decision (*b*) can be used without probabilities
(*c*) both a and b

7.5. The expected value of perfect information (EVPI)
(*a*) places two-sided bounds (upper and lower) on how much should be spent in gathering information
(*b*) can be determined without using probabilities
(*c*) refers to the utility of additional information
(*d*) equals the expected regret of the optimal decision under risk

7.6. The concept of utility is a way to
(*a*) measure the attractiveness of money
(*b*) take into account aversion to risk
(*c*) take into account inclination to take risk
(*d*) *a* and *b*
(*e*) *a*, *b*, and *c*

7.7. Which of the following does not apply to a decision tree ?
(*a*) A square node is a point at which a decision must be made.
(*b*) A circular node represents an encounter with uncertainty.
(*c*) One chooses a sequence of decisions which have the greatest probability of success.
(*d*) One attempts to maximize expected return.

7.8. The expected value of perfect information (EVPI)
(*a*) shows the cost necessary to produce perfect information about the future.
(*b*) shows the maximum possible increase in expected return with sample information

(*c*) shows the expected increase in information required to select the optimal decision

(*d*) all of the above

7.9. When computing the expected value of perfect information (EVPI) it is important that the payment is made

(*a*) in advance of receiving the information

(*b*) after receiving the information

(*c*) in an irrevocable way (*d*) both (*a*) and (*c*)

7.10. The criterion which selects the action for which maximum payoff is lowest is known as.

(*a*) Max-min criterion (*b*) Min-max criterion

(*c*) Max-max criterion (*d*) None of these

7.11. When decision are made sequentially in time,

(*a*) decision trees cannot be employed

(*b*) Bayes Theorem must be used

(*c*) the terminal value at the end of each sequence of branches is the net of the cash flows on that sequence

(*d*) the terminal value at the end of each sequence of branches is an expected net cash flow.

Answers

7.1. (*d*)	7.2. (*b*)	7.3. (*d*)	7.4. (*b*)	7.5. (*d*)	7.6. (*e*)	7.7. (*c*)
7.8. (*b*)	7.9. (*d*)	7.10. (*b*)	7.11. (*b*)			

Chapter 8

Game Theory

8.1. INTRODUCTION

Practical problems that frequently occur in Military operations, Business administration, Economics etc. require decision-making in competitive situation.

Conflicting or competitive situation: Situation where, there are two or more opposing parties with conflicting interests and where the action of one depends upon the action taken by the opponent, is known as Conflicting or competitive situation. For example, candidates in an election, countries involved in a war, advertisement and marketing campaign by competing firms. In a competitive situation the alternatives available to each competitor may be either finite or infinite.

8.1.1. Theory of Games

A body of knowledge that deals with making decisions when two or more intelligent and rational opponents are involved under condition of conflict and competition is known as theory of games. Game theory studies choice of optimal behaviour when costs and benefits of each option are not fixed, but depend upon the choices of other individuals.

The Game theory was developed in twentieth century. Von-Newmann is considered as father of Game theory. Von-Newmann's approach was based on the principle of 'best out of the worst'.

8.1.2. Game

A competitive situation is called a Game if it has following properties:

1. There are finite number of rational and intelligent participants (players)
2. Each player has a finite number of strategies (choices or alternatives) available to him.
3. All relevant information is known to each player in advance.
4. A play takes place when each player chooses one of his strategies simultaneously.
5. Every combination of strategies determines outcome, which results in loss or gain or draw (usually called payoff).

8.1.3. Applications

1. It is used by Chess players to win their games.

2. Economists have used game theory to analyze a wide array of economic phenomena, including auctions, bargaining, social network formation, and voting systems.
3. Army generals use this technique to plan war strategies.
4. In biology, game theory has been used to understand many different phenomena. It was used to explain the evolution and stability of the approximate 1:1 sex ratios. Additionally, biologists have used game theory to explain the emergence of animal communication, analyze fighting behaviour and territoriality.
5. Computer scientists have used games to model interactive computations.
6. A firm decides about Advertising media and its content based on decision taken by rivals using Game theory.
7. Managers make use of it to plan their Marketing strategies.
8. It can be used in Political negotiations.

8.2. BASIC DEFINITIONS

1. **Player:** The competitors in the game are known as players. A player may be an individual or group of individuals or an organization.
2. **Move:** The way in which the game progresses between states through exchange of information and pieces is called move. Moves are defined by the rules of the game and can be made in either alternating fashion, occur simultaneously for all players, or continuously for a single player until he reaches a certain state or declines to move further. Moves may be choice or by chance. For example, choosing a card from a deck or rolling a die is a chance move with known probabilities. On the other hand, asking for cards in blackjack is a choice move.
3. **Information:** A state of perfect information is when all players in a game know all moves. Games without chance elements like chess are games of perfect information, while games with chance involved like asking for blackjack are games of imperfect information.
4. **Strategy:** It is the decision rule a player uses to decide which course of action he should take from available alternatives. Strategy is of two kinds:
   (*a*) **Pure strategy:** A pure strategy is a decision always to select same course of action. In this, each player knows exactly what other player is going to do. Hence, a deterministic situation is obtained. It is simplest game; matrix contains saddle point.
   (*b*) **Mixed strategy:** Mixed strategy is a decision to choose atleast two of his course of actions with fixed probabilities. A probabilistic situation is obtained and there is no saddle point in matrix.

A player's strategy provides him with a move depending on what the other player has done, and/or chance events, as the game unfolds. A game with a finite number of stages and a finite number of moves at each stage is called a **finite game.** A finite game has a finite number of distinct strategies available to each player. The game of chess is a finite game, despite its complexity and vast number of strategies.

The choice of a particular strategy by each player determines the pay-off to each player.

5. **Optimal Strategy:** A course of action, which puts the player in the most preferred position, irrespective of the strategy of his competitors is called optimal strategy. The strategy for which, Maximin = Minimax, is optimal strategy.
6. **Payoff:** Payoff is the outcome of playing the game. Each player can choose among a number of actions to influence his pay-off. However, each player's pay-off depends on the other players' choices. Moreover, outcomes can also be influenced by **"chance occurrences."**
7. **Value of Game:** It is the expected payoff of the game when all the players of game follow their optimal (best) strategies. The game is called **fair** if the value of the game is zero and unfair if it is **non-zero**. Best strategy is determined on the basis of Maximin-Minimax criteria.
8. **Payoff matrix:** When the players select their particular strategies, the payoffs (gains or losses) represented in the form of a matrix is known as the payoff matrix. For positive entries, B makes payments to A i.e. for player A, it represents gains and for player B, it always represents losses. Player A is maximizing player and player B is minimizing player (like to minimize losses). For negative entries, A makes payments to B. The player A is called the maximizing player since he would like to maximize his gains, while the player B is called the minimizing player since he would like to minimize his losses.

		Player B				
		1	2	3	……	n
Player A	1	a_{11}	a_{12}	a_{13}	……	a_{1n}
	2	a_{21}	a_{22}	a_{23}	……	a_{2n}
	3	a_{31}	a_{32}	a_{33}	……	a_{3n}
	⋮	⋮	⋮	⋮	⋮	⋮
	m	a_{m1}	a_{m2}	a_{m3}	……	a_{mn}

In the payoff matrix, the number of rows is equal to the number of courses of action of the player A written on the left of the matrix, and the number of columns is equal to the number of courses of action of player B that is always written on the top of the matrix.

9. **Rationality:** In the language of Game Theory rationality implies that each player tries to maximize his/her payoff irrespective to what other players are doing. In essence each player has to decide a set of moves which are in accordance with the rules of the game and which maximize his/her rewards.
10. **Maximin Principle:** Best out of worst outcome. Maximizing player (A) decides to play that strategy which corresponds to the maximum of the minimum gains for his different options.
11. **Minimax Principle:** Minimizing player (B) selects that strategy which corresponds to the minimum of maximum losses for his different courses of options.
12. **Solving a game:** means determining the optimal strategies of both the players and the value of the game.

13. **Saddle or equilibrium point:** It is that point in the payoff matrix where the maximum of row minimum coincides with the minimum of column maximum. (Maximin = Minimax). The payoff at the saddle point is called the value of the game.

    A game is said to be fair if Maximin = Minimax = 0

1. A pure strategy game (possesses saddle point) can be solved by Maximin – Minimax criterion.
2. A mix strategy game (A game with no saddle point) can be solved by following methods
    (*a*) Algebric method (2 x 2)
    (*b*) Calculus method
    (*c*) Linear Programming method

8.3. CLASSIFICATION OF GAME THEORY

Game Theory can be classified in two branches:

1. ***Co-operative game theory:*** Here players may co-operate with one another. The players are allowed to form coalitions to share decisions, information and payoffs.
2. ***Non co-operative game theory:*** In this case the players work independently without assuming anything about what other players are doing. The players follow the rationality assumption and play to maximize their own payoffs irrespective of other players' payoffs.

    There are 2 forms of non-cooperative games.

    (*a*) **Extensive Form:** If players have some information about the choices of other players, the game is presented in extensive form.

    Here the players make their moves in turns. There is more than one move in the game and the previous moves of all opponents are known to everyone. An extensive form game is characterized by rules that dictate all possible moves in a state. It may indicate which player can move at which times, the payoffs of each chance determination, and the conditions of the final payoffs of the game to each player.

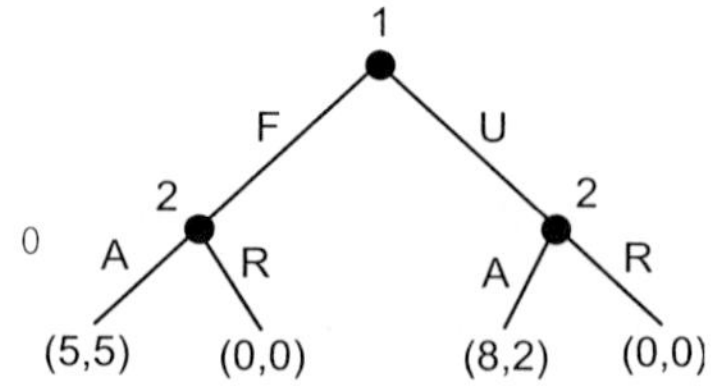

Fig. 8.1 An extensive form game

Games here are presented as trees (as shown above). Here each vertex (or node) represents a point of choice for a player. The player is specified by a number listed at the vertex. The lines out of the vertex represent a possible action for that player. The payoffs are specified at the bottom of the tree.

In the game pictured here, there are two players. *Player 1* moves first and chooses either *F* or *U*. *Player 2* sees *Player 1*'s move and then chooses *A* or *R*. Suppose that *Player 1* chooses *U* and then *Player 2* chooses *A*, then *Player 1* gets 8 and *Player 2* gets 2.

(*b*) Normal Form: Here the game is one shot and the players make simultaneous moves. A normal form game already has a complete list of all possible combinations of strategies and payoffs, thus removing the element of player choices. In short, in a normal form game, the best move is always known.

	Player 2 chooses left	*Player 2 chooses right*
Player 1 chooses top	4, 3	–1, –1
Player 1 chooses bottom	0, 0	3, 4

A normal form game

The normal (or strategic form) game is a matrix which shows the players, strategies, and payoffs (see the example). Here there are two players; one chooses the row and the other chooses the column. Each player has two strategies, which are specified by the number of rows and the number of columns. The payoffs are provided in the interior. The first number is the payoff received by the row player (*Player 1*); the second is the payoff for the column player (*Player 2*). Suppose that *Player 1* plays top and that *Player 2* plays left. Then *Player 1* gets 4, and *Player 2* gets 3.

8.4. TYPES OF GAMES

8.4.1. Zero Sum Games

Games in which algebric sum of gains and losses of all players is zero, are called zero sum games. The play does not cut or add a single paisa to the total capital of players. Which means if one player is able to improve his payoff by using some good strategy, the payoff of others is going to decrease *i.e.*, Only redistribution of money takes place. Zero-sum is a special case of a more general **constant sum game** where the benefits and losses to all players sum to the same value. Cutting a cake is zero- or constant-sum because taking a larger piece for yourself reduces the amount of cake available for others.

8.4.2. Constant Sum Games

In constant sum game, sum of payoff's of all the players for each outcome of the game, is a constant.

Example 8.1. Consider the game of Chess. There are two players, one playing with White pieces and other playing with Black pieces. There are three possible outcomes. O = {Black Wins, White Wins, Draw}.

Define payoffs as

	Black Wins	White Wins	Draw
U_b	1	0	½
U_w	0	1	½
Sum	1	1	1

This is a constant sum game, sum of payoffs is constant. If white increases his payoff by a win the payoff of black goes down and vice versa. Above type of game is called strictly competitive game. Win for one player is loss for the other.

8.4.3. Non-zero-sum Games

Situations where participants can all gain or suffer together, such as a country with an excess of bananas trading with another country for their excess of apples, where both benefit from the transaction, are known as **non-zero-sum games**. Most economic situations are non-zero-sum, since valuable goods and services can be created, destroyed, or badly allocated, and any of these will create a net gain or loss.

8.4.4. Two Person Zero Sum Games

Also known as rectangular games. Zero sum games with two players are called two person zero sum games. In this case, loss (gain) of one is exactly equal to the gain (loss) of other.

The **irreconcilable, conflicting interests** between the two players in a zero-sum game resemble **parlour games and military encounters** between enemy states. Players make moves and counter-moves, until the **rules of engagement** declare the game is ended. The rules of engagement determine what each player can or must do at each stage as the game unfolds. For example, games like **chess** require many moves to resolve the conflict.

Example 8.2.

Player 2

		A	B	C
Player 1	1	30, -30	-10, 10	20, -20
	2	10, -10	20, -20	-20, 20

A zero sum game

Consider the two-player zero-sum game shown in figure.

The first player chooses in secret one of the two actions 1 or 2; the second player, unaware of the first player's choice, chooses in secret one of the three actions A, B or C. Then, the choices are revealed and each player's points total is affected according to the payoff for those choices.

If the first player chooses action 2 and the second player chose action B. When the payoff is allocated the first player gains 20 points and the second player loses 20 points.

Now, in this game both players know the payoff matrix and attempt to maximize the number of their points. What should they do?

Player 1 could reason as follows: "with action 2, I could lose up to 20 points and can win only 20, while with action 1, I can lose only 10 but can win up to 30, so action 1 looks a lot better." With similar reasoning, player 2 would choose action C. If both players take these actions, the first player will win 20 points. But what happens if player 2 anticipates the first player's reasoning and choice of action 1, and deviously goes for action B, so as to win 10 points? Or if the first player in turn anticipates this devious trick and goes for action 2, so as to win 20 points after all?

8.5. PURE STRATEGY GAMES (Games with saddle point)

A pure strategy game can be solved, using following steps:

1. Set up the payoff matrix.
2. Find the row minimums and select the largest of these as player 1's Maximin strategy.
3. Find the column maximums and select the smallest of these as player 2's Minimax strategy.
4. If the Maximin value is equal to the Minimax value, the game is pure strategy game and that value is the saddle point.
5. Payoff at saddle point is the value of the game.

Example 8.3. Solve the game whose pay-off matrix is :

		B		
		I	II	III
A	I	– 3	– 2	6
	II	2	0	2
	III	5	– 2	– 4

(UPTU: 2004-2005)

Solution: (1) Search for saddle point using maximin-minimax criterion.

		B			
		I	II	III	Row minimum
A	I	(–3)	–2	[6]	–3
	II	2	[(0)]	2	[0] ← Maximin
	III	[5]	–2	–4	–4
Column maximum		5	(0) ↑ Minimax	6	

Here, Maximin = Minimax = 0 = value of game. Element enclosed both by circle and rectangle is saddle point. Hence, both players will adopt II strategy. Game is a pure strategy game.

8.5.1. Dominance

The size of game's payoff matrix can be reduced by eliminating a course of action, which is so inferior to another as never to be used. Such a course of action is said to be dominated by superior strategy(ies).

8.5.2. Principle of Dominance

For rows

(*a*) In matrix of player A, if all the entries in a row (x) are greater than or equal to the corresponding entries of another row (y), than row y is dominated by row x. Row y can be deleted.

(*b*) In matrix of player A, if sum of elements of any two rows (x and y) is greater than or equal to the corresponding elements of a third row (z), then row z is dominated by x and y. Row z can be deleted.

For columns

(*a*) In matrix of player A, if all the entries in a column (x) are less than or equal to the corresponding entries of another column (y), then column y is dominated by x. Column y can be deleted.

(*b*) In matrix of player A, if sum of elements of any two columns (x and y) is less than or equal to the corresponding elements of a third column (z), then column z is dominated by x and y. Column z can be deleted.

Example 8.4. Solve following game using principle of dominance.

		Player B				
		1	2	3	4	5
	1	2	3	3	8	4
Player A	2	5	6	3	7	8
	3	6	7	9	8	7
	4	4	2	8	4	3

Solution: (*i*) All the elements of column 1 are less than or equal to corresponding elements of column 4. Column 4 is dominated, hence, deleted.

		Player B			
		1	2	3	5
	1	2	4	3	4
Player A	2	5	6	3	8
	3	6	7	9	7
	4	4	2	8	3

(*ii*) Now, all the elements of row 4 are less than corresponding elements in row 3. Row 4 can be deleted.

		Player B			
		1	2	3	5
	1	2	4	3	4
Player A	2	5	6	3	8
	3	6	7	9	7

(*iii*) In above matrix, each element of column 1 is less than corresponding element of column 5. Delete column 5.

		Player B		
		1	2	3
	1	2	4	3
Player A	2	5	6	3
	3	6	7	9

(*iv*) Finally, each element of row 1 and row 2 is less than corresponding element of row 3. Row 1 and 2 can be deleted.

		Player B		
		1	2	3
Player A	3	6	7	9

Minimum value in above matrix is 6.

Therefore, value of game = 6

Optimal strategy for A is 3 and for B is 1.

Above example is for illustration purpose only, however, the principle of dominance is used when saddle point does not exist. (explained in example 8.6)

8.5.3. Nash Equilibrium

A Nash Equilibrium a set of strategies such that each player believes that it is doing the best it can, given the strategy of opponents. i.e. each player chooses a strategy that is his best choice, subject to what strategies the opponents chooses. The player has no incentive to deviate from the choosen strategy, since he is satisfied and feels that he has made the best decision, therefore Nash strategies are stable strategies.

In a game where pure strategy is used, Nash Equilibrium may or may not exist. Nash Equilibrium is possible only if we assume that all players understand the game and are rational.

Example 8.5. Consider a husband and wife who have different plans to spend the evening together and have fun. The husband is interested in watching a cricket match while the wife is interested in going out to watch a movie. We can model this as a game with husband as first player (P_1) and wife as the second player(P_2).

The payoff matrix for this game is given below. The pay-offs for both players in the four outcomes are explained here: -

(*a*) When P_1 and P_2 both go for cricket, P_1 has payoff of 10 because that option was his preference, while P_2 has payoff of 5 because cricket was not her preference.

(*b*) When P_1 goes for cricket while P_2 goes for movie, since they are separated from each other, the whole objective of having fun together gets defeated, so each ends up with a payoff of 1.

(*c*) When P_1 goes for movie and P_2 goes for cricket, they go for something they don't prefer and also go alone, so each gets a negative payoff of –10.

(*d*) When P_1 goes for movie and P_2 goes for movie, then since movie is P_2's preference she gets payoff of 10, while P_1 gets payoff of 5 because his preference was cricket.

On analyzing this matrix, one can see that there are 2 Nash Equilibrium: (10,5) and (5,10).

(10,5) is a Nash Equilibrium because if P_2 chooses cricket, then P_1 is better off choosing cricket compared to movie and if P_1 chooses cricket, then P_2 is better off choosing cricket compared to movie. A similar analysis proves that (5,10) is also Nash Equilibrium

Movie/Cricket Game (Outcome Matrix)

$P_1\downarrow / P_2 \rightarrow$	Cricket	Movie
Cricket	10 , 5	1 , 1
Movie	–10 , –10	5 , 10

8.6. MIXED STRATEGY GAMES (Games without saddle point)

A game, in which saddle point does not exist, is called as mixed strategy game. Here, each player adopts chance move and begins to play in random manner in a way that his average payoff over large number of plays of game is optimal.

Following methods can solve a mixed strategy game:

(*a*) Algebric method

(*b*) Graphical method

(*c*) Linear Programming method

8.6.1. Algebric Method

Steps to solve a 2x2 game are listed below:

1. Develop the payoff matrix.
2. Apply the Maximin-Minimax principle to see if the game has saddle point. If it does not have a saddle point, mixed strategy exists.
3. Subtract the smaller payoff in each row from larger one and the smaller payoff in each column from the larger one.
4. Interchange each of these pairs of subtracted numbers found above.
5. Put each of the interchanged numbers over the sum of the numbers of that pair.
6. Simplify the fraction to obtain the required strategies.
7. Value of game can be found either from A's point of view or B's point of view.

Illustration

Oddments in a pay off matrix are calculated as shown in matrix below

		Player B		Oddments
		1	2	
Player A	1	a	b	$\vert c-d \vert$
	2	c	d	$\vert a-b \vert$
Oddments		$\vert b-d \vert$	$\vert a-c \vert$	

$|b - d|$ is absolute value of $(b - d)$

$$\text{Probability of selecting alternative 1 by } A,\ p_1 = \frac{|c-d|}{|a-b|+|c-d|}$$

$$\text{Probability of selecting alternative 2 by } A,\ p_2 = \frac{|a-b|}{|a-b|+|c-d|}$$

$$\text{Probability of selecting alternative 1 by } B,\ q_1 = \frac{|b-d|}{|a-c|+|b-d|}$$

$$\text{Probability of selecting alternative 2 by } B,\ q_2 = \frac{|a-c|}{|a-c|+|b-d|}$$

$$\text{Value of game, } V = \frac{a|c-d|+c|a-b|}{|a-b|+|c-d|} = \frac{b|c-d|+d|a-b|}{|a-b|+|c-d|}$$

$$= \frac{a|b-d|+b|a-c|}{|a-c|+|b-d|} = \frac{c|b-d|+d|a-c|}{|a-c|+|b-d|}$$

Example 8.6. Solve following 4 × 4 game optimally

Playerr B

Player A	1	2	3	4
1	6	2	4	8
2	2	– 1	1	12
3	2	3	3	9
4	5	2	6	10

Solution: (*i*) Calculate maximin and minimax value

Player B

Player A	1	2	3	4	Row minimum
1	6	2	4	8	2
2	2	–1	1	12	– 1
3	2	3	3	9	2 → Maximin
4	5	2	6	10	2
Column maximum	6	3 (Minimax)	6	12	

Maximin value (2) is not equal to Minimax value (3). Therefore, game has no saddle point.

(*ii*) In above table, sum of the values in column 1 and column 2 is less than or equal to the corresponding value in column 4. Column 4 is dominated, hence deleted.

Player B

Player A	1	2	3
1	6	2	4
2	2	– 1	1
3	2	3	3
4	5	2	6

(*iii*) Now, the sum of values in row 1 and row 3 is greater than or equal to the corresponding value in row 2. Row 2 is dominated and deleted.

Player B

Player A	1	2	3
1	6	2	4
3	2	3	3
4	5	2	6

(*iv*) The values in column 2 of above matrix are less than or equal to the corresponding values in column 3. Column 3 is dominated and deleted.

Player B

Player A	1	2
1	6	2
3	2	3
4	5	2

(*v*) The values in row 1 are greater than or equal to the corresponding values in row 4. Row 4 is dominated and deleted.

		Player B 1	2	Oddments
Player A	1	6	2	|2 – 3| = 1
	3	2	3	|6 – 2| = 4
Oddments		|2 – 3| = 1	|6 – 2| = 4	

Probability of selection of alternative 1 by $A = p_1 = \dfrac{1}{1+4} = \dfrac{1}{5}$

Probability of selection of alternative 3 by $A = p_3 = \dfrac{4}{(1+4)} = \dfrac{4}{5}$

Probability of selection of alternative 1 by $B = q_1 = \dfrac{1}{(1+4)} = \dfrac{1}{5}$

Probability of selection of alternative 2 by $B = q_2 = \dfrac{4}{(1+4)} = \dfrac{4}{5}$

Value of game, $V = \dfrac{6\times1+2\times4}{1+4} = \dfrac{14}{5} = 2.8$

Optimal solution is $A\left(\dfrac{1}{5}, 0, \dfrac{4}{5}, 0\right), B\left(\dfrac{1}{5}, \dfrac{4}{5}, 0, 0\right)$

$V = 2.8$

8.6.2. Graphical Method for 2 x *n* Games

Games where one player has only two courses of actions while the other has more than two are called 2 x n games. 2 x 2 sub games, within which solution of game lies, are identified. Graphical method is helpful in finding out which of the optimal strategies of both the players assign non-zero probabilities to same number of pure strategies. Graphical method is useful for 2 x n and m x 2 games. It consists of two graphs:

1. The payoff (gains) available to player A versus his strategies options.
2. The payoff (losses) available to player B versus his strategies options.

Steps to solve a 2 *x n* game using graphical method are listed below:

(*i*) Reduce the size of the payoff matrix of player A applying the dominance property, if it exists.

(*ii*) Assuming x as probability of selection of alternative 1 by player A and $1 - x$ be the probability of selection of alternative 2 by player A. Calculate the expected gain function of player *A* with respect to each of the alternatives of player *B*.

(*iii*) For each of the gain functions which are obtained in step ii, find the value of the gain, when x = 0 as well as 1.

(*iv*) Plot the gain functions on a graph by assuming a suitable scale. Keep x on *X* - axis and the **gain** on *Y*-axis.

(*v*) Since the player A is Maximin player, find the highest intersection point in the lower boundary of the graph. It is Maximin point.

(*vi*) If the number of lines passing through the Maximin point is only two, form a 2 *x* 2 payoff matrix from the original problem by retaining only the columns corresponding to those two lines and go to step *viii*; otherwise go to step *vii*.

(*vii*) Identify any two lines with opposite slopes passing through that Maximin point. Then form a 2 *x* 2 payoff matrix from the original problem by retaining only the columns corresponding to those two lines, which are having opposite slopes.

(*viii*) Solve the 2 *x* 2 game using oddments and find the strategies for players A and B and also the value of the game following the steps listed in Algebric method.

Steps to solve *m x* 2 games using graphical method are listed below:

(*i*) Reduce the size of the payoff matrix of player A applying the dominance property, if it exists.

(*ii*) Assuming y as probability of selection of alternative 1 by player *B* and $1 - y$ be the probability of selection of alternative 2 by player *B*. Calculate the expected gain function of player *B* with respect to each of the alternatives of player *A*.

(*iii*) For each of the gain functions which are obtained in step ii, find the value of the gain, when $y = 0$ as well as 1.

(*iv*) Plot the gain functions on a graph by assuming a suitable scale. Keep y on *X* - axis and the **gain** on *Y*-axis.

(*v*) Since the player B is Minimax player, find the lowest intersection point in the upper boundary of the graph. It is Minimax point.

(*vi*) If the number of lines passing through the Minimax point is only two, form a 2 x 2 payoff matrix from the original problem by retaining only the rows corresponding to those two lines and go to step **viii**; otherwise go to step **vii**.

(*vii*) Identify any two lines with opposite slopes passing through that Minimax point. Then form a 2 x 2 payoff matrix from the original problem by retaining only the rows corresponding to those two lines, which are having opposite slopes.

(*viii*) Solve the 2 x 2 game using oddments and find the strategies for players A and B and also the value of the game following the steps listed in Algebric method.

Example 8.7. Solve following 2 × 3 game graphically

		Player B		
		1	2	3
	1	1	3	11
Player A	2	8	5	2

Solution: (*i*) Saddle point in the problem does not exists. Consider that player A plays mixed strategy against player B.

$$S_A = \begin{bmatrix} A_1 & A_2 \\ p_1 & p_2 \end{bmatrix} \text{ here, } p_2 = 1 - p_1$$

A's expected pay off aginst B's pure move are :

B's pure move	A's expected pay off, $E(p_1)$
B_1	$E_1(p_1) = 1p_1 + 8(1-p_1) = -7p_1 + 8$
B_2	$E_2(p_1) = 3p_1 + 5(1-p_1) = -2p_1 + 5$
B_3	$E_3(p_1) = 11p_1 + 2(1-p_1) = 9p_1 + 2$

(*ii*) Expected pay off equations are plotted as functions of p_1 (figure 8.2). Figure shows the pay off of each column represented as points on two vertical axes 1 and 2 at unit distance apart.

As player *A* wishes to maximise his minimum expected pay off, take highest point of intersection (x) on lower envelope. x is point of intersection of lines representing B_2 and B_3 (relevent moves). Hence, delete B_1

		Player B		
		B_2	B_3	Oddments
Player A	A_1	3	11	3
	A_2	5	2	8
Oddments		9	2	

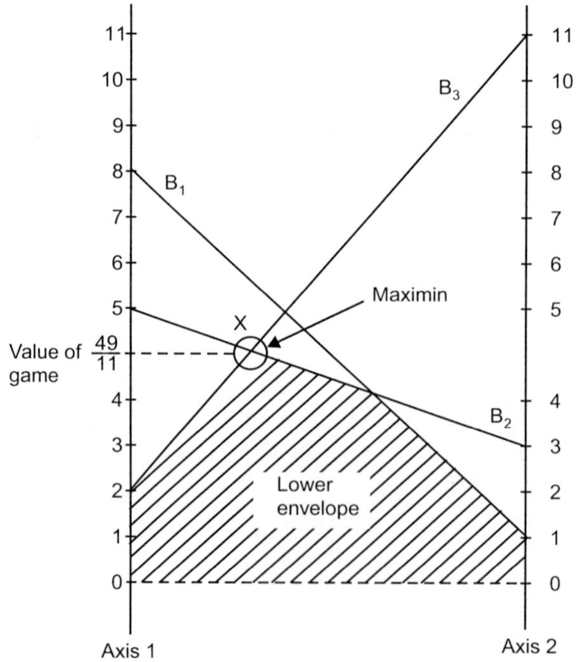

Fig. 8.2.

$$p_1 = \frac{3}{3+8} = \frac{3}{11}$$

$$p_2 = \frac{8}{3+8} = \frac{8}{11}$$

$$q_1 = \frac{9}{9+2} = \frac{9}{11}$$

$$q_2 = \frac{2}{9+2} = \frac{2}{11}$$

Value of game, $V = \dfrac{3\times 3+5\times 8}{3+8} = \dfrac{49}{11}$

Optimal solution is

$$A\left(\frac{3}{11},\frac{8}{11}\right), B\left(0,\frac{9}{11},\frac{2}{11}\right)\ V = \frac{49}{11}$$

8.6.3. Linear Programming Method

Example 8.8. Solve the following game by linear programming technique :

$$\text{Player } A \quad \overset{\text{Player } B}{\begin{pmatrix} 1 & -1 & 3 \\ 3 & 5 & -3 \\ 6 & 2 & -2 \end{pmatrix}}$$

Solution: Since some of the entries in the payoff matrix are negative, we add a suitable constant to such entries to ensure them all positive. Thus, adding a constant $c = 4$ to each element, we get the following revised payoff matrix :

$$\text{Player } A \quad \overset{\text{Player } B}{\begin{pmatrix} 5 & 3 & 7 \\ 7 & 9 & 1 \\ 10 & 6 & 2 \end{pmatrix}}$$

Let the strategies of the two players be

$$S_A = \begin{bmatrix} A_1 & A_2 & A_3 \\ p_1 & p_2 & p_3 \end{bmatrix} \quad S_B = \begin{bmatrix} B_1 & B_2 & B_3 \\ q_1 & q_2 & q_3 \end{bmatrix}$$

where $p_1 + p_2 + p_3 = 1$ and $q_1 + q_2 + q_3 = 1$.

The linear programming formulation for the problem is :

For Player A

Maximize u = Minimize $\dfrac{1}{u} = x_1 + x_2 + x_3$ subject to the constraints :

$$5x_1 + 7x_2 + 10x_3 \geq 1,\ 3x_1 + 9x_2 + 6x_3 \geq 1,$$
$$7x_1 + x_2 + 2x_3 \geq 1 : \text{ and } x_j \geq 0$$

For Player B

Minimize v = Maximize $\dfrac{1}{v} = y_1 + y_2 + y_3$.

$$5y_1 + 3y_2 + 7y_3 \leq 1,\ 7y_1 + 9y_2 + y_3 \leq 1.$$

$$10y_1 + 6y_2 + 2y_3 \leq 1; \text{ and } y_j \geq 0$$

where $x_j = p_j/u$ $(j = 1, 2, 3)$ and $y_j = \dfrac{q_j}{v}$ $(j = 1, 2, 3,)$: u = minimum expected gain to A and v = minimum expected loss to B.

Now solve the problem for player B. By introducing slack variables $y_4 \geq 0$, $y_5 \geq 0$ and $y_6 \geq 0$: the iterative simplex tables are :

Initial Iteration. Introduce y_3 and drop y_4.

C_B	B	X_B	y_1	y_2	y_3	y_4	y_5	y_6
0	y_4	1	5	3	7	1	0	0
0	y_5	1	7	9	1	0	1	0
0	y_6	1	10	6	2	0	0	1
	$1/v$	0	1	1	1	0	0	0

First Iteration. Introduce y_2 and drop y_5.

C_B	B	X_B	y_1	y_2	y_3	y_4	y_5	y_6
1	y_3	1/7	5/7	3/7	1	1/7	0	0
0	y_5	6/7	44/7	60/7	0	–1/7	1	0
0	y_6	5/8	60/8	36/7	0	–2/7	0	1
	$1/v$	1/7	2/7	4/7	0	1/7	0	0

Final Iteration. Optimum Solution.

C_B	B	X_B	y_1	y_2	y_3	y_4	y_5	y_6
1	y_3	1/10	2/5	0	1	3/20	–1/20	0
1	y_2	1/10	11/5	1	0	–1/60	7/60	0
0	y_6	1/5	24/5	0	0	–1/5	–3/5	1
	$1/v$	1/5	–2/5	0	0	–2/5	–1/5	0

The expected value of game, therefore, is

$$v^0 = v–4 = 5 –4 =1.$$

Optimum strategies for player B are

$$q_1^0 = 0,\ q_2^0 = \frac{1}{10} \times 5 = \frac{1}{2} \text{ and } q_3^0 = \frac{1}{10} \times 5 = \frac{1}{2}$$

Making use of duality, the optimum strategies for player A are

$$p_1^0 = \frac{2}{15} \times 5 = \frac{2}{3},\ p_2^0 = \frac{1}{15} \times 5 = \frac{1}{3} \text{ and } p_3^0 = 0.$$

Hence the optimum solution to the given problem is

$$S_A = \begin{bmatrix} A_1 & A_2 & A_3 \\ 2/3 & 1/3 & 0 \end{bmatrix} \quad S_B = \begin{bmatrix} B_1 & B_2 & B_3 \\ 0 & 1/2 & 1/2 \end{bmatrix} \text{ and } v^0 = 1.$$

8.7. SENSITIVITY ANALYSIS

Sensitivity analysis, also known as stability analysis, is a technique for determining how much an expected return will change in response to a given change in an input variable, all other things remaining unchanged.

8.7.1. Steps in Sensitivity Analysis

1. Begin with consideration of a nominal base-case situation, using the expected values for each input.
2. Calculate the base-case output.
3. Consider a series of "what-if" questions to determine by how much the output would deviate from this nominal level if input values deviated from their expected values.
4. Each input is changed by several percentage points above and below its expected value, and the expected payoff is recalculated.
5. The set of expected payoff is plotted against the variable that was changed.
6. The steeper the slope (i.e., derivative) of the resulting line, the more sensitive the expected payoff is to a change in the variable.

8.8. LIMITATIONS OF GAME THEORY

1. Analysis of game becomes complex and difficult as the number of players increases.
2. In situation of many players, some players may allow each other to share the secrets of business to work out collusion.
3. Assumption that players have complete knowledge of the strategies is not practical.

SOLVED EXAMPLES

Example 8.9. Solve the following game :

$$\text{Player } A \quad \begin{array}{c} \\ I \\ II \\ III \\ IV \end{array} \overset{\text{Player } B}{\begin{array}{cccc} I & II & III & IV \\ \end{array}} \begin{bmatrix} 3 & 2 & 4 & 0 \\ 3 & 4 & 2 & 4 \\ 4 & 2 & 4 & 0 \\ 0 & 4 & 0 & 8 \end{bmatrix}$$

Solution: From the above payoff matrix, we observe that first row is dominated by third row and first column is dominated by third column. The reduced payoff matrix is

$$\begin{array}{c} \\ II \\ III \\ IV \end{array} \begin{array}{c} \begin{array}{ccc} II & III & IV \end{array} \\ \begin{bmatrix} 4 & 2 & 4 \\ 2 & 4 & 0 \\ 4 & 0 & 8 \end{bmatrix} \end{array}$$

Now, none of the pure strategies of player B is inferior to any of his other strategies. However, a convex linear combination (average) of strategies III and IV dominates strategy II of player B, yielding the reduced payoff matrix

$$\begin{array}{c} \\ II \\ III \\ IV \end{array} \begin{array}{c} \begin{array}{cc} III & IV \end{array} \\ \begin{bmatrix} 2 & 4 \\ 4 & 0 \\ 0 & 8 \end{bmatrix} \end{array}$$

Again, we observe that none of the pure strategies of player A is inferior to any of his other strategies. However, a convex linear combination of strategies III and IV dominates strategy II of player A, resulting in the reduced payoff matrix

$$\begin{array}{ccc} & & \text{Player B} \\ & & \begin{array}{cc} \text{III} & \text{IV} \end{array} \\ \text{Player A} & \begin{array}{c} \text{III} \\ \text{IV} \end{array} & \begin{bmatrix} 4 & 0 \\ 0 & 8 \end{bmatrix} \end{array}$$

Now, let

$$S_A = \begin{bmatrix} A_1 & A_2 & A_3 & A_4 \\ 0 & 0 & p_1 & p_2 \end{bmatrix}, \quad S_B = \begin{bmatrix} B_1 & B_2 & B_3 & B_4 \\ 0 & 0 & q_1 & q_2 \end{bmatrix}$$

where $p_1 + p_2 = 1, q_1 + q_2 = 1$, and then using the method of solving 2×2 games, we can easily obtain the optimum strategies as

$$S_A = \begin{bmatrix} A_1 & A_2 & A_3 & A_4 \\ 0 & 0 & 2/3 & 1/3 \end{bmatrix}, \quad S_B = \begin{bmatrix} B_1 & B_2 & B_3 & B_4 \\ 0 & 0 & 2/3 & 1/3 \end{bmatrix}$$

The value of game is 8/3.

EXERCISE

8.1. What is game theory? How is it important to military operations?

8.2. State the properties, which a competitive situation should have, if it is to be called a game.

8.3. Explain the following terms:
 (*i*) Players, Strategy
 (*ii*) Two-person zero-sum game
 (*iii*) Principle of dominance
 (*iv*) Pure strategy in game theory
 (*v*) Mixed strategy
 (*vi*) Value of game
 (*vii*) Saddle point
 (*viii*) Payoff matrix
 (*ix*) Rectangular game.

8.4. Describe applications of game theory.

8.5. Describe 'Maximin principle' of game theory. Explain pure strategy and saddle point.

8.6. How is the concept of dominance used in simplifying the solution of a rectangular game?

8.7. Define mixed strategy and a fair game.

8.8. What are characteristics of game theory?

8.9. Explain difference between pure strategy and mixed strategy.

8.10. Explain the optimal strategy on the basis of Minimax criterion of optimality.

8.11. Explain the Minimax principle used in game theory.

8.12. Describe the rules for finding the saddle point.

8.13. Establish the relationship between game theory and linear programming with the help of an example.

8.14. Write steps of solving a two-person zero-sum game as a linear programming problem.

8.15. When do players apply mixed strategies?

8.16. What types of games are solved graphically?

8.17. When do you say the game is fair?

8.18. When do you say the game is stable?

8.19. What are major applications and limitations of game theory?

8.20. Find the optimum strategies of the players in the following games.

(*a*)

		B 1	2	3
A	1	30	20	40
	2	55	50	60
	3	60	30	40

(*b*)

		B 1	2	3
A	1	10	5	20
	2	65	50	40
	3	55	25	30

[**Ans.** *a* – 50, *b* – 40]

8.21. Players *A* and *B* play a game in which each player has three coins [25p, 50p and 100p (one rupee)]. Each of them selects a coin without the knowledge of the other person. If the sum of the values of the coins is an even number, *A* wins B's coin. If that sum is an odd number, *B* wins A's coin.

(*a*) Develop a payoff matrix with respect to Player A.

(*b*) Find the optimal strategies for the players.

[**Ans.** A (0, 1, 0), B(0, 1, 0), 50]

8.22. Consider the following payoff matrix with respect to Player *A* and solve it optimally.

		B 1	2
A	1	10	8
	2	6	12

[**Ans.** 9]

8.23. Consider the following payoff matrix with respect to Player *A* and solve it optimally.

		B 1	2
A	1	8	18
	2	16	9

[**Ans.** 216/17]

8.24. Solve the following games:

		B 1	2	3
A	1	–3	4	2
	2	7	8	5
	3	6	2	9

[**Ans.** 6.2]

8.25. Consider the following 4 × 4 game played by Players *A* and *B* and solve it optimally.

		Player B			
		1	2	3	4
Player A	1	12	4	8	16
	2	4	–2	2	24
	3	4	6	6	18
	4	10	4	12	20

[**Ans.** 5.6]

8.26. Solve the following 3×5 game using dominance property.

		Player B				
		1	2	3	4	5
Player A	1	6	15	30	21	6
	2	3	3	6	6	4
	3	12	12	24	36	3

[**Ans.** 6]

8.27. Consider the payoff matrix of Player A as shown below and solve it optimally using graphical method.

		Player B				
		1	2	3	4	5
Player A	1	4	2	5	–6	6
	2	7	–9	7	4	8

[**Ans.** – 46/21]

8.28. Consider the following payoff matrix of Player A and solve it optimally using the graphical method.

		Player B				
		1	2	3	4	5
Player A	1	7	8	4	6	8
	2	–8	6	1	9	6

[**Ans.** 4]

8.29. Consider the payoff matrix of player A as given below and solve it optimally using graphical method.

		Player B	
		1	2
Player A	1	5	3
	2	6	4
	3	2	–7
	4	9	–8

[**Ans.** 4]

8.30. The following table represents the payoff matrix with respect to Player A. Solve it optimally using linear programming method.

		Player B		
		1	2	3
Player A	1	2	3	–4
	2	5	–2	6
	3	2	6	3

[**Ans.** 3.09877]

8.31. Solve the following game whose payoff matrix is given below:

Player B

$$\text{Player A} \quad \begin{array}{c} \\ A_1 \\ A_2 \\ A_3 \end{array} \begin{array}{c} \begin{array}{cccc} B_1 & B_2 & B_3 & B_4 \end{array} \\ \begin{bmatrix} 1 & 7 & 3 & 4 \\ 5 & 6 & 4 & 5 \\ 7 & 2 & 0 & 3 \end{bmatrix} \end{array}$$

[**Ans.** 4]

8.32. Use dominance propertyto reduce the size of the following game to 2 × 2 game and hence find the optimal strategies and value of the game.

(*a*) Player B

$$\text{Player } A \quad \begin{array}{c} \\ A_1 \\ A_2 \\ A_3 \end{array} \begin{array}{c} \begin{array}{ccc} B_1 & B_2 & B_3 \end{array} \\ \begin{bmatrix} 5 & 7 & 11 \\ 2 & -1 & 8 \\ 18 & -6 & 10 \end{bmatrix} \end{array}$$

(*b*) Player B

$$\text{Player } A \quad \begin{array}{c} \\ A_1 \\ A_2 \\ A_3 \\ A_4 \end{array} \begin{array}{c} \begin{array}{ccccc} B_1 & B_2 & B_3 & B_4 & B_5 \end{array} \\ \begin{bmatrix} 2 & 4 & 3 & 8 & 4 \\ 5 & 6 & 3 & 7 & 8 \\ 6 & 7 & 9 & 8 & 7 \\ 4 & 2 & 8 & 4 & 3 \end{bmatrix} \end{array}$$

[**Ans.** (A_3, B_1), 6]

8.33. In zero-sum two person children's game of stone paper and scissors, both players simultaneously call out stone, paper or scissors. The combination of paper and stone is a win of one unit for player calling paper (paper covers stone); stone and scissors is a win for stone (stone breaks scissors), and scissors and paper is a win for scissors (scissors cut paper). A call of the same item represents no pay-off. Write the pay-off matrix and the equivalent linear programming problem to the above game. Find the optimal strategy for both the players and the value of the game.

8.34. For the following payoff table, transform the zero-sum game into an equivalent linear programming problem and solve it by simplex method :

Player B

$$\text{Player } A \quad \begin{array}{c} \\ A_1 \\ A_2 \\ A_3 \end{array} \begin{array}{c} \begin{array}{ccc} B_1 & B_2 & B_3 \end{array} \\ \begin{bmatrix} 0 & 2 & 2 \\ 3 & -1 & 3 \\ 4 & 4 & -2 \end{bmatrix} \end{array}$$

[**Ans.** 17/11]

8.35. Firm X is fighting for its life against the determination of firm Y to drive it out of the industry, Firm X has the choice of increasing price, leaving it unchanged or lowering it. Firm Y has the same three options. Firm X's gross sales in the event of each of the pairs of choices are shown below :

		Firm Y's pricing strategies		
		Increase price	Do not change	Reduce price
Firm X's pricing Strategies	Increase price	90	80	110
	Do not change	110	100	90
	Reduce price	120	70	80

Assuming firm X as the maximizing one and solve the problem as a linear programming problem.

8.36. The following matrix represents the payoff to P_1 in a rectangular game between two persons P_1 and P_2 :

$$P_1 \overset{P_2}{\begin{bmatrix} 8 & 15 & -4 & -2 \\ 19 & 15 & 17 & 16 \\ 0 & 20 & 15 & 5 \end{bmatrix}}$$

By the property of dominance, reduce the game to a 2 × 4 game and solve it graphically. **[Ans.** 245/16]

8.37. Solve the following '2-person, zero-sum' game:

		Player B		
		B_1	B_2	B_3
Player A	A_1	10	5	7
	A_2	6	7	5
	A_3	7	6	7

[Ans. 19/3]

OBJECTIVE TYPE QUESTIONS

8.1. Two-person zero-sum game that the
(*a*) sum of losses to one player equals the sum of gains to other.
(*b*) sum of losses to one player is not equal to the sum of gains to other.
(*c*) both (*a*) and (*b*). (*d*) none of the above.

8.2. Game theory models are classified by the
(*a*) number of players (*b*) sum of all playoffs
(*c*) number of strategies (*d*) all of the above.

8.3. A game is said to be fair, If
(*a*) both upper and lower values of the game are same and zero.
(*b*) upper and lower values of the game are not equal
(*c*) upper value is more than lower value of the game
(*d*) none of the above.

8.4. What happens when maximin and minimax values of the game are same?
(*a*) No solution exists (*b*) Solution is mixed
(*c*) Saddle point exists (*d*) None of the above

8.5. A mixed strategy game can be solved by
(*a*) algebraic method (*b*) matrix method
(*c*) graphical method (*d*) all of the above.

8.6. The size of the payoff matrix of a game can be reduced by using the principle of
(*a*) game inversion (*b*) rotation reduction
(*c*) dominance (*d*) gram transpose.

8.7. The payoff value for which each player in a game always selects the same strategy is called the
(*a*) saddle point (*b*) equilibrium point
(*c*) both (*a*) and (*b*) (*d*) none of the above

8.8. Games which involve more than two players are called
(*a*) conflicting games (*b*) negotiable games
(*c*) n-person games (*d*) all of the above

8.9. When the sum of gains of one player is equal to the sum of losses to another player in a game, this situation is known as
(*a*) biased game (*b*) zero-sum game
(*c*) fair game (*d*) all the above

8.10. When no saddle point is found in a payoff matrix of game, the value of the game is then found by
(*a*) knowing joint probabilities of each row and column combination to calculate expected payoff for that combination and adding all such values.
(*b*) reducing size of the game to apply algebraic method
(*c*) both (*a*) and (*b*) (*d*) none of the above.

Answers

8.1. (*a*) 8.2. (*d*) 8.3. (*a*) 8.4. (*c*) 8.5. (*d*) 8.6. (*c*) 8.7. (*c*)
8.8. (*c*) 8.9. (*b*) 8.10. (*c*)

❏❏❏

Chapter 9

Sequencing Problem

9.1. INTRODUCTION

Sequencing problem is concerned with an appropriate selection of series of jobs to be done on a finite number of service facilities (e.g. machines), in some well defined order so as to optimize some effectiveness measure (such as overall cost). A practical situation of sequencing problem is to determine the sequence in which number of products should be processed on different machines to optimize some measure of effectiveness.

Let, there are n jobs $(1, 2, 3,..., n)$, to be performed one at a time, each of which requires processing on some or all of m different machines $(A, B, C, ...)$. The order of processing each job through machines is given (for example, job I is processed through machines A, C, B-in this order). The time that each job must require on each machine is known. The effectiveness (*i.e.* cost, time or mileage, etc.) can be measured for any given sequence of jobs at each machine, and the most suitable sequence is to be selected (which optimizes the effectiveness measure) among all $(n!)^m$ theoretically possible sequences. Although, theoretically, it is always possible to select the best sequence by testing each one, but it is practically impossible because of large number of computations. In particular, if $m = 5$ and $n = 5$, the total number of possible sequences will be $(5!)^5 = 25{,}000{,}000{,}000$. Hence, the effectiveness for each of $(5!)^5$ sequences are to be computed before selecting the most suitable one. But, this approach is practically impossible to adopt.

Mathematically, let

A_i = time for job i on machine A,

B_i = time for job i on machine B, etc.

T = time from start of first job to completion of the last job.

Then, the problem is to determine for each machine a sequence of jobs $i_1, i_2, i_3, ..., i_n$, where

$(i_1, i_2, i_3, ..., i_n)$ is the permutation of the integers which will minimize T.

9.2. BASIC TERMS USED

The following terms and notations will be used in sequencing :

1. **Number of Machines:** It means the service facilities through which a job must pass before it is completed. For example, a book to be published has to be processed through typing, editing, printing, binding, etc. In this example, the book constitutes the *job* and the different processes constitute the *number of machines*.

2. **Processing Order:** It refers to the order (sequence) in which various machines are required for completing the job.
3. **Processing Time:** It means the time required by each job on each machine. The notation T_{ij} will denote the processing time required for i^{th} job by j^{th} machine $(i = 1, 2, ..., n; j = 1, 2, ..., m)$.
4. **Idle Time on a Machine:** This is the time for which a machine remains idle during the total elapsed time. The notation X_{ij} shall be used to denote the idle time of machine j between the *end* of the $(i - 1)^{th}$ job and the start of the ith job.
5. **Total Elapsed Time:** This is the time interval between starting the first job and completing the last job in a particular order by given set of machines. This also includes *idle time,* if exists. It will be denoted by the symbol T.
6. **No Passing Rule:** This rule means that passing is not allowed, *i.e.* the same order of jobs is maintained over each machine. If each of the n-jobs is to be processed through two machines A and B in the order AB, then this rule means that each job will go to machine A first and then to B.

9.3. GENERAL ASSUMPTIONS

1. No machine can process more than one job at a time.
2. Each operation, once started, must be performed till completion.
3. A job is an entity, *i.e.* even though the job represents a lot of individual parts, no lot may be processed by more than one machine at a time.
4. Each operation must be completed before any other operation, which it must precede, can begin.
5. Time intervals for processing are known and independent of the order in which operations are to be performed. Processing times do not change.
6. There is only one of each type of machine.
7. A job is processed as soon as possible subject to ordering requirements.
8. All jobs are known and are ready to start processing before the period under consideration begins.
9. The time required to transfer jobs from one machine to another is negligible and taken equal to zero.
10. The order of completion of the jobs is independent of the sequence of jobs.

9.4. SOLUTION OF SEQUENCING PROBLEMS

At present, solutions of following cases are available:

1. n jobs and two machines A and B, all jobs processed in the order AB.
2. n jobs and three machines A, B and C, all jobs processed in the order ABC.
3. Two jobs and m machines. Each job is to be processed through the machines in a prescribed order (which is not necessarily the same for both the jobs).
4. Problems with n jobs and m-machines.

9.5. PROCESSING n JOBS THROUGH TWO MACHINES

The problem can be described as: (*i*) only two machines A and B are involved, *(ii)* each job is processed in the order AB, and *(iii)* the exact or expected processing times $A_1, A_2, A_3, ..., A_n; B_1, B_2, B_3, ..., B_n$ are known

Processing Times	Job (1)				
	1	2	3	...	n
A_i	A_1	A_2	A_3	...	A_n
B_i	B_1	B_2	B_3	...	B_n

The problem is to sequence (order) the jobs so as to minimize the total elapsed time T. The solution procedure adopted by ***Johnson*** (1954) is given below.

Solution Procedure

Step 1. Select, the least processing time occurring in the list $A_1, A_2, A_3,..., A_n$ and $B_1, B_2, B_3, ..., B_n$. If there is a tie, either of the smallest processing time should be selected.

Step 2. If the least processing time is A_r select r^{th} job *first*. If it is B_s, do the sth job *last* (as the given order *is AB)*.

Step 3. There are now n - 1 jobs left to be ordered. Again repeat steps 1 and n for the reduced set of processing times obtained by deleting processing times for both the machines corresponding to the job already assigned. Continue till all jobs have been ordered. The resulting ordering will minimize the elapsed time T.

Proof: Since passing is not allowed, all n jobs must be processed on machine A without any *idle time* for it. On the other hand, machine B *is* subject to its remaining idle time at various stages. Let Y_i be the time for which a machine B remains idle after completing $(i - 1)^{th}$ job and before starting processing the ith job $(i = 1, 2, ..., n)$. Hence, the total elapsed time T is given by

$$T = \sum_{i=1}^{n} B_i + \sum_{i=1}^{n} Y_i$$

where some of the Y_i's may be zero.

Now we wish to minimize T. However, since $\sum_{i=1}^{n} B_i$ is the total time for which machine B has to work and thus being constant, it does not form a part or minimizing T. So the problem is reduced to that of minimizing $\sum_{i=1}^{n} Y_i$. A very convenient procedure for obtaining a sequence of performing jobs so as to minimize $\sum_{i=1}^{n} Y_i$ is well explained by the following Gantt Chart.

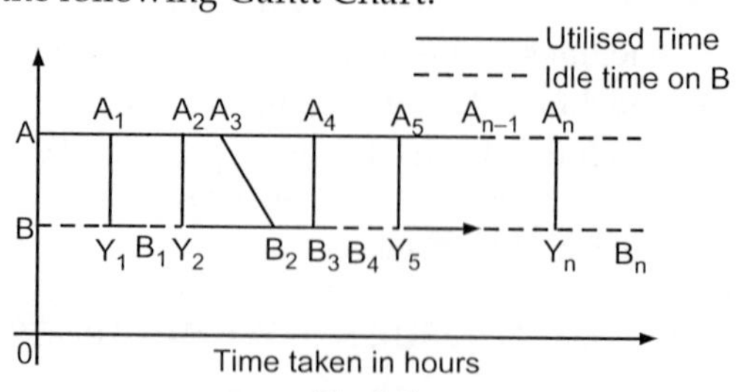

Fig. 9.1.

From this chart, it is clear that

$$Y_1 = A_1$$

$$Y_2 = \begin{cases} A_1 + A_2 - Y_1 - B_1, & \text{if } A_1 + A_2 > Y_1 + B_1 \\ 0 & \text{otherwise} \end{cases}$$

The expression for Y_2 may be rewritten as
$Y_2 = \max \{A_1 + A_2 - Y_1 - B_1, 0\}$
Thus, $Y_1 + Y_2 = \max \{A_1 + A_2 - B_1, A_1\}$, since $Y_1 = A_1$.
Similarly, $Y_3 = \max \{A_1 + A_2 - A_3 - B_1 - B_2 - Y_1 - Y_2, 0\}$

$$\therefore \qquad Y_1 + Y_2 + Y_3 = \max \left\{ \left(\sum_{i=1}^{3} A_i - \sum_{i=1}^{2} B_i \right) . \sum_{i=1}^{2} Y_i \right\}$$

$$= \max \left\{ \left(\sum_{i=1}^{3} A_i - \sum_{i=1}^{2} B_i \right) . \left(\sum_{i=1}^{2} A_i - B_i \right), A_i \right\}$$

In general, we get

$$\sum_{i=1}^{n} Y_i = \max \left\{ \left(\sum_{i=1}^{n} A_i - \sum_{i=1}^{n-1} B_i \right), \left(\sum_{i=1}^{n-1} A_i - \sum_{i=1}^{n-2} B_i \right), \ldots A_i \right\}$$

$$= \max_{1 \le r \le n} \left\{ \sum_{i=1}^{r} A_i - \sum_{i=1}^{r-1} B_i \right\}$$

Now, if we denote $\sum_{i=1}^{n} Y_i$ by $D_n(S)$, then the problem becomes that of finding the sequence $< S^* >$ for processing the jobs 1, 2 ..., n so as to have the inequality $D_n(S^*) \le D_n(S_0)$ for any sequence $<S_0>$ other than $<S^*>$. In other words, we have to find the optimal sequence $< S >$ so as to minimize $D_n(S)$. This can be done iteratively by successively interchanging the consecutive jobs. Each such interchange of jobs gives a value of D_n(S) less than or equal to its value before the change.

Johnson's Algorithm for n Jobs 2 Machines

The Johnson's iterative procedure for determining the optimal sequence for an n-jobs 2-machine sequencing problem can be outlined as follows:

Step 1: Examine the A_i's and B_i's for $i = 1,2, ..., n$ and find out min (A_i, B_i)

Step 2: *(i)* If this minimum be A_k for some $i = k$, do (process) the kth job first of all.

(ii) If this minimum be B_r *for* some $i = r$, do (process) the rth job last of all.

Step 3: *(i)* If there is a tie for minima $A_k = B_r$ process the kth job first of all and r^{th} job in the last.

(ii) If the tie for the minimum occurs among the A_i's, select the job corresponding to the minimum of B_i's and process it first of all.

(iii) If the tie for minimum occurs among the B_i's, select the job corresponding to the minimum of A_i's and process it in the last. Go to next step.

Step 4: Cross-out the jobs already assigned and repeat steps 1 to 3 arranging the jobs next to first or next to last, until all the jobs have been assigned.

Step 5: Calculate idle time for machine A and B.

Step 6: Calculate total elapsed time to process all jobs through 2 machines.

Example 9.1: There are five jobs each of which must go through the two machines A and B in the order AB. Processing times are given below:

	Processing time (hours)				
Job	1	2	3	4	5
Time for *A*	5	1	9	3	10
Time for *B*	2	6	7	8	4

Determine a sequence for five jobs that will minimize the elapsed time T. Calculate the total idle time for the machines in this period.

Solution : Apply steps 1 and 2 of solution procedure. It is seen that the smallest processing time is one hour for job 2 on the machine A. So list the job 2 at first place as shown below.

2				

Now, the reduced list of processing times becomes

Job	*A*	*B*
1	5	[2]
3	9	7
4	3	8
5	10	4

Again, the smallest processing time in the reduced list is 2 for job 1 on the machine *B*. So place job 1 last.

2				1

Continuing in the like manner, the next reduced list is obtained

Job	*A*	*B*
3	9	[7]
4	3	8
5	10	4

Leading to sequecne

2	4			1

and the list

Job	A	B
3	9	[7]
5	10	4

gives rise to sequence

2	4		5	1

Finally, the optimal sequence is obtaied

2	4	3	5	1

Further, it is also possible to calculate the minimum elapsed time corresponding to the optimal sequencing, using the individual processing time given in the statement of the problem. The details are given in table.

Job sequence	Machine *A*		Machine *B*	
	Time in	Time out	Time in	Time out
2	0	1	1	7
4	1	4	7	15
3	4	13	15	22
5	13	23	23	27
1	23	28	28	30

Thus, the minimum time, *i.e.*, the time for starting of job 2 to completion of the last job 1, is 30 hrs only. During this time, the machine *A* remains idle for 2 hrs (from 28 to 30 hrs) and the machine *B* remains idle for 3 hrs only (from 0-1, 22-23, and 27-28 hrs). The total elapsed time can also be calculated by using Gantt chart as follows :

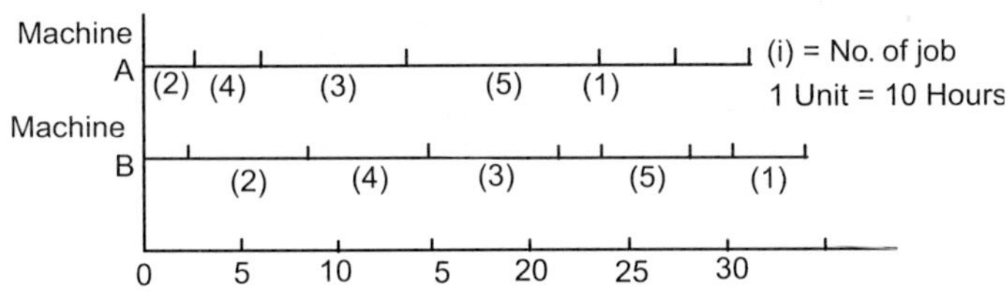

Fig. 9.2.

From the Fig. 9.2, it can be seen that the total elapsed time is 30 hrs, and the idle time of the machine B is 3 hrs. In this problem, it is observed that job may be held in inventory before going to the machine. For example, 4th job will be free on machine A after 4th hour and will start on machine B after 7th hr. Therefore, it will be kept in inventory for 3 hrs. Here it is assumed that the storage space is available and the cost of holding the inventory for each job is either same or negligible. For short duration process problems, it is negligible. Second general assumption is that the order of completion of jobs has no significance, *i.e.* no job claims the priority.

9.6. PROCESSING n JOBS THROUGH THREE MACHINES

The problem can be described as: *(i)* Only three machines *A*, *B* and C are involved, *(ii)* each job is processed in the prescribed order *ABC*, *(iii)* transfer of jobs is not permitted, *i.e.* adhere strictly to the order over each machine, and *(iv)* exact or expected processing times are given in table.

Job	Machine *A*	Machine *B*	Machine *C*
1	A_1	B_1	C_1
2	A_2	B_1	C_2
3	A_3	B_3	C_3
:	:	:	:
n	A_n	B_n	C_n

So far no general procedure is available for obtaining an optimal sequence in this case.

However, the earlier method adopted by *Johnson* (1954) can be extended to cover the special cases where *either one* or *both* of the following conditions hold:

(*i*) The minimum time on machine $A \geq$ the maximum time on machine B.

(*ii*) The minimum time on machine $C \geq$ the maximum time on machine B.

The procedure explained here is to replace the problem with an equivalent problem, involving n jobs and two fictitious machines denoted by G and H, and corresponding time Gj and Hj are defined by

Here... stands for other machines, if any. Applying the rules to this example, it is observed by taking A as X, and D as Y (1 rule), that delete the programs containing *ad*. Such a program is 16th only. Again by n rule taking A as X and C as Y, all those programs are deleted which contain *lie, i.e.,* the 5th program. Other rules are not applicable to our problem. Thus we have only following five programs.

$$G_i = A_i + B_i,\ H = B_i + C_i$$

If this problem with prescribed ordering GH is solved, the resulting optimal sequence will also be optimal for the original problem.

Rules for deleting the programs which cannot be optimal.

Rule no.	**Technological ordering for**		**Delete programs containing**
	Job 1	**Job 2**	
I	X ... Y	Y ...	xy
II	X ... Y	... XY	$\bar{x}y$
III	... X ... Y	... XY ...	$\bar{x}y$
IV	... X Y ...	X ... Y ...	xy
V	... X Y ... Z ...	... X ... YZ	$x\bar{y}z$
VI	... X ... YZ ...	... XY ... Z	$\bar{x}y\bar{z}$

Here... stands for other machines, if any. Applying the rules to this example, it is observed by taking A as X, and D as Y (I rule), that delete the programs containing *ad*. Such a program is 16thonly. Again by n rule taking A as X and C as Y, all those programs are deleted which contain *lie, i.e.,* the 5th program. Other rules are not applicable to our problem. Thus we have only following five programs.

Program No.

1	2	4	6	8
$\bar{a}$	a	a	a	a
$\bar{b}$	$\bar{b}$	b	$\bar{b}$	b
$\bar{c}$	$\bar{c}$	$\bar{c}$	c	c
$\bar{d}$	$\bar{d}$	$\bar{d}$	$\bar{d}$	d

Now finally we enumerate all these programs one by one using Gantt Chart as shown in Fig. 9.3.

From these charts it is clear that optimum program is 6th and the minimum elapsed time is 18 hours.

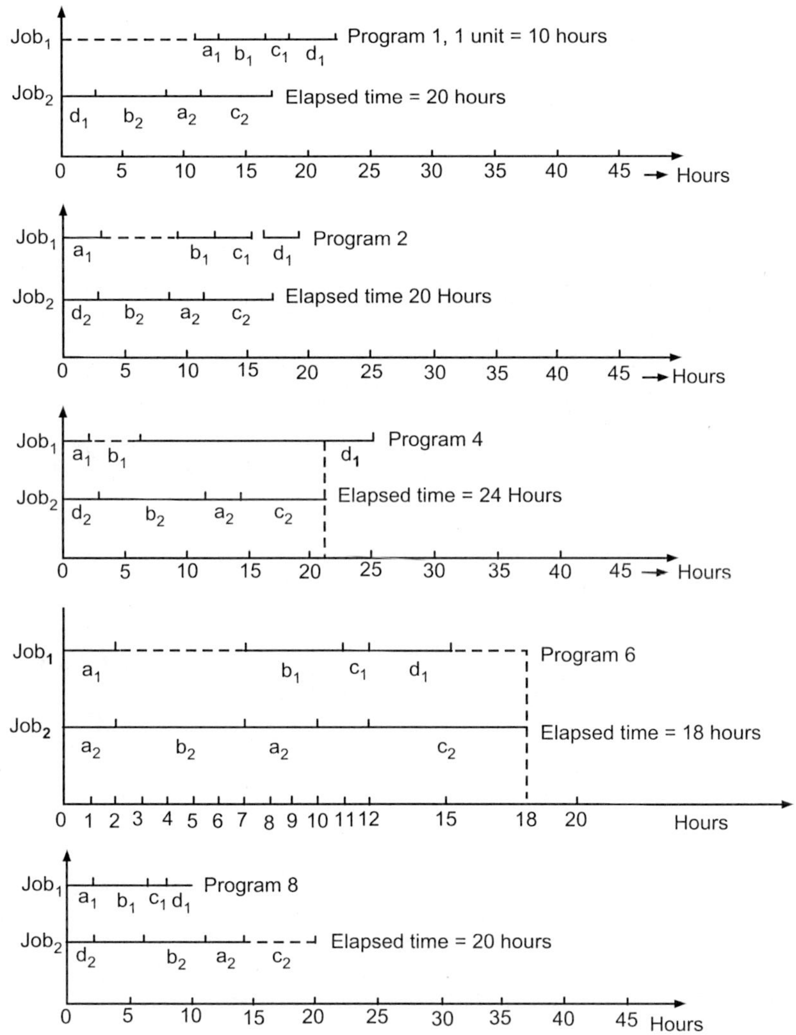

Fig. 9.3.

Example 9.2. There are five jobs, each of which must go through machines, A, B and C in the order ABC. Processing times are given in Table.

Job 1	Processing Times		
	A_1	B_1	C_1
1	8	5	4
2	10	6	9
3	6	2	8
4	7	3	6
5	11	4	5

Determine a sequence for five jobs that will minimize the elapsed time T.

Solution: Here min $A_i = 6$, max $B_i = 6$, min $C_i = 4$.

Since one of two conditions is satisfied by min A_i =max B_i, so the procedure adopted in *Example 1* can be followed.

The equivalent problem, involving five jobs and two fictitious machine G and H, becomes:

Job 1	Processing Times	
	$G_i = (A_i + B_i)$	$H_i = (B_i + C_i)$
1	13	9
2	16	15
3	8	10
4	10	9
5	15	9

This new problem can be solved by the procedure described earlier. Because of ties, possible optimal sequences are :

(i) | 3 | 2 | 1 | 4 | 5 | (ii) | 3 | 2 | 4 | 1 | 5 |

(ii) | 3 | 2 | 4 | 5 | 1 | (iv) | 3 | 2 | 5 | 4 | 1 |

(v) | 3 | 2 | 1 | 5 | 4 | (vi) | 3 | 2 | 5 | 1 | 4 |

It is possible to calculate the minimum elapsed time for first sequence as shown in Table.

Job	Machine A		Machine B		Machine C	
	Time in	Time out	Time in	Time out	Time in	Time out
3	0	6	6	8	8	16
2	6	16	16	22	22	31
1	16	24	24	29	31	35
4	24	31	31	34	35	41
5	31	42	42	46	46	51

Thus, any of the sequence from (*i*) to (*vi*) may be used to order the jobs through machines A, B and C. They all will give a minimum elasped time of 51 hrs. Idle time for machine A is 9 hrs, for B 31 hrs, for C 19 hrs.

9.7 PROCESSING n JOBS THROUGH m MACHINES

Let each of the n jobs be processed through m machines, say MI, $M2$, $M3$, ..., M $m$in the order M_1, M_2, M_3 ... M_m, and T_{ij} denote the time taken by the ith machine to complete the jth job. The step-by-step procedure for obtaining an optimal sequence is as follows:

Step 1. First find, (*i*) min, (T_{ij}) (*ii*) min (T_{mj}) and (*iii*) max $(T_{2j}, T_{3j}, \dots T_{m-1})$ for $j = 1, 2, \dots, n$.

Step 2. Then check whether

(*i*) min $(T_{1j}) \geq$ max (T_{ij}) for $i = 2, 3, \dots m - 1$ or (*ii*) min $(T_{mj}) \geq$ max (T_{if}) for $i = 2, 3, \dots m - 1$.

Step 3. If inequalities of step 2 are not satisfied, this method fails. Hence, go the next step.

Step 4. Convert the m-machine problem into 2-machine problem considering two fictious machines G and H. So that,

$$T_{Gj} = T_{1j} + T_{2j} + \dots + T_{(m-1)} \text{ and } T_{Hj} = T_{2j} + T_{3j} + \dots T_{mj}.$$

Now determine the optimal sequence of n jobs through 2 machines by using the optimal sequence algorithm.

Step 5. In addition to conditions given in step 4, if $T_{2j} + T_{3j} + \dots + T_{(m-1)} = C$ (a fixed positive constant) for all $j = 1, 2 \dots n$, then determine the optimal sequence for n jobs and two machines M_1 and M_m in the order $M_1 M_m$ by using the optimal sequence algorithm.

Example 9.3 Solve the following sequencing problem giving an optimal solution when passing is not allowed.

		Job (*i*)				
		A	*B*	*C*	*D*	*E*
Machine (*i*)	M_1	11	13	9	16	16
	M_2	4	3	5	2	6
	M_3	6	7	5	8	4
	M_4	15	8	13	9	11

Solution : In this example,

$$\min_j T_{1j} = 9 = T_{13}, \ \min_j T_{4j} = 8 = T_{42}\ ; \ \max_j T_{2j} = 6 = T_{25}, \ \min_j T_{3j} = 8 = T_{34}.$$

Since the conditions

$$(\min_j T_{1j} \geq T_{ij}, i = 2, 3) \text{ and } (\min_j T_{4j} \geq \max_j T_{ij}, i = 2, 3)$$

are satisfied convert this problem into five jobs and two-machine problem.

Also, $M_{2j} + M_{3j} = 10$, a fixed positive constant for all $j = 1, 2, \dots 5$, therefore the problem will reduce to an optimal sequence for five jobs and two machines, M_1 and M_4 in the order M_1M_4 meaning thereby M_2 and M_3 have no effect on the optimality of the sequences. Following the usual optimal sequence algorithm, obtain the optimal sequences $C \to A \to E \to D \to B$. Therefore, the total clasped time may be calculated as follows :

		Machine			
		M_1	M_2	M_3	M_4
Job	*C*	0 – 9	9 – 14	14 – 19	19 – 32
	A	9 – 20	20 – 24	24 – 30	32 – 45
	E	20 – 36	36 – 42	42 – 46	46 – 57
	D	36 – 52	52 – 54	54 – 62	62 – 71
	B	52 – 65	65 – 68	68 – 75	75 – 83

This table shows that the total elapsed time is 83 hrs.

Example 9.4 There 4 jobs each of which has to go through the machines M_i, $i = 1, 2, \dots,$ in the order $M_1, M_2 \dots M_6$. Processing times are given.

		Machine					
		M_1	M_2	M_3	M_4	M_5	M_6
job (*j*)	A	20	10	9	4	12	27
	B	19	8	11	8	10	21
	C	13	7	10	7	9	17
	D	22	6	5	6	10	14

Determine a sequence of thes four jobs, which minimizes the total elapsed time T.

Solution : In this example,

$(\min_j T_{1j} = 13,\ \min_j T_{6j} = 14)$ and $\max_j T_{2j} = 10,\ \max_j T_{3j} = 11,\ \max_j T_{4j} = 8,$

$\max_j = T_{5j} = 1).$

Since, the conditions :

$(\min_j T_{1j} \geq \max_j T_{ij})$ and $(\min_j T_{6j} \geq \max_j T_{ij})$ for $i = 2, 3, 4, 5$

are satisfied, the problem can be converted into 4-job and 2-machine problem. Thus, if G and H are two fictitious machines such that

$$T_{aj} = \sum_{i=1}^{m-1} T_{ij} \text{ and } T_{Hj} = \sum_{i=2}^{m} T_{ij}$$

then the problem can be reformulated as 4-job and 2-machine problem.

		Jobs			
		A	B	C	D
Machines	G	55	56	46	49
	H	62	58	50	41

Using the optimal sequence algorithm, an optical sequence is obtained as $C \sim A \sim B \sim D$. The total elapsed time may be calculated as:

		M_1	M_2	M_3	M_4	M_5	M_6
	C	0 – 13	13 – 20	20 – 30	30 – 37	37 – 46	46 – 63
	A	13 – 33	33 – 43	43 – 52	52 – 56	56 – 68	68 – 95
Job	B	33 – 52	52 – 60	60 – 71	71 – 79	79 – 89	95 – 116
	D	52 – 74	74 – 80	80 – 85	91 – 101	91 – 101	116 – 130

This table shows that the total elapsed time is 130 hrs.

9.8. PROCESSING TWO JOBS THROUGH *m*-MACHINES : GRAPHICAL METHOD

In the two job m-machine problem, there is a graphical procedure, which is rather simple to apply and usually provides good (though not necessarily optimal) results. The following example will make the graphical procedure clear.

Example 9.5 Use graphical method to minimize the time needed to process the following jobs on the machines shown below, i.e. for each machine find the job, which should be done first.

Also calculate the total time needed to complete both the jobs.

Job 1	Sequence of Machines	A	B	C	D	E
	Time	2	3	4	6	2
Job 2	Sequence of Machines	C	A	D	E	B
	Time	4	5	3	2	6

Solution: Step 1. First, draw a set of axes, where the horizontal axis represents processing time on job 1 and the vertical axis represents processing time on job 2

Step 2. Layout the machine time for two jobs on the corresponding axes in the given technological order Machine A takes 2 hrs for job 1 and 5 hrs for job 2. Construct the rectangle PQRS for the machine A. Similarly, other rectangles for machines *B*, *C*, *D* and E are constructed as shown.

Step 3. Make programme by starting from the origin 0 and moving through various states of completion (points) until the point marked Graphical solution for the 2-job 5-machine sequencing problem. 'finish' is obtained. Physical interpretation of the path thus chosen involves the series of segments, which are horizontal or vertical or diagonal making an angle of 45° with the horizontal. Moving to the right means that job 1 is proceeding while job 2 is idle, and moving upward means that job 2 is proceeding while job 1 is idle, and moving diagonally means that both the jobs are proceeding simultaneously. Further, both the jobs cannot be processed simultaneously on the same machine. Graphically, diagonal movement through the blocked-out (shaded) area is not allowed, and similarly for other machines too.

Step 4. To find an optimal path. An optimal path (programme) is one that minimizes idle time for job (horizontal movement). Similarly, an optimal path is one that minimizes idle time for job 2 (vertical movement). Choose such a path on which diagonal movement is as much as possible. According to this, choose a good path by inspection as shown by arrows.

Step 5. To find the elapsed time: The elapsed time is obtained by adding the idle time for either of the job to the processing time for that job. In this problem, the idle time for the chosen path is 3 hrs for the job 1 and zero for the job 2. Thus, the total elapsed time, 17 + 3 = 20 hrs is obtained.

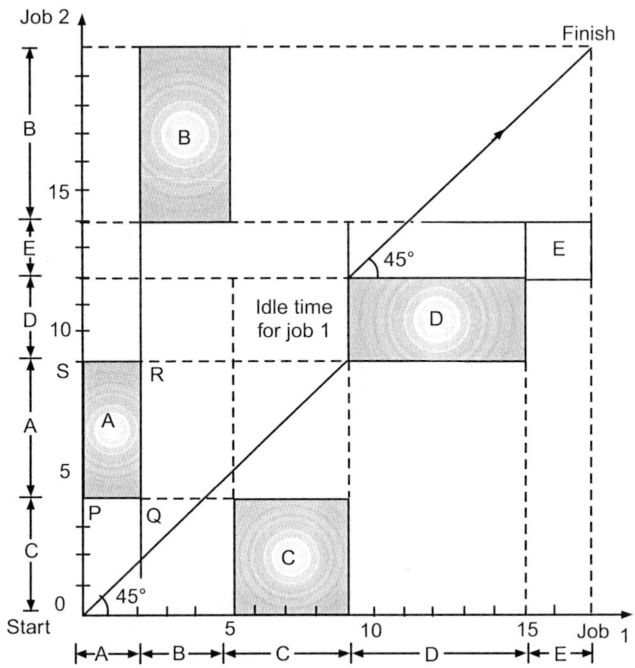

Fig. 9.4.

EXERCISE

9.1. What is sequencing problem? Explain and illustrate.

9.2. What is 'no passing' rule in sequencing algorithm?

9.3. Write the important assumptions made in sequencing problems.

9.4. Describe the Johnson's procedure for determining an optimal sequence for processing '*n*' jobs through two machines.

9.5. Explain how to process '*n*' jobs through 3 machines. **(UPTU 2008-09)**

9.6. Explain the elements that characterize a sequencing problem.

9.7. How does the sequencing technique help the production manager?

9.8. Explain the procedure of obtaining an optimal solution to a three machine n job problem. State assumptions made.

9.9. Give an example of sequencing model and describe it.

9.10. Give Johnson's procedure for determining an optimal sequence for processing '*n*' items on two machines. Give justification of the rule used in the procedure. **(UPTU 2008-09)**

9.11. Write in brief on sequencing of jobs.

9.12. How would you use Gantt chart for solving sequencing problems?

9.13. Explain the procedure of obtaining an optimal solution to a two jobs *m* machines sequencing problem using Graphical Method.

9.14. A company has six jobs in hand *A, B, C, D, E* and *F*. All jobs have to go through two machines M_1 and M_2. The time required for each job on each machine, in hours, is given below :

	A	*B*	*C*	*D*	*E*	*F*
M_1	3	12	18	9	15	6
M_2	9	18	24	24	3	15

9.15. Five jobs are to be performed, first on machine X and then on machine Y. The time taken, in hours by each job on each machine, is given below:

Job	*A*	*B*	*C*	*D*	*E*
Machine *X*	12	4	20	14	22
Machine *Y*	6	14	16	18	10

Determine the optimal sequence of jobs that minimizes the total elapsed time to complete the jobs. Also calculate the minimum time.

9.16. Find the sequence that minimizes the total time required in performing the following jobs on three machines in the order ABC, processing time in hours is given below:

Job	1	2	3	4	5	6
Machine *A*	8	3	7	2	5	1
Machine *B*	3	4	5	2	1	6
Machine C	8	7	6	9	19	9

9.17. A foreman wants to process four different jobs on three machines – shaping, drilling and tapping machine, the sequence of operations being shaping-drilling-tapping. Determine the optimal sequence for the four jobs to minimize the time elapsed (time from the start of first job to the end of last job) if the process times are:

Job	Shaping (minutes)	Drilling (minutes)	Tapping (minutes)
1	13	3	18
2	18	8	4
3	8	6	13
4	23	10	8

9.18. Use graphical method to obtain total minimum elapsed time to process the following jobs on the machines *A*, *B*, *C*, *D* and *E*.

Job 1 sequence	*A*	*B*	*C*	*D*	*E*
Time	3	4	2	6	2
Job 2 sequence	*B*	*C*	*A*	*D*	*E*
	5	4	3	2	6

9.19. Find an optimal sequence for the following sequencing problem of four jobs and five machines when passing is not allowed, having the processing time (in hours) as given below:

Job	1	2	3	4
Machine *A*	6	5	4	7
Machine *B*	4	5	3	2
Machine *C*	1	3	4	2
Machine *D*	2	4	5	1
Machine *E*	8	9	7	5

❑❑❑

Unit IV

Chapter 10

Stochastic Inventory Models

10.1. INTRODUCTION

Inventory is defined as an idle resource of an organization. It is a physical stock of goods kept for future use. The inventory can be in the form of raw materials, parts, semi-finished goods, furniture, machinery etc.

Inventory is a stock of anything held to meet some future demand. It is created when the rate of receipts exceeds the rate of disbursements.

Inventory can also be defined as the stock of goods, commodities, other economic resources that are stored or reserved to ensure smooth and efficient running of business operations.

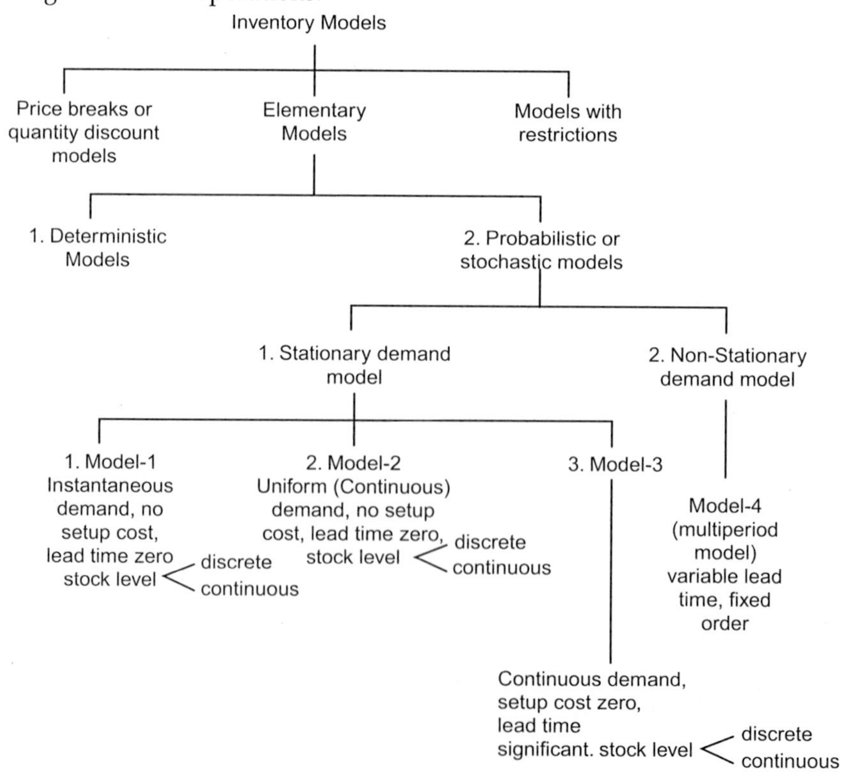

Fig. 10.1. Classification of Inventory Models.

Inventory Control is designed to support the requisition processing, inventory management, purchasing, and physical inventory reconciliation functions of inventory management through a set of highly interactive capabilities.

Inventory control is the function of directing the movement of goods through the entire manufacturing cycle from the requisitioning of raw materials to the inventory of finished goods orderly arranged to meet the objectives of maximum customer satisfaction with minimum investment and low cost plant operations.

An Inventory System is made up of a set of policies and controls designed to monitor the levels of inventory to answer the following questions:

1. What levels should be maintained?
2. When stock should be replenished? and
3. How large orders should be? i.e. what is the optimal size of the order?

10.2. REASONS FOR MAINTAINING INVENTORIES

Adequate inventories facilitate production activities and help to assure customers of good service. On the other hand, carrying inventories holds up working capital on goods that are idle (i.e. not earning any return on investment). Hence, the major problem of inventory management is to maintain adequate but not excess inventories. Some of the major reasons for holding inventories are:

1. It helps in smooth and efficient running of business.
2. It provides adequate service to the customers with variable (immediate and seasonal) demands.
3. It reduces the possibility of duplication of orders.
4. It helps in maintaining economy when demand of an item is seasonal or fluctuating.
5. It helps in minimizing the loss due to deterioration, damage etc.
6. It protects against supply errors, shortages and stock-outs.
7. It acts as a buffer stock when raw materials are received late and many items are rejected.
8. Takes advantage of price discounts by bulk purchasing.
9. It helps in leveling production activities, stabilizing employment and improving labour relations.
10. To decouple successive stages in operations so that breakdowns do not stop the entire system.
11. To protect against future price and delivery uncertainties, such as strikes, price increases and inflation.
12. It facilitates the production of different products on the same facilities.
13. It provides means of obtaining and handling materials in economic lot sizes.
14. It allows for changes and maintenance of the production system.

10.3. FORMS OF INVENTORIES

In general, there are three forms of inventories held by a firm:

1. **Raw material inventory:** These are those basic materials, which have not undergone any conversion since their receipt from the suppliers. These materials are used for manufacturing of product. Different raw materials are used for different industries. For example cotton used in textile mills,

fruits in canning industries, timber for furniture makers, steel for automobile firms, are raw materials.

2. **Work-in-process inventory:** Work-in-process inventory comprises items or materials in semi-finished condition of manufacturing or goods stored during production process. Raw materials become work-in-process at the end of first operation and remain in that category until they become finished products. Work-in-process inventory can be found on conveyors, trucks, and pallets or in temporary area of storage waiting to be worked upon or assembled.
3. **Finished goods inventory:** Finished goods are final products ready to be shipped from the factory. Products usually leave "work-in-process" category and enter "finished goods" category at a point of final inspection when they are ready for delivery to the customer or stock room.

10.4. TYPES OF INVENTORIES

There are five types of inventories, according to purpose for which these are used:

1. **Transportation inventories:** The amount of transportation inventories depends upon the time consumed in transportation and the nature of demand.
2. **Buffer inventories:** These are maintained to meet the uncertainty of demand and supply. In real-life problems, sales and production times cannot be predicted accurately. So to meet these fluctuations, safety stocks are carried.
3. **Anticipation inventories:** These are built in advance by anticipating or foreseeing the future demand. Anticipation inventories are built up for the season of large sales, promotion programme or a plant shut down period. For example inventory of crackers is built up before Deepawali season.
4. **De-coupling inventories:** The inventories used to reduce the interdependence of various stages of production system are known as De – coupling inventories.
5. **Lot-size inventories:** Practically, it rarely happens that the rate of consumption is same as the rate of production. Hence, items are produced in large quantities in batches, as the demand and supply are generally different.

10.5. INVENTORY COSTS

In determining an optimum inventory level, the criterion most often used is cost function. The classical inventory analysis identifies three major cost components.

1. Inventory carrying costs or stockholding costs:

These costs arise on account of maintaining the inventory in stocks and the interest paid on the capital tied up with the stocks. Holding cost vary directly with the size of the inventory as well as the time for which the item is held in stock. It includes the cost of storage, maintenance and handling. Various components of stockholding costs are:

(*a*) **Cost of money or capital tied up in inventories:** This is one of the most important components. Money borrowed from the banks may cost interest. But usually the problem is viewed slightly different way i.e. how much the organization would have earned, had the capital been invested in an alternative project such as developing a new product, etc.

(*b*) **Cost of storage space:** This consists of rent for space. Besides space expenses, this will also include heating, lighting and other atmospheric control expenses.

(*c*) **Depreciation and deterioration costs:** These costs are especially important for fashion items or items undergoing chemical changes during storage. Fragile items such as crockery are liable to damage, breakage etc.

(*d*) **Pilferage cost:** It depends upon the nature of the item. Valuables such as gunmetal bushes and expensive tools may be more tempting, while there is hardly possibility of heavy casting or forging being stolen. While the former must be kept under lock and key, the letter may be simply dumped in the stockyard.

(*e*) **Obsolescence cost:** It depends upon the nature of the item in stock. Electronic and computer components are likely to be fast outdated. Changes in design also lead to obsolescence. It may be possible to quantify the percentage loss due to obsolescence.

(*f*) **Handling costs:** Handling cost includes all costs associated with movement of stocks, such as cost of labour, overhead cranes, gantries and other machinery used for this purpose.

(*g*) **Record keeping and administrative cost:** There is no use of keeping stocks unless one can easily know whether or not the required item is in stock. This signifies the need of keeping funds for record keeping and necessary administration.

(*h*) **Taxes and insurance:** Most organizations have insurance cover and this costs around 1% to 2% of the invested capital.

2. Procurement costs or Setup costs:

These costs include the fixed cost associated with placing of an order or setting up machinery before starting the production. This also includes costs of purchase, requisition, follow up, receiving the goods, quality control, etc. Also called order costs or replenishment cost, these costs are assumed to be independent of the quantity ordered or produced but directly proportional to the number of order placed. At times, however, these costs may not bear any relationship to the number of orders. More than one stock item may be ordered on one set of documents, the clerical is not divisible and without the existing staff increasing or decreasing, there may be considerable scope for changing the number of orders. In such a case, the acquisition cost relationship may be quadratic or stepped instead of the straight line.

3. Shortage Costs or Stock-out costs:

The cost of running out of stock is known as shortage or stock out cost. These costs are associated with either a delay in meeting demands or the inability to meet it at all. Therefore, shortage costs are usually interpreted in two ways. First, when unfulfilled demand has been met at later stage, which is referred, as backlog case. This results in **loss of goodwill** and the cost of idle equipment. These are proportional to the quantity that is short as well as the delay time. Secondly, when demand is lost, these costs are proportional to the quantity only, which is lost. This may cause the cancellation of orders, loss of sales and profit.

10.6. ELEMENTS OF INVENTORY CONTROL

1. Demand: The number of units required per period is called demand. The demand pattern of a commodity may be either deterministic or probabilistic. Deterministic demand can be constant or variable. The Probabilistic demand occurs when the demand over a certain period of time is not known with certainty. It is described by a probability distribution function. A probabilistic demand can be either stationary or non-stationary over a period of time.

Types of Demand

(*i*) **Independent demand:** It pertains to finished or end products type of demand. In other words, the demands for various items are unrelated to each other and needed quantities are determined separately or independently.

(*ii*) **Dependent demand:** With this type of demand the need for any item is a direct result of the need for some other item, usually a higher-level item of which it is a component or subassembly. As an example, the need for tires is dependent on the need for cars. Material Requirements Planning (MRP) deals with this type of demand.

2. Lead Time: The time gap between placing of an order and its actual arrival in the inventory is known as lead-time (LT). Lead time has two components, namely the administrative lead time - from initiation of procurement action until the placing of an order, and the delivery lead time - from placing of an order until the delivery of the ordered material. If the lead time is a variable and known stochastically, the amount and timing of replenishment is determined by considering the expected costs of holding and shortage over lead time.

3. Order Cycle: The time period between placements of two successive orders is known as an order cycle. The order may be placed on the basis of following two types of inventory review system:

(*a*) **Continuous review:** The record of the inventory level is checked continuously until a specified point (called re-order point) is reached when a new order is placed. This is often called as the two-bin system. This divides the inventory into two parts and places it physically, or on paper, in two bins. Items are drawn from only one bin and when it is empty, a new order is placed. Demand is then satisfied from the second bin until the order is received. Upon receipt of the order, enough items are placed in the second bin to make up the earlier total. The remaining items are placed in the first bin. This procedure is then repeated.

(*b*) **Periodic review:** In this system the inventory levels are viewed at equal interval of time and orders are placed at such intervals. The quantity ordered each time depends on the available level at the time of review.

4. Time horizon: The time period over which the inventory level will be controlled is called the time horizon. This horizon may be finite or infinite depending upon the nature of the demand for the commodity.

5. Re-order level (ROL): The level between maximum or minimum stock, at which purchasing (or manufacturing) activities start for replenishment, is known as re-order level. It is that level of inventory when fresh order should be placed with suppliers for procuring additional inventory equal to Economic Order Quantity (EOQ).

ROL = Average daily use x Lead time in days

Maximum level = (ROL + Re-ordered quantity) – (Minimum consumption x minimum re-order period)

Minimum level = ROL – (Normal consumption x Normal re-order period)

Also, ROL = Expected demand during lead time + safety stock

6. Stock replenishment: Although an inventory problem may operate with lead-time, the actual replacement of stock may occur instantaneously or uniformly.

7. Buffer stock / safety stock: Buffer stock is extra inventory maintained in addition to the inventory required corresponding to normal consumption level. Buffer / Safety stock is maintained to avoid the stock out condition. Greater is buffer stock maintained; less is the risk of stock outs.

The purpose of maintaining the buffer stock is to maintain a balance between the cost of shortages and the extra inventory cost. If the buffer stock is maintained at a low level, the inventory holding cost will be low but shortages will occur frequently and the cost of shortages will be high. On the other hand, if the buffer stock is at high level, cost of shortages will be low but inventory cost would be high.

Safety stock is also defined as the minimum additional inventory to serve as a safety margin or cushion to meet an unanticipated increase in usage.

Safety stock = (Maximum lead time – Average lead time) x consumption rate

8. Types of Inventory Systems

1. Fixed-order quantity: Here order quantity or size remains constant, however the time between orders varies.
2. Fixed-time period: Here the time between orders remains constant but the order size varies.

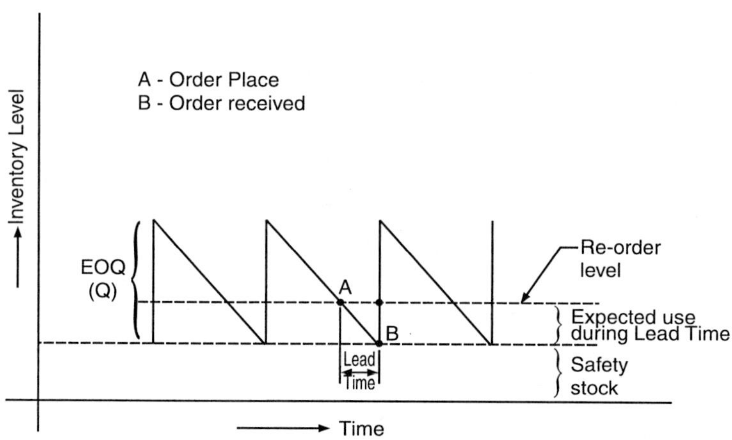

Fig. 10.2. Inventory Order Cycle.

10.7. ADVANTAGES OF INVENTORY CONTROL

The main advantages of inventory control are:

1. Inventory control ensures an adequate supply of items to customers and avoids the shortages as far as possible at the minimum cost.

2. It ensures a smooth and efficient running of the business organization.
3. It eliminates the possibility of duplicate ordering.
4. It makes use of available capital and space in a most effective manner and avoids an unnecessary expenditure on high inventory.
5. The risk of loss due to change in prices of items is reduced
6. It provides advantages of quantity discount on bulk purchases.
7. It helps to minimize the loss due to deterioration, obsolescence, damages or pilferage etc.
8. It helps in maintaining the economy by absorbing some of the fluctuations when the demand for an item fluctuates or is seasonal.
9. Inventory serves as a buffer stock required due to delay in supply of product.
10. Inventory management minimizes and controls accumulation and build up of surplus stock – and removes the dead movable surplus stock as far as possible.

10.8. ECONOMIC ORDER QUANTITY (EOQ)

The concept of Economic Order Quantity (EOQ) or Economic Lot Size was first given by Ford W Harris in 1913 in order to balance cost of holding too much stock against that of ordering in small quantities too frequently. Economic Order Quantity (EOQ) is an optimal order quantity where total annual stocking costs are at a minimum.

Economic Order Quantity (EOQ) is the size of an optimal order for which the aggregate of the costs of procuring the inventory and costs of holding the inventory is minimum.

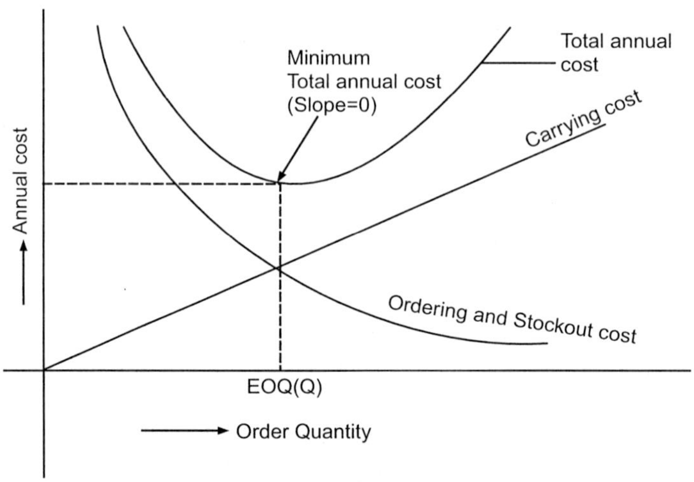

Fig. 10.3. Economic Order Quantity (EOQ)

It is clear from the above figure that the minimum total cost occurs at the point where the ordering cost and inventory carrying costs are equal.

1. Economic order quantity is determined with following five **assumptions:**

(*a*) The demand rate is known with certainty and constant over time. Therefore the depletion of inventory results in a straight line with slope equal to the negative of the demand rate.

(*b*) Replenishments (order quantity) arrive in a batch equal to the order quantity all at once rather than piecemeal. There are no limitations on lot size.

The replenishment results in a *vertical* line on the graph, rather than a line with a positive finite slope.

(*c*) Annual holding cost and annual ordering cost are the *only* costs that are relevant to the *order quantity* decision.

There are no quantity discounts, so per-unit price is irrelevant.

(*d*) Replenishment decisions for one item (say doughnuts) are made independently from replenishment decisions for other items (say coffee).

Model enhancements are required to consider situations where several items are purchased from the same supplier, or several items belong to a product family that can share the same setup.

(*e*) There is no uncertainty in lead-time or supply.

Therefore, replenishment orders can be timed so that no stock-outs occur. Because none occur, stock-out costs are irrelevant to the decision.

The minimum inventory equals zero, the maximum inventory equals the *EOQ*, and the average cycle inventory equals *EOQ*/2.

T These five assumptions describe an unrealistic situation. However, there are other models, and the *EOQ* often is a reasonable first approximation when the actual situation is only somewhat similar to the one considered by the assumptions.

2. Calculating the EOQ

(*a*) Annual holding cost

$$\text{Annual holding cost} = \frac{Q}{2}(C_1)$$

where $Q/2$ = average cycle inventory, and

C_1 = cost of holding one unit in inventory for a year.

(*b*) Annual ordering cost

With Q in the denominator, the annual ordering cost varies inversely with Q.

$$\text{Annual ordering cost} = \frac{D}{Q}(C_3)$$

where D = annual demand

D/Q = number of orders placed in one year

C_3 = average cost of placing one order

(In manufacturing, this cost is called setup cost.)

Because these are the only relevant costs, the annual total of the relevant costs:

(*c*) Total inventory costs (*C*)

$$C = \frac{Q}{2}(C_1) + \frac{D}{Q}(C_3)$$

(*d*) EOQ formula

The minimum total cost occurs when the ordering cost is approximately equal to holding cost.

$$\frac{Q}{2}(C_1) = \frac{D}{Q}(C_3)$$

$$Q^* = \sqrt{\frac{2DC_3}{C_1}} = EOQ$$

(e) Time between orders (TBO): Sometimes policies are expressed in terms of the time between replenishments. *TBO* for a lot size is the elapsed time between receiving orders of Q units. If $Q = EOQ$, the *TBO* expressed in terms of months, is:

$$TBO_{EOQ} = \frac{EOQ}{D}\text{(12 months/year)}$$

(*f*) Optimum number of orders placed per year, $n^* = D / Q^* = \sqrt{DC_1 / 2C_3}$

Minimum total annual inventory cost, $TC^* = ½\ Q^*C_1 + (DC_3 / Q^*)$ $= \sqrt{2DC_1 C_3}$

(g) Effect of changes

Change in the demand rate

Q. What happens to cycle inventory if the demand rate increases?

A. D is in the numerator of the *EOQ* equation. *EOQ* varies directly as the square root of D. If D quadruples, *EOQ* doubles.

Change in the setup costs

Q. What happens to lot sizes if setup costs decrease?

A. S is in the numerator of the *EOQ* equation. *EOQ* varies directly as the square root of S. If S decreases by a factor of 4, *EOQ* is halved.

Change in the holding costs

Q. What happens if interest rates drop?

A. Interest charged on inventory investment is a component of holding cost. H is in the denominator of the *EOQ* equation. If the interest rate drops, *EOQ* increases.

Errors in estimating D, H, and S.

Q. How critical are errors in estimating D, H, and S?

A. Errors such as overestimating ordering cost may be offset by other errors such as overestimating holding cost. The square root also reduces the effect of errors. If one misses a cost or demand estimate by 10%, the effect on total annual cost is often undetectable.

Example 10.1. A manufacturing company purchases 9000 parts of a machine for its annual requirements, ordering one month usage at a time. Each part costs Rs.20. The ordering cost per order is Rs. 15 and the carrying charges are 15 % of the average inventory per year. Suggest a more economical purchase policy for the company. How much would your advice save the company per year?

Solution : Annual demand, $D = 9000$ parts per year, ordering cost per order, C_3 = Rs. 15. Carrying cost, C_1 = 15% of the average inventory cost per year

$$C_1 = 20 \times \frac{15}{100} = 3 \text{ per part per year}$$

$$Q^*\ (EOQ) = \sqrt{\frac{2DC_3}{C_1}} = \sqrt{\frac{2 \times 9000 \times 15}{3}} = 300 \text{ units.}$$

Optimum time between orders,

$$t^* = \frac{Q^*}{D} = \frac{300}{9000}$$

$$t^* = \frac{1}{30} \text{ year}$$

Minimum total annual inventory cost, T_C^*

$$T_C^* = \sqrt{2DC_1C_3}$$

$$= \sqrt{2 \times 9000 \times 3 \times 15} = 900.$$

If the company follows the policy of ordering every month, the annual ordering cost is = 12 × 15 = 180

and lot size of inventory each month

$$= \frac{9000}{12} = 750 = Q.$$

Average inventory at any time = $\frac{Q}{2} = \frac{750}{2} = 375.$

Storage cost at any time = 375 C_1 = 375 × 3 = 1125

Total annual cost = 1125 + 180 = 1305

Hence, the company is advised to purchase 300 parts at time interval of 1/30 year instead of ordering 750 parts each month.

Net saving of company = 1305 – 900 = Rs. 405 per year.

10.9. DETERMINATION OF OPTIMUM SAFETY STOCK

The safety stock depends upon the service level required by the company. It is determined such that the probability of running out of stock during lead time does not exceed a given value. To explain this, it is assumed here that the demand, x_L, during lead time L is normally distributed with mean μ_L and standard deviation σ_L *i.e.*, $N(\mu_L, \sigma_L)$ consider following :

L = Lead time between placing and receiving an order

μ_L = Average demand during lead time.

σ_L = Standard deviation of demand during lead time.

x = Random variable representing demand

B = Buffer stock

α = Maximum allowable probability of running out of stock during lead time.

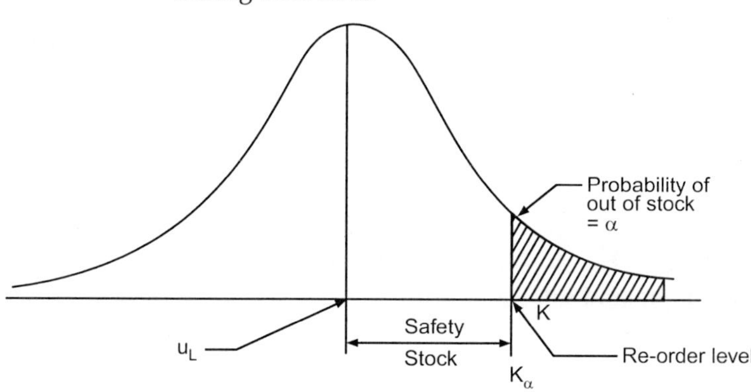

Fig. 10.4.

Probability of running out of stock,

$$P\{x \geq B + \mu_L\} \leq \alpha$$

converting random variable x into standard random variable (z), using

$$z = \frac{x - \mu_L}{\sigma_L}$$

Therefore, we have

$$P\left\{z \geq \frac{B}{\sigma_L}\right\} \leq \alpha$$

This condition is shown in figure by K_α which is determined from standard normal tables :

$$\therefore \qquad P\{z \geq K_\alpha\} = \alpha$$

Hence, optimum buffer stock,

$$B = K_\alpha \cdot \sigma_L$$

For example 95% service level means that chances of running out of stock are 5%. Corresponding to this, the value of K_α from Normal table is equal to 1.645. Therefore, the safety stock = 1.645 σ_L.

If d is demand per unit time, and σ – standard deviation,

$$\mu_L = dL$$

and

$$\sigma_L = \sqrt{\sigma^2 L}$$

$$\text{Reorder level} = \mu_L + K_\alpha \ \sigma_L$$
$$= dL + K_\alpha \ \sigma_L$$

Example 10.2. Daily demand for a product is normally distributed with mean, 60 units and a standard deviation of 6 units. The lead-time is constant at 9 days. The cost of placing an order is Rs.200 and the annual holding cost are 20 % of the unit price of Rs.50. A 95 % service level is desired for the customers, who place orders during the record period. Determine the order quantity and the order level for the item in question, assuming that there are 300 working days during a year.

Solution : Daily demand, d = 60 units; σ = 6 units ; Lead time, L = 9 days.

Annual demand of the production, $D = 60 \times 300 = 18{,}000$ units.

Cost of placing an order = 200 = C_3

$$\text{Annual holding cost} = 20 \times \frac{50}{100} = 10 = C_1$$

Order quantity,

$$Q^* = \sqrt{\left(\frac{2DC_3}{C_1}\right)} = \sqrt{\frac{2 \times 18000 \times 200}{10}}$$
$$= 848.528 \approx 849.$$

Expected demand during lead time,

$$\mu_L = \text{average demand per day} \times \text{lead time} = d \times L$$
$$\mu_L = 60 \times 9 = 540 \text{ units.}$$

Variance of demand during lead time,

$$\sigma^2 = \sum_{i=1}^{9} \sigma_i^2$$
$$\sigma^2 = 9 \times (6)^2 = 324.$$

Standard deviation during lead time,

$$\sigma_L = \sqrt{324} = 18.$$

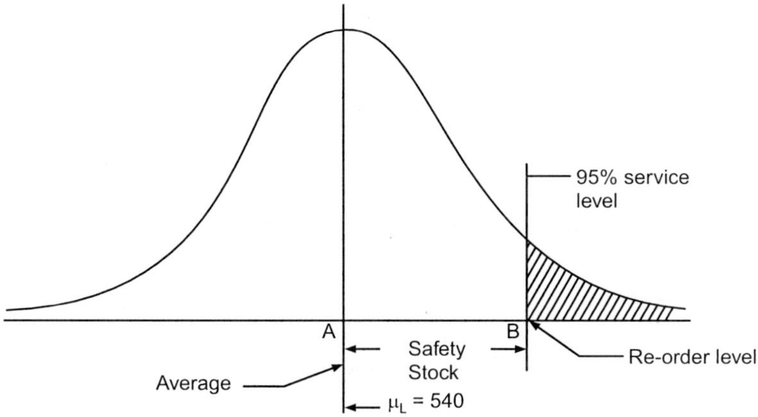

Fig. 10.5.

For 95% service level, point B in the diagram is 1.645 standard deviation away from the mean (from normal table).

Hence, safety stock = 1.645 σ_L =1.645 × 18

Reorder level, X is calculated as follows

$$z = \frac{X - \mu_L}{\sigma_L} \Rightarrow 1.645 = \frac{X - 540}{18}$$

[z = 1.645 for 95% service level]

or X = 1.645 × 18 + 540 = 569.61

$\approx$ 570 units.

Example 10.3. A music store offers a best-selling compact disc. The daily demand for the disc is approximately normally distributed with mean 200 discs and a standard deviation of 20 discs. The cost of keeping the discs in the store is Rs. 0.04 per disc per day. It costs the store Rs. 100 to place a new order. The supplier normally specifies a 7-day lead-time for delivery. Assuming that the store wants to limit the probability of running out of discs during the lead-time to no more than .02, determine the store's optimal inventory policy.

Solution : Daily demand, d = 200 discs

Standard deviation, σ = 20 discs

Cost of holding discs in store, C_1 = .04 per disc per day.

Cost of placing an order, C_3 = 100

Lead time, L = 7 days,

Probability of running out of stock, α = 2%

Average demand during lead time, $\mu_L = dL$ = 200 × 7 = 1400

Standard deviation during lead time, $\sigma_L = \sqrt{\sigma^2 L}$

$$\sigma_L = \sqrt{(20)^2 \times 7} = 52.9$$

Economic Order Quantity,

$$Q^* = \sqrt{\frac{2dC_3}{C_1}}$$

$$Q^* = \sqrt{\frac{2 \times 200 \times 100}{.04}} = 1000 \text{ discs.}$$

($D \rightarrow d$, daily demand is taken)

Buffer stock, $B = \sigma_L K_\alpha$

Value of K_α for $\alpha = 2\%$, from Normal table, is equal to 2.055.

Therefore, $B = 52.9 \times 2.055 = 108.7$

≈ 109 discs.

Reorder level $= B + \mu_L$

$= 109 + 1400 = 1509$ discs.

Hence, store's optimal inventory policy is to order 1000 discs whenever the inventory level drops to 1509 discs.

10.10. SINGLE -PERIOD INVENTORY MODEL

This model deals with perishable items or short-lived items such as newspapers or goggles during the swimming season. Orders are placed to cover one demand period and ordering costs are factored into the unit cost charged by the vendor. For example, fashion items become obsolete at the end of the season. Under this model products must have the following characteristics:

1.The product is only viable during a single time period,

2.The demand for the product is highly variable, but follows a known or assumed probability distribution,

3.The scrap value of the product (salvage value) after the demand period has elapsed is less than the initial cost of the product.

10.10.1. Concept and Derivation

The optimal stocking decision, using marginal analysis, occurs at the point where the benefits derived from carrying the next unit are less than the costs for that unit. Therefore, we should stock that quantity where the profit from the sale of the last unit is equal to or greater than the losses if the unit remains unsold.

Symbolically: MP > ML

where, MP = marginal profit, profit from the nth unit if sold,

ML = marginal loss, loss from n^{th} unit if it is not sold.

Since we do not know the outcomes with certainty, we must use the probabilities associated with selling and not selling the nth unit as follows: Assume p is the probability of the nth unit being sold and therefore $(1 - p)$ is the probability of the nth unit not sold. Optimal ordering occurs when the expected marginal profit for the nth unit is equal to or greater than the expected value of the marginal loss for that unit:

$$p(MP) > (1 - p)(ML),$$

Rearranging terms and solving for p *(critical probability) we get:*

Critical Probability, $p > ML/(MP + ML)$

Illustration: Given that the: selling price = Rs.100/unit, vendor's price = Rs. 70/unit, salvage value = Rs. 30/unit. Also demand during period is expected to range between 35 and 40 units. The demand probabilities and associated cumulative probabilities are as follows:

Number of units Demanded	*Probability of demand (P)*	*Number of units*	*Probability of selling this unit (p)*
35	0.10	1 to 35	1.00
36	0.15	36	0.90
37	**0.25**	**37**	**0.75** ⟵
38	0.25	38	0.50
39	0.15	39	0.25
40	0.10	40	0.10
41	0.00	41 or more	0.00

$$p = ML/(ML+MP) = 40/(40+30) => 0.57$$

Net Profit Calculations

N	*P*	*p*	*MP* (*p*)(100-70)	*ML* (1 – *p*)(70-30)	*Net* MP-ML	*Total Profit*
35	0.10	1.00	(1)(30) = 30	(1– 1)(40) = 0	30 – 0 = 30	30 × 35 = 1050
36	0.15	0.90	27	4	23	1073
37	0.25	0.75	(.75)(30) = 22.50	(1-.75)(40)=10	12.50	1085.50
38	0.25	0.50	15	20	– 5.00	1080.50
39	0.15	0.25	7.50	30	– 22.50	1068
40	0.10	0.10	3.00	36	– 33.00	1035
41	0.00	0.00	0.00	40	– 40.00	995

Therefore, optimal policy is to order 37 units.

Different Single period inventory models are described in following paragraphs.

10.11. MODEL –1: INSTANTANEOUS DEMAND, NO SET – UP COST

Instantaneous Demand, No Set-Up Cost : It is assumed in this model that all demand distributions are stationary and independent over time.

Case (*i*) Discrete Stock Units :

Assumptions of this model are :

(*i*) '*t*' is the constant interval between orders. It may also be considered as unity *e.g.*, daily, monthly etc.

(*ii*) *z* is stock in discrete units for time '*t*'.

(*iii*) *d* is the estimated demand with probability $p(d)$.

(*iv*) C_1 is the holding cost per item per unit time.

(*v*) C_2 is the shortage cost per item per unit time.

(*vi*) Lead time is zero to determine the optimum order level.

(*vii*) Total demand is filled at the beginning of the period.

Depending upon the amount demanded (*d*), the inventory position just after the demand occurs may be either positive (surplus) or negative (shortage) (shown in figure)

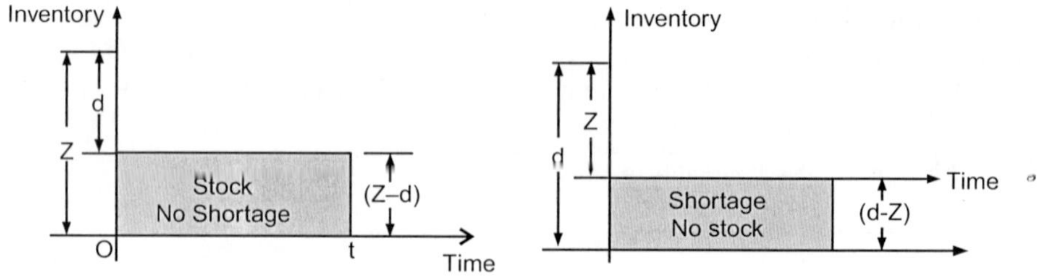

(a) $d < z$ (over supply) (b) $d > z$ (shortage)

Figure 10.6 Instantaneous demand with discrete stock

case (a) when demand 'd' does not exceeds stock 'z' i.e., $d \leq z$.

In this case, total demand is satisfied and there is no shortage.

$$\therefore \text{ Holding cost per unit time} = \begin{cases} (z-d)C_1 & d \leq z \\ C_1 \times 0 & d > z \end{cases}$$

Case (*b*) when demand '*d*' exceeds the stock *i.e.*, $d > z$

Here, demand is not satisfied at all

$$\therefore \text{ Shortage cost per unit time} = \begin{cases} C_2 \times 0 & d \leq z \\ (d-z)C_2 & d > z \end{cases}$$

To get the expected cost, cost is multiplied by its probability $p(d)$ and to get the total expected cost, all the expected costs are added up.

Total expected cost,

$$C(z) = \sum_{d=0}^{z} (z-d)\, C_1\, p(d) + \sum_{d=z+1}^{\infty} C_1 \times 0 \,.\, p(d) + \sum_{d=0}^{z} C_2 \,.\, 0 \,.\, p(d)$$

$$+ \sum_{d=z+1}^{\infty} (d-z)C_2\; p(d).$$

$$C(z) = \sum_{d=0}^{z} (z-d)\, C_1\, p(d) + \sum_{d=z+1}^{\infty} (d-z)C_2\; p(d) \qquad \ldots (1)$$

For minimum value of C(z), following condition must be satisfied :

$$\Delta C(z-1) < 0 < \Delta C(z)$$

Difference of equation (1) can be written if

$$C_1[(z+1)-d]\, p(d) = C_2[d-(z+1)]\, p(d)$$

Taking first difference of equation (1)

$$\Delta C(z) = C_1 \sum_{d=0}^{z} [\{(z+1)-d\} - (z-d)]\, p(d)$$

$$+ C_2 \sum_{d=z+1}^{\infty} [\{d-(z+1)\} - (d-z)]\, p(d)$$

Simplifying and substituting $\sum_{d=z+1}^{\infty} p(d) = 1$ and satisfying the condition $\Delta C(z) > 0$, the optimum value of stock level z can be obtained by relationship

$$\sum_{d=0}^{z-1} p(d) < \frac{C_2}{C_1 + C_2} < \sum_{d=0}^{z} p(d) \qquad \text{... (2)}$$

* If z_0 is such that

$$\sum_{d=0}^{z_0-1} p(d) < \frac{C_2}{C_1 + C_2} = \sum_{d=0}^{z_0} p(d) \qquad \text{... (3)}$$

then $\quad C(z_0 + 1) = C(z_0)$

and optimum value of z is either z_0 or $(z_0 + 1)$.

* If z_0 is such that

$$\sum_{d=0}^{z_0-1} p(d) = \frac{C_2}{C_1 + C_2} < \sum_{d=0}^{z_0} p(d) \qquad \text{... (4)}$$

This leads to $C(z_0 - 1) = C(z_0)$, optimum value of z is either $(z_0 - 1)$ or z_0.

Example 10.4. A newspaper boy buys papers for 30 paise each and sells them for 70 paise. He cannot return unsold newspapers. Daily demand has the following distribution.

No of customers	23	24	25	26	27	28	29	30	31	32
Probability	.01	.03	0.06	0.10	0.20	0.25	0.15	0.10	0.05	0.05

If each day's demand is independent of the previous day's how many papers should he order each day?

Solution: $\quad C_1 = 0.30, C_2 = 0.70 - 0.30 = 0.40$

	23	24	25	26	27	28	29	30	31	32
$P(d)$	0.01	0.03	0.06	0.10	0.20	0.25	0.15	0.10	0.05	0.05
$\Sigma P(d)$	0.01	0.04	0.10	0.20	0.40	0.65	0.80	0.90	0.95	1.00

$$\frac{C_2}{C_1 + C_2} = \frac{40}{30 + 40} = 0.5714$$

Now, $0.40 < 0.5714 < 0.65$

$\therefore$ No of papers ordered = 28.

Case (*ii*) Continuous Stock Level :

Here, stock levels are assumed to be continuous. Hence, the probability $p(d)$ is replaced by $f(x)\,dx$, where $f(x)$ is probability density function $\int_{x_1}^{x_2} f(x)\,dx$ = probability of an order within the range x_1 and x_2.

Cost equation is :

$$C(z) = C_1 \int_0^z (z - x) f(x)\,dx + C_2 \int_z^\infty (x - z) f(x)\,dx \qquad \text{... (5)}$$

For optimum value of $C(z)$

$$\frac{dC(z)}{dz} = 0$$

$$\therefore \qquad \frac{dC(z)}{dz} = (C_1 + C_2) \int_0^z f(x)\,dx - C_2 = 0 \qquad \left(\int_0^{\infty} f(x)\,dx = 1\right)$$

or
$$\int_0^z f(x)\,dx = \frac{C_2}{(C_1 + C_2)} \qquad \text{... (6)}$$

and
$$\frac{d^2C(z)}{dz^2} = (C_1 + C_2)\left[f(x)\cdot\frac{dx}{dz}\right]_0^z = (C_1 + C_2) f(z) > 0.$$

Therefore, optimum value of z can be obtained which satisfies equation (6).

Example 10.5. An ice cream company sells one of its type of ice creams by weight. If the product is not sold on the day it is prepared, it can be sold for a loss of 50 paise per pound. But there is an unlimited market for one day old ice creams. On the other hand the company makes a profit of Rs 3.20 on every pound of ice creams sold on the day it is prepared. If daily order form a distribution with $f(x) = 0.02 - 0.0002x$, $0 \le x \le 100$, how many pounds of ice creams should the company prepare every day.

Solution : Given $\quad C_1$ = Rs. 0.50, C_2 = Rs. 3.20

Let Q be the amount of ice cream prepared every day

Now
$$\int_0^Q f(x)\,dx = \frac{C_2}{C_1 + C_2}$$

$$\int_0^Q (0.02 - 0.0002x)\,dx = \frac{3.2}{3.2 + 0.5}$$

$$\left(0.02x - \frac{0.0002x^2}{2}\right)_0^Q = 0.865$$

$0.0002Q^2 - 0.04Q + 1.730 = 0$

On solving, $Q = 136.7$ and 63.5

But $Q = 136.7$ is not possible ($\because x \le 100$)

Hence, $Q = 63.5$ pounds is the optimum quantity.

Case (*iii*) Re-order Lead Time Significant :

In this case, re-order lead time is also considered. Given that :

(*a*) n is member of order cycle periods in the re-order lead time.

(*b*) z_0 is stock level present before placing the order.

(*c*) $q_1, q_2, q_3, \ldots, q_{n-1}$ are quantities already ordered and due to be received on 1st, 2nd, 3rd... and $(n-1)^{\text{th}}$ days.

(*d*) $f(x') = f(x_1 + x_2 + x_3 + x_4 + \ldots + x_n)$, where x' is the demand during the lead time.

We have to determined q_n for which the total expected cost of nth order cycle is minimum. Stock at the end of 1st period, z_1 = [stock level already present before placing the order + quantity already ordered and due to be received in 1st period] — demand during first period or $z_1 = (z_0 + q_1) - x_1$

Similarly, $z_2 = (z_1 + q_2) - x_2 = z_0 + (q_1 + q_2) - (x_1 + x_2)$ and

$$z_n = (z_{n-1} + q_n) - x_n = z_0 + (q_1 + q_2 + \ldots + q_n) - (x_1 + x_2 + \ldots + x_n)$$

$$= \left[z_0 + \sum_{i=1}^{n-1} q_i + q_n \right] - \sum_{i=1}^{n} x_i$$

$$z_n = z' - x'$$

here, $$z' = \left[z_0 + \sum_{i=1}^{n-1} q_i + q_n \right] \text{ and } x' = \sum_{i=1}^{n} x_i$$

$\therefore$ $$dz' = d\, q_n \qquad \left(dz_0 = 0,\ d \sum_{i=1}^{n-1} q_i = 0, \text{ both are constant} \right)$$

Now replacing z by z' and x by x' in cost equation (5)

$$C(z') = C_1 \int_0^{z'} (z' - x') f(x')\, dx' + C_2 \int_{z'}^{\infty} (x' - z') f(x')\, dx'$$

Optimum value of z' is determined which satisfies

$$\int_0^{z'} f(x')\, dx' = \frac{C_2}{(C_1 + C_2)} \qquad \ldots(7)$$

and the optimum value of q_n is

$$q_n^* = z^* - \left(z_0 + \sum_{i=1}^{n-1} q_i \right) \qquad \ldots(8)$$

Here, z^* is optimum value of z'.

Example 10.6. A shop owner places orders daily for goods which will be delieverd 7 days later (*i.e.*, the reorder lead time is 7 days). On a certain day, the owner has 10 items in stock. Furthermore, on the 6 previous days, he has already placed orders, for the delivery of 2, 4, 1, 10, 11 and 5 items in that order, over each of the next 6 days. Assume C_1 = Rs. 0.15, C_2 = Rs. 0.95. and the distribution requirement over a 7-day period (x') is $f(x') = 0.02 - 0.000\,2x'$.

Solution : Substituting the values of C_1, C_2 and $f(x')$ in the expression

$$\int_0^{z'} f(x')\, dx' = \frac{C_2}{C_1 + C_2}$$

we get, $$\int_0^{z'} (0.02 - 0.0002x')\, dx' = \frac{0.95}{0.15 + 0.95}$$

or $$\left[0.02x' - 0.0002\left(\frac{x'^2}{2} \right) \right]_0^{z'} = 0.8636$$

or $\qquad 0.02z' - 0.0001\, z'^2 = 0.8636$

or $0.0001z'^2 - 0.02z' + 0.8636 = 0$

$$z'^2 - 200z' + 8600 = 0$$

Solving this quadratic equation, $z' = 63$ or 136 items.

Since $z' = z_0 + \sum_{i=1}^{6} q_i + q_7$, we have $63 = 10 + (2 + 4 + 1 + 10 + 11 + 5) + q_7$

or $\qquad q_7 = 63 - 10 - 33 = 20$ items

Thus, the optimum order quantity is 20 items.

10.12. MODEL –2: CONTINUOUS DEMAND, NO SET–UP COST

In this model, the withdrawals from stock are continuous and rate of withdrawals is assumed to be constant. Demand distributions are dependent over time.

Case (*i*) Continuous Units :

(*i*) t is scheduling period

(*ii*) z is stock level to which the stock is raised at the end of period t.

(*iii*) $f(x)$ is probability density function for known demand x.

(*iv*) C_1 is the carrying cost per item per unit time.

(*v*) C_2 is the shortage cost per item per unit time.

(*vi*) Lead time is zero.

(*vii*) Demand rate is constant.

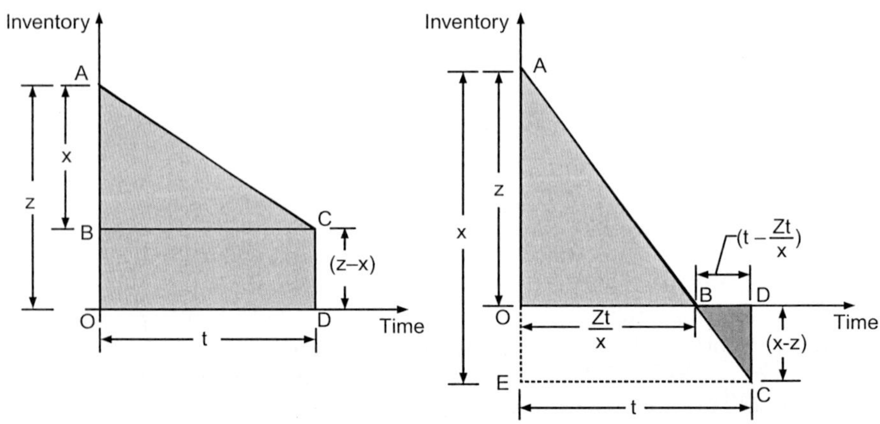

(*i*) $x \le z$ (*ii*) $x > z$

Fig. 10.7. Continuous demand with continuous units.

There are two possibilities :

(*i*) $x \le z$: No shortage, only carrying cost is involved.

$$\text{Holding cost} = C_1 \times \text{Inventory area } AODC$$

$$= C_1 \times \frac{1}{2}\,[CD + AO] \times BC$$

$$= C_1 \times \frac{1}{2}\,[2z - x]\,.\,t$$

(*ii*) $x > z$: Shortage cost is also involved.

Area of inventory, $\Delta AOB = \frac{1}{2}\, z \,.\, \frac{zt}{x} = \frac{z^2 t}{2x}$

(From similar ΔAEC and ΔAOB, $\frac{AO}{AE} = \frac{OB}{EC}$ or $\frac{z}{x} = \frac{OB}{t}$ or $OB = \frac{zt}{x}$)

Area of shortage, $\Delta BDC = \frac{1}{2}\, BD \times CD$

$$= \frac{1}{2}\left(t - \frac{zt}{x}\right)(x - z) = \frac{1}{2x}\,(x - z)^2\, t$$

Total average cost $= C_1 \frac{z^2 t}{2x} + C_2 \,.\, \frac{1}{2x}(x - z)^2\, t$

Therefore, cost equations is

$$C(z) = \int_0^z \left[\frac{C_1}{2}(2z - x)\right] f(x)\, dx + \int_z^\infty \left[\frac{C_1 z^2}{2x} + \frac{C_2 (x - z)^2}{2x}\right] f(x)\, dx \qquad \text{...(9)}$$

For optimum value of $C(z)$,

$$\frac{dC(z)}{dz} = 0$$

or $$\frac{dC(z)}{dz} = (C_1 + C_2)\int_0^z f(x)\, dx + (C_1 + C_2)\int_z^\infty \frac{z\, f(x)}{x}\, dx - C_2 = 0$$

$$\left[\int_0^\infty f(x)\, dx = 1\right]$$

or $$\int_0^z f(x)\, dx + z\int_z^\infty \frac{f(x)\, dx}{x} = \frac{C_2}{(C_1 + C_2)} \qquad \text{... (10)}$$

$C(z)$ is optimum provided $\frac{d^2 C(z)}{dz^2}$ is positive.

Case (*ii*) Discrete Units :

Replacing $f(x)\, dx$ by probability $P(x)$ in equation (9).

Cost equation becomes

$$C(z) = \sum_{x=0}^{z} C_1\left[z - \frac{x}{2}\right] P(x) + \sum_{x=z+1}^{\infty}\left[\frac{C_1 z^2}{2x} + \frac{C_2 (x - z)^2}{2x}\right] P(x) \qquad \text{...(11)}$$

Proceeding in similar manner as in model-1, case (*i*), finding $\Delta C(z)$.

The range of optimum value of z can be obtained by following relationship

$$\sum_{x=0}^{z-1} P(x) + \left(z - \frac{1}{2}\right)\sum_{x=z}^{\infty}\frac{P(x)}{x} < \left[\frac{C_2}{C_1 + C_2}\right] < \sum_{x=0}^{z} P(x) + \left(z + \frac{1}{2}\right)\sum_{x=z+1}^{\infty}\frac{P(x)}{x} \qquad \text{... (12)}$$

Example 10.7. The probability distribution of monthly sales of a certain item is as follows :

Monthly sales	0	1	2	3	4	5	6
Probability	0.02	0.05	0.30	0.27	0.20	0.10	0.06

The cost of carrying inventory is Rs. 10.00 per unit per month. The current policy is to maintain a stock of four items at the beginning of each month. Assuming that the cost of shortage is proportional to both time and quantity short, obtain the imputed cost of a shortage of one item for one time unit **(UPTU : 2003–2004)**

Solution : Given that

1. optimum stock, $z = 4$ items,
2. carrying cost C_1 = Rs. 10.00 per item per month,
3. the probability $p(x)$ for sale x in each month is as follows:

$p(0)$	$p(1)$	$p(2)$	$p(3)$	$p(4)$	$p(5)$	$p(6)$
0.02	0.05	0.30	0.27	0.20	0.10	0.06,

(4) the shortage cost C_2 is to be determined,

(5) the range of monthly sales x is given from 0 to 6 times in discrete units (not from 0 to ∞ here),

$$\sum_{x=0}^{z-1} p(x) + (z - \tfrac{1}{2}) \sum_{x=z}^{\infty} \frac{p(x)}{x} < \frac{C_2}{C_1 + C_2} < \sum_{x=0}^{z} p(x) + (z + \tfrac{1}{2}) \sum_{x=z+1}^{\infty} \frac{p(x)}{x}$$

Now least value of C_2 can be determined by considering,

$$\frac{C_2}{C_1 + C_2} = \sum_{x=0}^{z} p(x) + (z - \tfrac{1}{2}) \sum_{x=z}^{\infty} \frac{p(x)}{x}$$

Therefore, substituting the given values, we get

$$\frac{C_2}{10 + C_2} = \sum_{x=0}^{3} p(x) + (4 - \tfrac{1}{2}) \sum_{x=4}^{6} \frac{p(x)}{x}$$

$$= [p(0) + p(1) + p(2) + p(3)] + \frac{7}{2}\left[\frac{p(4)}{4} + \frac{p(5)}{5} + \frac{p(6)}{6}\right]$$

$$= (0.02 + 0.05 + 0.30 + 0.27) + \frac{7}{2}\left[\frac{0.20}{4} + \frac{0.10}{5} + \frac{0.06}{6}\right] = 0.92$$

$\therefore$ Least value of $C_2 = \dfrac{9.2}{0.08}$ = Rs. 115.

Similarly, the greatest value of C_2 can be determined by considering

$$\frac{C_2}{C_1 + C_2} = \sum_{x=0}^{z} p(x) + (z + \tfrac{1}{2}) \sum_{x=z+1}^{\infty} \frac{p(x)}{x}$$

Substituting the given values, we get

$$\frac{C_2}{10 + C_2} = \sum_{x=0}^{4} p(x) + (4 + \tfrac{1}{2}) \sum_{x=5}^{6} \frac{p(x)}{x} = 0.84 + \frac{9}{2} \times 0.03 = 0.975$$

$\therefore$ Greatest value of $C_2 = \frac{9.75}{0.025}$ = Rs. 390.

Hence, the required range of values for the imputed cost C_2 can be written as Rs. 115 < C_2 < Rs. 390.

10.13. MODEL-3: CONTINUOUS DEMAND, NO SET–UP COST WITH CONSTANT LEAD TIME

In this model, the re-order lead time is also taken into account.

Condition : (*i*) Discrete Units :

Let (*i*) t_p is the prescribed interval between orders

(*ii*) x is the demand during the period t_p with probability $F(x)$

(*iii*) y is the demand during lead time $L = T$, with probability $G(y)$.

The amount in stock during the specific period depend on the order level z, the demand y during the lead time and the demand x during t_p. There are three possible situations depending upon the values of x, y and z (shown in figure). The period t_p begins when the lead time T is elapsed. Here, we have to minimize the expected cost during t_p in three different situations—A, B and C.

Case (*A*) : The period *tp starts* with stock – $(z - y)$ and ends with remaining stock – $[(z - y) - x]$.

$$\text{Total cost} = C_1 \times \text{inventory area } ABCD$$

$$= C_1 \cdot \frac{1}{2}\,[(z-y) + (z-y-x)]\, t_p$$

$$= C_1\left(z - y - \frac{x}{2}\right) t_p \qquad ...(13)$$

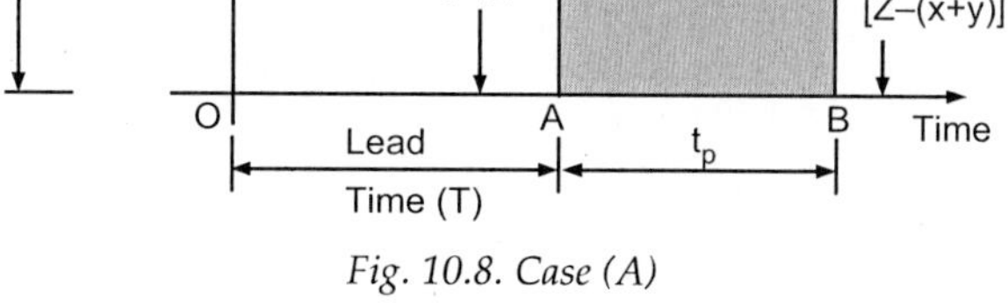

Fig. 10.8. Case (A)

Case (*B*) : The period t_p begins with stock $(z - y)$ and ends with shortage $(x + y - z)$

$$\text{Area representing inventory cost} = \frac{1}{2}\,(z-y)\left[\frac{(z-y)}{x}\right] t_p$$

$$= \frac{1}{2x}\,(z-y)^2\, t_p$$

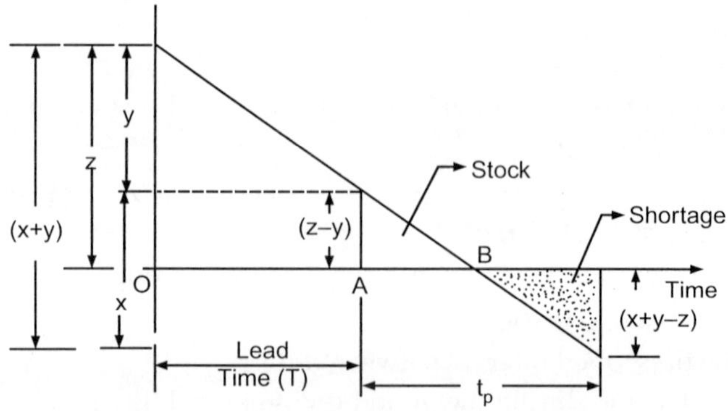

Fig. 10.9. Case (B).

$$\text{Area representing shortage cost} = \frac{1}{2}(x + y - z)\left[t_p - \frac{(z-y)}{x} t_p\right]$$

$$= \frac{1}{2x}(x + y - z)^2 t_p$$

$$\text{Total cost} = C_1 \frac{1}{2x}(z - y)^2 t_p + C_2 \frac{1}{2}(x + y - z)^2 t_p \qquad \text{...(14)}$$

Case (*C*) : The period t_p begins with a shortage $(y - z)$ and ends with a shortage $(x + y - z)$.

The shortage cost = C_2 × shortage area *ABCD*

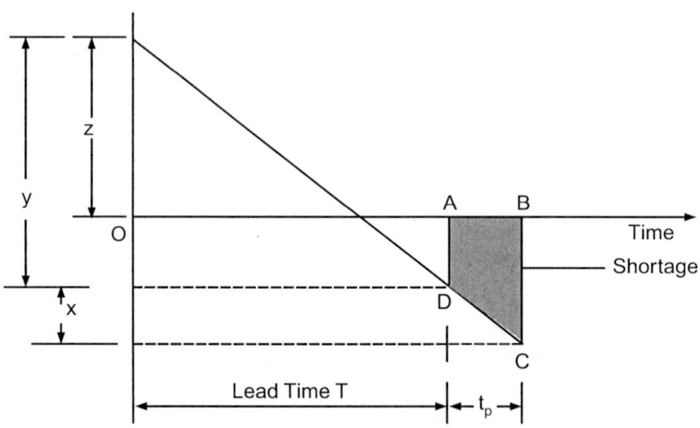

Fig. 10.10. Case (C).

$$= C_2 \times \frac{1}{2}[(y - z) + (x + y - z)] t_p$$

$$= C_2 \left(\frac{x}{2} + y - z\right) t_p \qquad \text{...(15)}$$

The total expected cost C(z) per unit time is given by following cost equation :

$$C(z) = \sum_{y=0}^{z} \sum_{x=0}^{z-y} C_1\left(z - y - \frac{x}{2}\right) F(x)\, G(y) + \sum_{y=0}^{z} \sum_{x=z-y+1}^{\infty} \frac{1}{2x}[C_1(z - y)^2$$

$$+ C_2(x + y - z)^2]\, F(x)\, G(y) + \sum_{y=z+1}^{\infty} \sum_{x=0}^{\infty} C_2 \left(\frac{x}{2} + y - z\right) F(x)\, G(y) \quad \text{... (16)}$$

Using the concept of differences, we can find the optimum value of z for which $\Delta C(z) \geq 0$ from following relationship.

$$\sum_{y=0}^{z} G(y) \left[\sum_{x=0}^{(z-y)} F(x) + \left(z - y + \frac{1}{2}\right) \sum_{x=z-y+1}^{\infty} \frac{F(x)}{x} \right] \geq \frac{C_2}{(C_1 + C_2)} \quad \text{... (17)}$$

Condition (*ii*) Continuous Units :

In cost equation (16), $F(x)$ and $G(y)$ are replaced by probability functions $f(x)\,dx$ and $g(y)\,dy$ respectively and $\sum \sum$ by $\int \int$.

Now cost equation is

$$C(z) = \int_0^z g(y)\,dy \int_0^{z-y} C_1 \left(z - y - \frac{x}{2}\right) f(x)\,dx + \int_0^z g(y)\,dy \int_{(z-y)}^{\infty}$$

$$\left[C_1 \frac{(z-y)^2}{2x} + C_2 \frac{(x - z + y)^2}{2x} \right] f(x)\,dx + \int_z^{\infty} g(y)\,dy \int_0^{\infty} C_2 \left(y - z + \frac{x}{2}\right) f(x)\,dx \text{... (18)}$$

For optimum value of z,

$$\frac{dC(z)}{dz} = 0$$

and

$$\frac{d^2C(z)}{dz^2} > 0$$

$$\frac{dC(z)}{dz} = (C_1 + C_2) \int_0^z g(y)\,dy \left[\int_0^{z-y} f(x)\,dx + (z-y) \int_{(z-y)}^{\infty} \frac{f(x)}{x}\,dx \right] - C_2 = 0$$

The optimum value of z can be determined from following relationship.

$$\int_0^z g(y)\,dy \left[\int_0^{z-y} f(x)\,dx + (z-y) \int_{(z-y)}^{\infty} \frac{f(x)}{x}\,dx \right] = \frac{C_2}{(C_1 + C_2)} \quad \text{... (19)}$$

10.14. MODEL –4: MULTI-PERIOD MODEL WITH CONSTANT LEAD TIME

In this model, the lead-time is constant with assumption of no set-up cost. A shortage occurs when the consumption of goods is more than that it was expected. To avoid such situation of stock out the only alternative is to raise the reorder level, so that sufficient safety stock is maintained. Therefore, to determine an optimal level of safety stock, the cost of carrying this safety stock is balanced by the shortage (stock-out) cost.

Fixed order System

In this system, the inventory position is reviewed continuously and maintained up-to a prescribed level. As and when the inventory level reaches the reorder level, an order is placed for fixed quantity Q, which is nearly equal to the economic order quantity. This ordered quantity will be received at the end of the lead-time period (*L*). As soon as the new supply is received, the cycle becomes complete. Three such cycles are shown below in figure 10.11.

1. **First cycle:** In first cycle, the demand rate and the lead-time both have their average values such that the replenishment is received just when the inventory level touches the safety stock level.
2. **Second cycle:** In this cycle, the lead-time (L) is slightly longer and the demand cuts into the safety stock before the replenishment is received.
3. **Third cycle:** In this cycle, the lead time (L) is much longer which results not only in a cutting into the safety stock but also in a shortage (stock-out) just before the receipt of replenishment.

Since the demand is uncertain there is possibility of either shortages or surpluses. If the demand during the lead time (*L*) is more than the re-order level quantity (r), then a shortage of $(d_L - r)$ quantity units exist, otherwise, a surplus of $(r - d_L)$ quantity units remain in the inventory system.

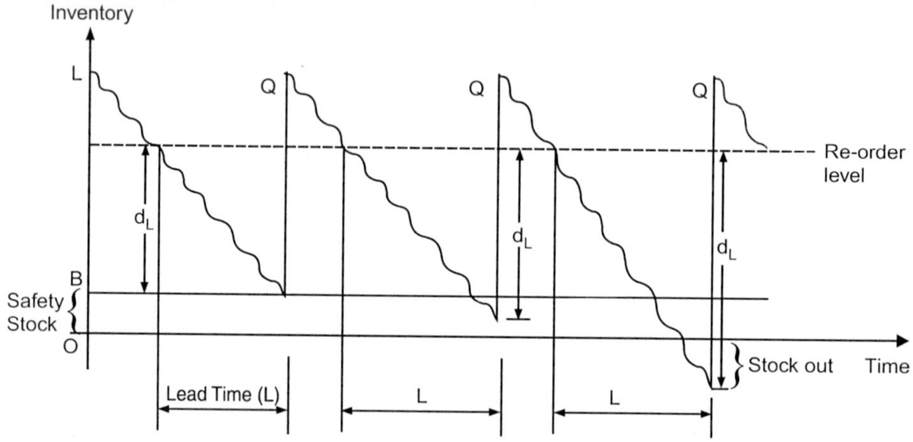

Fig. 10.11.

Total cost = Ordering cost (C_3) + (actual + additional) Carrying cost (C_1) + shortage cost (C_2) ...(20)

$$\text{Ordering cost} = \frac{D}{Q} C_3 \quad ...(21)$$

$$\text{Carrying cost} = \frac{Q}{2} C_1 \quad ...(22)$$

$$\text{Total shortage cost} = (d_L - r)\, C_2 \quad ...(23)$$

Due to probabilistic demand, the expected shortage cost per cycle

$$= C_2 \sum (d_L - r)\, p(d) \quad ...(24)$$

where $p(d)$ is the probability of possible demand level.

$$\text{Expected annual cost} = C_2 \left(\frac{D}{Q}\right) \sum (d_L - r)\, p(d) \quad ...(25)$$

$$\text{Safety stock,} \quad B = (r - d_L) \quad ...(26)$$

Safety stock is made up by increasing the re-order level quantity (r).

Increased carrying cost for maintaining a safety stock against shortages

$$= C_1 (r - d_L) \quad ...(27)$$

$$\text{Total carrying cost} = \frac{1}{2} Q C_1 + C_1 (r - d_L) = C_1 \left[\frac{Q}{2} + (r - d_L)\right] \qquad ...(28)$$

From equation (21), (25) and (28)

$$\text{Total cost} = C_3 \left(\frac{D}{Q}\right) + C_1 \left[\frac{Q}{2} + (r - d_L)\right] + C_2 \left(\frac{D}{Q}\right) \sum (d_L - r)\, p(d) \qquad ...(29)$$

Here, the decision policy is to find the values of the decision variables Q and r so that the total cost is lowest.

For the simplicity, it is assumed that Q and r are independent of each other. Approximated optimal order quantity (Q^*) can be calculated by using EOQ formula.

$$Q^* = EOQ = \sqrt{\left(\frac{2DC_3}{C_1}\right)}$$

The value of Q^* can be used to determine the value of r to minimize safety stock costs. The alternative values of 'r' can be obtained from following relation :

$$\text{Total cost, } C = C_1 (r - d_L) + C_2 \left(\frac{D}{Q^*}\right) \sum (d_L - r)\, p(d)$$

Example 10.8. A certain item has annual demand of 2000 units. The cost of placing an order is Rs. 400 and the annual carrying cost is Rs. 10 per unit. The costs of stock out are estimated to average Rs. 10. The demand during lead time tends to be randomly distributed throughout the year, so that a Poisson distribution can be assumed. There are 250 working days per year and lead time is 5 working days.

Demand during lead time	70	75	80	85	90	95	100
Probability	0.02	0.14	0.23	0.24	0.21	0.12	0.04

Determine the optimal order quantity and reorder level.

Solution : $D = 2000$ units/year, $C_3 =$ Rs. 400/order

$C_1 =$ Rs. 10 per unit/per year,

$C_2 =$ Rs. 10 per unit, and lead time (L) = 5 days.

The optimal value of order quantity, Q^*

$$Q^* = \sqrt{\left(\frac{2DC_3}{C_1}\right)} = \sqrt{\frac{2 \times 2000 \times 400}{10}} = 400 \text{ units}$$

The expected number of units *demanded* (d_L) during the lead time is given by

$$\begin{aligned} D_L &= d_{L_1} p(d_{L_1}) + d_{L_2} p(d_{L_2}) + d_{L3} p(d_{L_3}) + d_{L_4} p(d_{L4}) + d_{L_5} p(d_{L_5}) \\ &\quad + d_{L_6} p(d_{L_6}) + d_{L_7} p(d_{L_7}) \\ &= (70 \times 0.02) + (75 \times 0.14) + (80 \times 0.23) + (85 \times 0.24) \\ &\quad + (90 \times 0.21) + (95 \times 0.12) + (100 \times 0.04) \\ &= 85 \text{ expected units.} \end{aligned}$$

If the safety stock is not provided, then shortage will occur whenever the demand during lead time exceeds 85 units. The expected number of shortage units per reorder cycle are computed in the following table.

Reorder level (r)	*Safety stock level* (B)	*Lead time demand* (d_L)	*Shortages during lead time* ($d_L - r$)	*Prob. of demand during* $L[p(d_L)]$	*Expected shortage during* $L(d_L - r)\, p(d_L)$	*Expected units short per reorder cycle* $\Sigma(d_L - r)\, p(d_L)$
85	0	85	0	0.24	0	2.85
		90	5	0.21	1.05	
		95	10	0.12	1.20	
		100	15	0.04	0.60	
90	5	90	0	0.21	0	1.0
		95	5	0.12	0.6	
		100	10	0.04	0.4	
95	10	95	0	0.12	0	0.2
		100	5	0.04	0.2	
100	15	100	0	0.04	0	0

From above table it is observed that whenever there is no safety stock, the expected number of units short per reorder cycle are 2.85. But, when the buffer stock of 5, 10. and 15 units is provided, the expected shortage is 1.0, 0.2 and 0, respectively.

The annual carrying costs due to safety stock is 0, 50, 100 and 150 at the rate of Rs. 10 per unit for safety stock ranging from 0 to 15 units. These costs and the expected annual shortage cost is calculated below :

Safety stock level (B)	*Expected shortage reorder cycle* $\Sigma(d_L - r)\, p(d_L)$	*Per unit shortage cost* (C_2)	*Reorder cycle per year* (D/q^*)	*Expected annual shortage cost* $C_2(D/q^*)\, \Sigma(d_L - r)\, p(d_L)$
0	2.85	10	5	142.5
5	1.00	10	5	50.0
10	0.20	10	5	10.0
15	0	10	5	0

The carrying costs of various safety stock levels and expected annual shortage cost are combined together in the following table to show the inventory trade-off between the two.

Safety stock level	*Safety stock carrying cost*	*Expected annual shortage cost*	*Total annual expected cost*
0	0	142.5	142.5
5	50	50.0	100.0
10	100	10.0	110.0
15	150	0	150.0

From this table we observe that total annual expected cost reaches a minimum level at the safety stock level of 5 units. Hence, optimal reorder level, r = B + dL = 5 + 85 = 90 units.

10.15. SELECTIVE INVENTORY MANAGEMENT

Every organization consumes several items of sore. Since all the items are not of equal importance, a high degree control on inventories of each item is neither applicable nor useful. Therefore, it is necessary to classify the items in groups depending upon their utility/ importance. Such type of classification is known as selective inventory management. Some common analysis are described below.

10.15.1. Always Better Conrol (ABC) Analysis

Only a small percentage of items held in inventory deserve management's closest attention. ABC analysis is a very effective tool for selective inventory management. ABC analysis helps the management to place their efforts where results would be useful to the greatest possible extent.

The ABC analysis was first used by the General Electric Company in USA

The ABC analysis is an application of the **Pareto's Principle** – *a few high usage value items constitute a major part of the capital invested in inventories, where as bulk of items in inventory having low usage value constitute insignificant part of the capital.*

In ABC analysis the items are classified in three main categories based on their respective net usage value.

Class A items: More costly and valuable items are classified as class A. Such items have large investment but not much in number. More careful and closer control is needed for these items.

Class B items: The items having average consumption value are classified as class B. These items have less importance than class A items and require lesser degree of control than those in class A.

Class C items: The items having low net usage value are put in category C.

Procedure for 'ABC' analysis:

Following are the steps followed for classification of items into 'A', 'B' or 'C' category:

Step 1: Determine the number of units sold or used in the past over a period of time (over a period of month, a quarter or one year).

Step 2: Obtain the unit cost for each manufactured or purchased item in inventory.

Step 3: Calculate the net usage value (in Rs.) of each item consumed over the selected period of time by multiplying the unit cost by the usage.

Step 4: Arrange the items in a descending order according to their respective net usage value calculated in step 3.

Step 5: Prepare a table showing unit cost, consumption and usage value for each item.

Step 6: Calculate the cumulative sum of the number of the items and the usage value for each item obtained in step3.

Step 7: Calculate the percentage of the values obtained in step 6 with respect to the grand total of the corresponding columns.

Step 8: Draw a graph by taking percentage of items on X-axis and the associated percentage of net usage value on Y-axis. After plotting various points on the graph, draw a smooth curve.

Step 9: Mark three points on the curve roughly where the slope changes sharply. These points are called the points of inflection.

Step 10: The net usage value and the percentage of items at these points classify the items into three groups **A** – items of high usage value, **B** – items of medium usage value and **C** – items of low usage value.

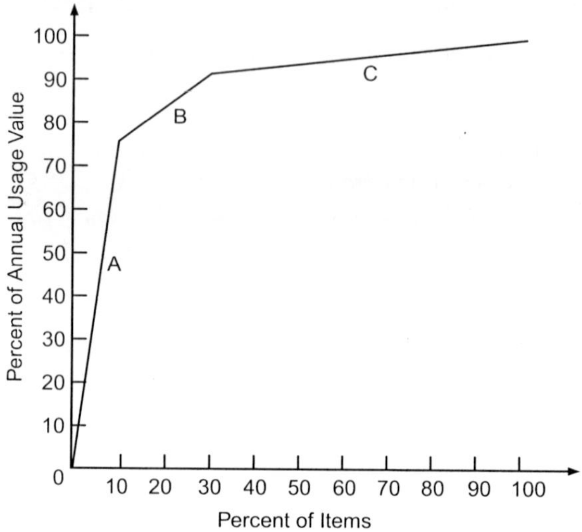

Fig. 10.12. ABC classification of items

Disadvantages and limitations of ABC analysis

1. It is very difficult to classify all items in A, B and C category where large number and variety of inventories are present.
2. ABC analysis is based on usage value. Functional importance of items is ignored.
3. By reducing the number of orders for class C items huge inventory is piled up causing problems of storage, pilferage and loss due to deterioration.
4. Casual approach towards C class items results in production hold up due to non-availability of these items.
5. The ABC analysis is based on prevailing market prices of product that keeps on fluctuating. The purpose of ABC analysis is defeated if it is not reviewed and updated periodically.
6. ABC analysis ignores many technical and functional aspects of items during classification.

Example 10.9. From the following details, perform ABC analysis of selective management.

Item	*Units*	*Unit cost (in Rs.)*
1	7,000	5.00
2	24,000	3.00
3	1,500	10.00
4	600	22.00
5	38,000	1.50
6	40,000	0.50
7	60,000	0.20
8	3,000	3.50
9	300	8.00
10	29,000	0.40
11	11,500	7.10
12	4,100	6.20

Solution : First rank the items according to their usage value:

Item	*Units*	*Unit cost*	*Total cost (Rs) Usage value*	*% of total cost*	*Ranking*
1	7,000	5.00	35,000	9.8	4
2	24,000	3.00	72,000	20.2	2
3	1,500	10.00	15,000	4.2	7
4	600	22.00	13,200	3.7	8
5	38,000	1.50	57,000	16.0	3
6	40,000	0.50	20,000	5.6	6
7	60,000	0.20	12,000	3.4	9
8	3,000	3.50	10,500	3.0	11
9	300	8.00	2,400	0.7	12
10	29,000	0.40	11,600	3.3	10
11	11,500	7.10	81,650	23.0	1
12	4,100	6.20	25,420	7.1	5
			Total 3,55,770	100.0	

The cut-of points have been determined from the following graph.

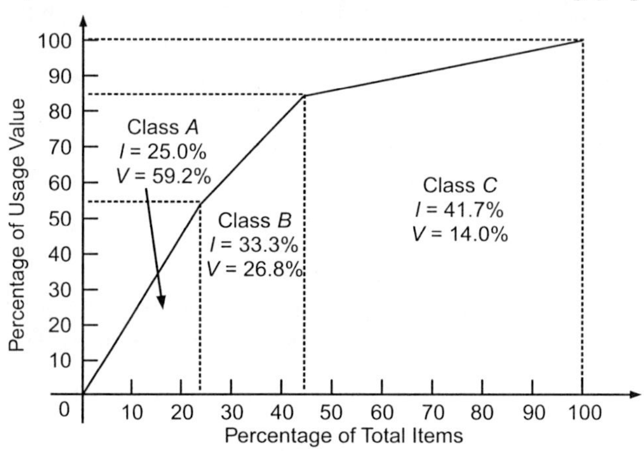

Fig. 10.13. The "ABC" classification.

Items ranked (1), (2) and (3) have large usage value, those ranked (4), (5), (6) and (7) a *moderate* usage value and those ranked (8) to (12) a *small* usage value. Therfore, it is appropriate to classify these as "*A*", "*B*" and "*C*" items respectively. The percentage of items and total usage value in each class is calculated in table below:

Original item No.	Order of ranking	Number of items	Percentage of items	Value	Cumulative value	Cumulative %	% of totall value	Cate-gory
11	(1)	3	25%	Rs. 81,650	Rs. 81,650	23.0	59.2%	A
2	(2)			72,000	1,53,650	43.2		
5	(3)			57,000	2,10,650	59.2		
1	(4)	4	33.3%	35,000	2,45,650	69.0	26.8%	B
12	(5)			25,420	2,71,070	76.2		
6	(6)			20,000	2,91,070	81.8		
3	(7)			15,000	3,06,070	86.0		
4	(8)	5	41.7%	13,200	3,19,270	89.7	14%	C
7	(9)			12,000	3,31,270	93.1		
10	(10)			11,600	3,42,870	96.4		
8	(11)			10,500	3,53,370	99.3		
9	(12)			2,400	3,55,770	100.0		
Total		12	100%	3,55,770			100	

10.15.2. V.E.D. Analysis

Some of the disadvantages of ABC analysis are taken care of in VED analysis. In VED analysis, items are classified according to their criticality (technical and functional). There are three categories:

V – Vital items

E – Essential items

D – Desirable items

Vital items are those whose availability must be ensured at all the times otherwise whole system becomes non-operative.

Essential items are those items which when demanded and not available reduce the efficiency of the system. For example telephone is an essential item. In case it is not operative, system will not be halted.

Desirable items are those whose non-availability neither stop the system nor reduce its efficiency.

The VED analysis is useful in capital-intensive process industries and in controlling the spare parts required for maintenance.

External forces influence the vitality of items. For example, some items become critical due to their short supplies. Proper planning of these items can prevent their short falls and hold up of system.

10.15.3. Hybrid Analysis

In this analysis the advantages of both ABC and VED analysis are combined thereby eliminating their disadvantages. Here, inventory is classified in the following five groups:

Group I – class A and Vital

Group II – class A and Essential
– class B and Vital

Group III – class A and Desirable
– class B and Essential
– class C and Vital

Group IV – class B and Desirable
– class C and Essential

Group V – class C and Desirable

Hybrid analysis ensures reduced overall inventory costs and prevents possible stock outs.

10.15.4. XYZ Analysis

XYZ analysis is based on closing inventory value of items.

X – Items with high inventory value

Y – Items with moderate inventory value

Z – Items with low inventory value

This analysis is generally performed once in a year during the annual stock taking. It helps in identifying items being stocked extensively.

10.15.5. FNSD Analysis

In this analysis, items are classified according to their usage rate or movement in the system. Four such categories are:

F – Fast moving items

N – Normal moving items

S – Slow moving items

D – Dead items

Fast moving and slow moving inventories are of great concern to the management because fast moving inventories might lead to stock out condition and slow moving inventories may lead to Dead stock. FNSD analysis is particularly useful in departmental stores. This analysis ensures the following:

1. Periodic review of all categories.
2. Take corrective measures to increase the supply of fast moving items.
3. Close monitoring of slow moving items.
4. Identify alternate/substitute uses of slow moving items.
5. Take appropriate steps to dispose of dead stock.

Example 10.10. The cost of holding an item in stock is Rs. 2 per unit and the shortage cost is Rs. 8. If Rs. 2 is purchasing cost per unit, determine the optimal order level of inventory. Following probability distribution is given

D	0	1	2	3	4	5
P_D	0.05	0.25	0.20	0.15	0.20	0.15

Solution : Given, C_1 = Rs. 2

C_2 = Rs. 8

Purchasing cost is Rs. 2

D	0	1	2	3	4	5
P_D	0.05	0.25	0.20	0.15	0.20	0.15
ΣP_D	0.05	0.30	0.50	0.65	0.85	1.00

$$\frac{C_2}{C_1 + C_2} = \frac{8}{10} = 0.8$$

$$0.65 < \frac{C_2}{C_1 + C_2} < 0.85$$

Therefore, optimal order level is 4 units.

Example 10.11. The demand for a certain product has a rectangular distribution between 4000 and 5000. Find the optimal order quantity, if the storage cost is Rs. 1 per unit and shortage cost is Rs. 7 per unit.

Solution : Given C_1 = Rs. 1 ; C_2 = Rs. 7

$$\int_0^Q f(x)\,dx = \frac{C_2}{C_1 + C_2}$$

Since the distribution is rectangular the p.d.f is given by $\frac{1}{b-a}$

Hence $$\int_{4000}^{Q} \frac{1}{5000-4000}\,dx = \frac{7}{8}$$

$$\Rightarrow \quad \int_{4000}^{Q} \frac{1}{1000}\,dx = \frac{7}{8}$$

$$\frac{1}{1000}(x)_{4000}^{Q} = \frac{7}{8}$$

$$\frac{1}{1000}(Q - 4000) = \frac{7}{8}$$

$$Q - 4000 = \frac{7000}{8}$$

$$Q = \frac{7000}{8} + 4000 = \frac{39,000}{8} = 4875$$

Q = 4875 units

Example 10.12. Let the probability density of a demand of a certain item during a day be

$$f(x) = \begin{cases} 0.1 & 0 \le x \le 10 \\ 0 & x > 10 \end{cases}$$

The demand is assumed to occur at a uniform pattern during the whole day. Assuming that the unit carrying cost of the item in inventory is Rs. 0.5 per day, and unit shortage cost is Rs. 4.5 per day. If the purchasing cost per unit is Rs. 0.5. Determine the optimum level of inventory.

Solution : Given, C_1 = Rs. 0.50, C_2 = Rs. 4.5, C_2 = Rs. 0.5

Let Q be the amount of item prepared every day

$$\int_0^Q f(x)\,dx = \frac{C_2 - C}{C_1 + C_2}$$

$$\int_0^Q 0.1\,dx = \frac{4.5 - 0.5}{4.5 + 0.5}$$

$$((0.1)\,x)_0^Q = 0.8$$

$$0.1\,Q = 0.8$$

$$\Rightarrow \qquad Q = 8 \text{ units}$$

10.16. EOQ PROBLEMS WITH PRICE BREAKS

Here, consider the class of inventory in which cost is a variable factor when items are purchased in bulk; some discount price is usually offered by the supplier. Such discounts are referred to as quantity discounts or price breaks.

If the price discount is available then total cost per unit of inventory system and items would be

$$TC = RK_1 + \frac{R}{Q}C_3 + \frac{1}{2}QK_1 \times I$$

where K_1 is the cost of manufacturing or purchasing per unit and I denotes the holding cost per unit, R is consumption rate during lead time.

$$\frac{dTC}{dQ} = 0 \Rightarrow \sqrt{2C_3R/K_1I}$$

$$\frac{d^2TC}{dQ^2} > 0 \text{ at } Q = \sqrt{2C_3R/K_1I}$$

$$\therefore \text{ The optimum value of } Q,\ Q^* = \sqrt{\frac{2C_3R}{K_1I}}$$

Now, consider EOQ problem when purchase cost is subjected to price breaks.

10.16.1. Case I: EOQ Problems with One Price Break

The procedure for obtaining EOQ for a single discount is given as follows.

Order quantity	Purchase cost per unit
$0 \le Q_1 < b$	K_{11}
$b \le Q_2$	K_{12}

where b is the quantity at K_{11} and beyond which the quantity discount applies and $K_{12} < K_{11}$.

Step 1: Compute Q^*_2 i.e., optimum order quantity for the lowest price (highest discount) and compare it with quantity b.

If $Q^*_2 \geq b$ then place the orders for quantities of size Q^*_2 and obtain the discount. Otherwise go to next step.

Step 2: If $Q^*_2 < b$, we cannot place order at the reduced price K_{12}. Therefore, in order to maintain the optimum order quantity we need only to compare the total inventory cost for $Q = Q^*_1$ (for price K_{11}) with $Q = b$.

The values of $TC\ (Q_1^*)$ and $TC(b)$ may be determined as follows.

$$TC(Q_1^*) = RK_{11} + \frac{R}{Q_1^*}C_3 + \frac{1}{2}Q_1^* \times K_{11} \times I$$

$$TC(b) = RK_{12} + \frac{R}{b}C_3 + \frac{1}{2}b \times K_{12} \times I$$

If $TC\ (Q_1^*) > TC(b)$ then $Q^* = \text{b}$ otherwise $Q^* = Q_1^*$.

Example 10.13. For the optimum order quantity for a product for which the price break are as follows :

Quantity	Unit cost (Rs.)
$0 \leq Q_1 < 500$	10.00
$500 \leq Q_2$	9.25

The monthly demand for the product is 200 units, the cost of storage is 2% of the unit cost and the cost of ordering is Rs. 350.00.

Solution : Given : C_3 = Rs. 350

R = 200 units per month

$I = 2/100 = 0.02$

K_{11} = Rs 10.00, K_{12} = Rs. 9.25

Step 1 : The highest discount available is 9.25

So we compute $Q_2^* = \sqrt{\frac{2RC_3}{K_{12}I}}$

$$= \sqrt{\frac{2 \times 350 \times 200}{9.25 \times 0.02}} = 870 \text{ units}$$

Now Q_2^* = 870 units and $b = 500$.

We have $Q_2^* > b$, $\therefore$ the optimum purchase quantity is given by $Q^* = Q_2^* = 870$ units.

Example 10.14. Find the optimum order quantity for a product, the price breaks of which are as follows.

Quantity	Unit cost (Rs.)
$0 \leq Q_1 < 800$	Re. 1.00
$800 \leq Q_2$	Re. 0.98.

The yearly demand for the product is 1600 units per year, cost of placing an order is Rs. 5, the cost of storage is 10% per year.

Solution : Given: R = 1600 units per year

$$C_3 = \text{Rs. 5 per order}$$
$$I_{C4} = 10\% = \text{Rs. } 0.10.$$
$$K_{11} = \text{Rs } 1.00,\ K_{12} = \text{Rs } 0.98$$

Step 1: The highest discount available is Rs. 0.98 = K_{12}. So we compute Q_2 at K_{12}

$$Q_2^* = \sqrt{\frac{2 \times R \times C_3}{K_{12} \times I}} = \sqrt{\frac{2 \times 1600 \times 5}{0.98 \times 0.1}}$$
$$= 404 \text{ units}$$

Now $Q_2^* = 404$ units, $b = 800$ units. In this case $Q_2^* < b$.

Step2 : Considering K_{11} = Rs. 1.00, we find the optimum order quantity

$$Q_1^* = \sqrt{\frac{2\,RC_3}{K_{11} \times I}} = \sqrt{\frac{2 \times 5 \times 1600}{1 \times 0.1}} = 400 \text{ units}$$

Given :
$$R = 10{,}000 \text{ units per year}$$
$$C_3 = \text{Rs. 5 per order}$$
$$I = \text{Rs. } 0.1,\ K_{11} = 100,\ K_{12} = \text{Rs. } 95$$

Since $Q_1^* = 400$ and $b = 800$, we get $Q_1^* < b$. We compare the optimum cost of procuring the least quantity which will entitle us a price break

$$TC(Q_1^*) = 1600 \times 1 + \frac{1600}{400} \times 5 + \frac{1}{2} \times 400 \times 1 \times 0.10$$
$$= \text{Rs. } 1640$$

$$TC(b) = 1600 \times 0.98 + \frac{1600}{400} \times 5 + \frac{1}{2} \times 800 \times 0.98 \times 0.10$$
$$= \text{Rs. } 1617.20$$

$$TC(Q_1^*) > TC(b)$$

$\therefore$ Optimum purchase quantity is given by $Q^* = b = 800$ units.

10.16.2. Case II: EOQ Problems with two Price Breaks

When there are two price breaks the situations is illustrated as follows.

Order quantity	Unit price (Rs)
$0 \le Q_1 < b_1$	K_{11}
$b_1 \le Q_2 < b_2$	K_{12}
$b_2 \le Q_3$	K_{13}

Where b_1 and b_2 are the quantities which determine the price breaks.

The procedure for finding the optimum order quantity is given in the following steps.

Step 1 : Compute Q_3^* and compare it with b_2

Step 2 : If $Q_3^* \ge b_2$ the optimum order quantity is Q^*_3. If $Q_3^* < b_2$ go to next step.

Step 3 : Compute Q_2^*, since $Q_3^* < b_2$, $Q_2^* < b_2$ because $Q_1^* < Q_2^* < \ldots < Q_n^*$

Thus either $Q_2^* < b_1$ or $b_1 \le Q_2^* < b_2$

Step 4 : If $Q_2^* < b_2$, and $b_1 \le Q_2^* < b_2$ then the same procedure as in the case of one price break is to be followed. *i.e.*, compare $TC(Q_2^*)$ and $TC(b_2)$ and then determine the optimum quantity.

Step 5 : If $Q_2^* < b_2$ and $Q_2^* < b_1$ then compute Q_1^* which will now satisfy the inequality $Q_1^* < b_1$. Compare $TC(Q_1^*)$ with $TC(b_1)$ and $TC(b_2)$ so as to get the optimum purchase quantity.

Example 10.15. Find the optimal order quantity for a product for which the price breaks are as follows.

Quantity	Unit cost (Rs.)
$0 \le Q_1 < 500$	10.00
$500 \le Q_2 < 750$	09.25
$750 \le Q_3$	08.75

The monthly demand for the product is 200 units, cost of storage is 2% of the unit cost and the cost of ordering is Rs. 350.

Solution : Given, $C_3 =$ Rs. 350, $I =$ Rs. 0.02,

$R =$ 200 units per month

$$Q_3^* = \sqrt{\frac{2RC_3}{K_{13} \times I}} = \sqrt{\frac{2 \times 200 \times 350}{(8.75)(0.02)}} = 894 \text{ units}$$

Since $\quad Q_3^* = 894 > b_2 = 750$

The optimum order quantity is given by $Q = Q_3^* = 894$ units.

Example 10.16. Find the optimum order quantity for a product for which the price breaks are as follows.

Quantity	Unit cost (Rs.)
$0 \le Q_1 < 100$	Rs. 20 per unit
$100 \le Q_2 < 200$	Rs. 18 per unit
$200 \le Q_3$	Rs. 16 per unit

The monthly demand for the product is 400 units. The storage cost is 20% of the unit cost of the product and the cost of ordering is Rs. 25. **(UPTU : 2003–2004)**

Solution : Given, $R =$ 400 units

$I =$ Rs. 0.20

$C_3 =$ Rs. 25.00

$$\text{calculate } Q_3^* = \sqrt{\frac{2RC_3}{K_{13} \times I}} = \sqrt{\frac{2 \times 400 \times 25}{16 \times 0.2}}$$

$$= 79 \text{ units}$$

$\therefore \quad Q_3^* = 79 < b_2 = 200$

$$Q_2^* = \sqrt{\frac{2RC_3}{K_{12} \times I}} = \sqrt{\frac{2 \times 400 \times 25}{18 \times 0.2}} = 75 \text{ units}$$

Again, since $Q_2^* < b_2$ and $Q_2^* < b_1 = 100$, we calculate Q_1^*

$$Q_1^* = \sqrt{\frac{2RC_3}{K_{11} \times I}} = \sqrt{\frac{2 \times 400 \times 25}{20 \times 0.2}} = 70 \text{ units}$$

Compare the $TC(Q_1^*)$ with $TC(b_1)$ and $TC(b_2)$. In order to find the optimum order quantity.

$$TC(Q_1^*) = 400 \times 20 + \frac{400}{70} \times 25 + \frac{1}{2} \times 70 \times 20 \times 0.2$$

$$= \text{Rs. } 8282.86$$

$$TC(b_1) = 400 \times 18 + \frac{400}{100} \times 25 + \frac{1}{2} \times 100 \times 18 \times 0.2$$

$$= \text{Rs. } 7480.00$$

$$TC(b_2) = 400 \times 16 + \frac{400}{200} \times 25 + \frac{1}{2} \times 200 \times 16 \times 0.2 = \text{Rs. } 6770.00$$

Since $TC(Q_1*) > TC(b_1) > TC(b_2)$ the optimum order quantity is given by $Q^* = b_2 = 200$ units.

Example 10.17. Find the optimal order quantity for a product for which the price breaks are as follows

Quantity	Unit cost (Rs.)
$0 \leq Q_1 < 500$	100
$500 \leq Q_2 < 750$	92.50
$750 \leq Q3$	87.50

The monthly demand of the product is 200 units, the holding cost is 2% of the unit cost and the ordering cost is Rs. 1000.

Solution : Given, $R = 200$ units per month

$I = 0.02$, $C_3 = \text{Rs. } 1000$

$K_{11} = \text{Rs. } 100$, $K_{12} = \text{Rs. } 92.150$, $K_{13} = 87.50$

We calculate $$Q_3^* = \sqrt{\frac{2RC_3}{K_{13} \times I}}$$

$$= \sqrt{\frac{2 \times 200 \times 1000}{87.5 \times 0.02}} = 478 \text{ units}$$

since $Q_3^* = 478 < b_2 = 750$

$$Q_2^* = \sqrt{\frac{2RC_3}{K_{12} \times I}} = \sqrt{\frac{2 \times 200 \times 1000}{92.5 \times 0.02}}$$

$$= 465 \text{ units}$$

Also, since $Q_2^* < b_2$ and $Q_2^* < b_1 = 500$

We calculate $$Q_1^* = \sqrt{\frac{2RC_3}{K_{11} \times I}} = \sqrt{\frac{2 \times 200 \times 1000}{100 \times 0.02}}$$

$$= 447 \text{ units}$$

Next we compare $TC(Q_1^*)$ with $TC(b_1)$ and $TC(b_2)$ to get the optimum order quantity.

$$TC(Q_1^*) = RK_{11} + \frac{R}{Q_1^*}C_3 + \frac{1}{2}Q_1^* \times K_{11} \times I$$

$$= 200 \times 100 + \frac{200}{447} \times 1000 + \frac{1}{2} \times 447 \times 0.02 \times 100 = \text{Rs. } 20894$$

$$TC(b_1) = RK_{12} + \frac{R}{b_1} C_3 + \frac{1}{2} b_1 \times K_{12} \times I$$

$$= 200 \times 92.5 + \frac{200}{500} \times 1000 + \frac{1}{2} \times 500 \times 92.5 \times 0.02$$

$$= \text{Rs. } 19362.50$$

$$TC(b_2) = RK_{13} + \frac{R}{b_2} C_3 + \frac{1}{2} b_2 \times K_{13} \times I$$

$$= 200 \times 87.5 + \frac{200}{750} \times 1000 + \frac{1}{2} \times 750 \times 87.5 \times 0.02$$

$$= \text{Rs. } 18422.90$$

Since, $Tc(Q_1^*) > Tc(b_1) > Tc(b_2)$ the optimal order quantity is $Q = b_2 = 750$ units. The optimal order quantity corresponds to the lowest total inventory cost.

Example 10.18. A shop keeper has a uniform demand of an item at the rate of 50 item per month. He buys from a supplier at a cost of Rs. 6 per item and the cost of ordering is Rs. 10 for each order. If the stock holding costs are 20% per year of stock value, how frequently should he replenish his stocks.

Now suppose the supplier offers a 5% discount on orders between 200 and 999 items and a 10% discount on orders exceeding or equal to 1000 items. Can the shop keeper reduce his costs by taking advantage of either of these discounts?

Solution: Given, $R = 100$ items per year

$C_3 =$ Rs. 10 per order

$C_1 =$ Rs. 1.20 (6×0.2)

$$Q^* = \sqrt{\frac{2C_3R}{C_1}} = \sqrt{\frac{2 \times 600 \times 10}{1.2}} = 100 \text{ items}$$

$$t^* = \frac{Q^*}{R} = \frac{100}{600} = \frac{1}{6} \text{ year}$$

$= 2$ months

The total annual cost includes the fixed cost, set-up cost and holding costs.

Fixed cost = Rs. 3600, (*i.e.*, 6×600)

Total 100 items are ordered each time. This means that a total of 6 orders are placed in a year. Hence, the replenishment cost is Rs. 60.00.

$\therefore$ Average inventory holding cost $= 50 \times 0.2 \times 6$

= Rs. 60.00

Hence the total cost = Rs. 3600 + 60 + 60 = Rs. 3720.00

In the case of discounts we have the following formulation

Quantity	Unit cost (Rs.)
$0 \le Q_1 < 200$	6.00
$200 \le Q_2 < 1000$	5.70 (5 % discount)
$100 \le Q_3$	5.40 (10% discount)

$$\therefore \quad Q_3^* = \sqrt{\frac{2C_3R}{K_{13}I}} = \sqrt{\frac{2\times 10\times 600}{5.4\times 0.20}} = 110 \text{ units}$$

$$\because \quad Q_3^* < b_2 = 1000,$$

$$\therefore \quad Q_2^* = \sqrt{\frac{2C_3R}{K_{12}I}} = \sqrt{\frac{2\times 10\times 600}{5.7\times 0.20}} = 105 \text{ units}$$

Again since $Q_2^* < b_1$

$$Q_1^* = \sqrt{\frac{2C_3R}{K_{11}I}} = \sqrt{\frac{2\times 10\times 600}{6\times 0.20}} = 100 \text{ units}$$

To find the optimum order quantity, we compare $Tc(q_1)$ with $Tc(b_1)$ and $Tc(b_2)$

$$TC(Q_1) = 10\times\frac{600}{100} + 600\times 6 + 0.2\times 6\times\frac{100}{2} = \text{Rs } 3720$$

$$TC(b_1) = 10\times\frac{600}{200} + 600\times 5.7 + 0.2\times\frac{5.7\times 200}{2}$$

$$= \text{Rs } 3564$$

$$TC(b_2) = 10\times\frac{600}{1000} + 600\times 5.4 + 0.2\times\frac{0.2\times 5.4\times 200}{2}$$

$$= \text{Rs } 3786$$

$$\because \quad TC(b_1) < TC(q_1) < TC(b_2)$$

Hence, the optimum purchase quantity is $q^* = b_1 = 200$ units. The shopkeeper should accept the offer at 5 % discount only as by doing this he is able to save Rs. 3720 – 3564 = Rs. 156 during the year.

10.16.3. Case III: EOQ Problems With n Price Breaks

When there are n price breaks the situations may be illustrated as follows.

Quantity	Unit cost (Rs.)
$0 \le Q_1 < b_1$	K_{11}
$b_1 \le Q_2 < b_2$	K_{12}
$\vdots$	$\vdots$
$b_{n-1} \le Q_n$	K_{1n}

where b_1, b_2, b_{n-1} are those quantities which determine the price breaks.

The procedure for obtaining the optimum order quantity in this case is as follows

Step 1: Compute Q_n^*. If $Q_n^* \ge b_{n-1}$. The optimum order quantity is reached *i.e.*, Q_n^*.

Step 2: If $Q_n^* < b_{n-1}$, compute Q_{n-1}. If $Q_{n-1}^* \ge b_{n-2}$, proceed as in the case of one price break. The optimum order quantity is determined by comparing $Tc(Q_{n-1}^*)$ with $Tc(b_{n-1})$.

Step 3: If $Q_n^* < b_{n-2}$, compute, Q_{n-2}. If $Q_{n-2}^* \ge b_{n-3}$, proceed as in the case of two price breaks *i.e.*, the optimum order quantity is determined by comparing $Tc(Q_{n-2}^*)$ with $Tc(b_{n-2})$ and $Tc(b_{n-1})$.

Sep 4 : If $Q_{n-2}^* < b_{n-2}$ compute Q_{n-3}^*. If $Q_{n-3} \ge b_{n-4}$ then compare $Tc(Q_{n-3}^*)$ with $Tc(b_{n-3})$, $Tc(b_{n-2})$ and $TC(b_{n-1})$.

Step 5 : Compute in this way unit $Q_{n-j}^* \geq (b_{n-j})$, $0 \leq j \leq n-1$. Then compare $Tc(Q_{n-j}^*)$ with $Tc(b_{n-j-2})$ until $Q_{n-j}^* \geq b_{n-j+1} \geq Tc(b_{n-1})$.

Example 10.19. Find the optimum order quantity for a product for which the price breaks are as follows.

Quantity	Unit cost (Rs.)
0–499	25
500–1499	24.80
1500–2999	24.60
Over 3000	24.40

Ordering cost is Rs. 180 per order, carrying cost is 0.10. Demand is 500 units per year. Also find the minimum inventory cost.

Solution : Given,

$$D = 500 \text{ units/year}$$
$$C_3 = \text{Rs. } 150/\text{order}$$
$$I = 0.10$$
$$K_{11} = 25,\ K_{12} = 24.80,$$
$$K_{13} = 24.60,\ K_{14} = 24.40$$
$$b_1 = 550,\ b_2 = 1500,\ b_3 = 3000$$

$$Q_4^* = \sqrt{\frac{2RC_3}{K_{14} \times I}} = \sqrt{\frac{2 \times 500 \times 180}{0.1 \times 24.4}} = 272 \text{ units}$$

As $Q_4^* = 272 < b_3 = 3000$, next calculate Q_3^*

$$Q_3^* = \sqrt{\frac{2RC_3}{IK_{13}}} = \sqrt{\frac{2 \times 500 \times 180}{0.10 \times 24.60}} = 270 \text{ units}$$

$\because$ $Q_3^* = 270 < b_2 = 1500$, next

Calculate
$$Q_2^* = \sqrt{\frac{2RC_3}{IK_{12}}} = \sqrt{\frac{2 \times 500 \times 180}{0.1 \times 25}}$$
$$= 269 \text{ units}$$

next we compute

$$Q_1^* = \sqrt{\frac{2RC_3}{IK_{11}}}$$
$$= \sqrt{\frac{2 \times 500 \times 180}{0.1 \times 24.8}}$$
$$= 268 \text{ units}$$

Now
$$TC(Q_1^*) = RK_{11} + \frac{R}{q_1{}^*} \times C_3 \frac{q_1{}^*}{2} (I \times K_{11})$$
$$= \text{Rs } 13170.82$$

$$TC(b_1) = RK_{12} + \frac{R}{b_1} \times C_3 + \frac{b_1}{2} (I \times K_{12})$$
$$= \text{Rs } 13200.00$$

$$TC(b_2) = RK_{13} + \frac{R}{b_2} \times C_3 + \frac{b_2}{2}(I \times K_{13})$$

$$= \text{Rs } 14205.00$$

$\because TC(q_1^*) > TC(b_1) > TC(b_2), Q_1^* = 268$ units
is the optimum order quantity.
Minimum inventory cost Rs. 13170.82

EXERCISE

10.1. Why the items are stocked in inventory?

10.2. Explain different types of costs in inventory system.

10.3. What are advantages and disadvantages of having inventory?

10.4. Explain different models of probabilistic inventory control.

10.5. Enlist various inventory models available. Explain any one of them in detail. **(UPTU: 2004-2005)**

10.6. Describe the basic characteristics of an inventory system.

10.7. Name various inventory models available with continuous and discrete demands. Discuss, in detail, any one such model. Give applications. **(UPTU: 2003-2004)**

10.8. Describe various types of inventory models.

10.9. Explain the following terms:
(*i*) Lead time (*ii*) Re-order point
(*iii*) Stock-out cost (*iv*) Set-up cost
(*v*) Holding cost (*vi*) Shortage or penalty cost
(*vii*) Safety stock (*viii*) Buffer stock

10.10. Distinguish between deterministic and stochastic inventory models.

10.11. Discuss any one stochastic model of inventory management. Derive the formula of optimum level of the inventory.

10.12. What are the objectives of inventory control?

10.13. Discuss the continuous case of a probabilistic inventory model with instantaneous demand and no set-up cost.

10.14. What are functions of inventory control?

10.15. Explain clearly the different costs that are involved in inventory problems with suitable examples. How are they inter-related?

10.16. Derive the relationship for EOQ for any one-inventory model.

10.17. What is purpose of safety stock in inventory decisions?

10.18. What is ABC analysis? Why is it necessary? What are the basic steps involved in implementing it?

10.19. What is selective inventory control? Write different selection techniques.

10.20. Discuss the problem of inventory control when the stochastic demand is uniform, production of commodity is instantaneous and lead-time is negligible. (Discrete case)

10.21. A shipbuilding firm uses rivets at a constant rate of 20,000 numbers per year. Ordering costs are Rs. 30 per year. The rivets cost Rs. 1.50 per number. The holding cost of rivets is estimated to be 12.5 % of unit cost per year. Determine the EOQ. [**Ans.** 1005 numbers]

10.22. A manufacturer of electric motors is required to purchase 2400 castings per year. These are subject to quantity discount of Rs. 0.75 from a quoted price of Rs. 10, if quantity purchased is 3000 at any given time. If the ordering cost is Rs. 100/- and inventory-carrying cost is 24 % per year, find:

(*i*) What should be the ordering quantity?

(*ii*) Number of orders placed in a year. **(UPTU: 2004-2005)**

10.23. A shopkeeper has to decide how much quantity of bread he should stock every week. The quantity of bread demanded in any week is assumed to be a continuous random variable with a given function $f(x)$. Let A be the unit cost of purchasing bread, B unit sale price, C refund on unit stale bread and finally D is the unit penalty cost. Find the optimum quantity of bread to be stocked. If $A = 8$, $B = 20$, $C = 2$, $D = 5$, and the demand is regular between 1000 and 2000, show that the optimum quantity is approximately 1739 breads.

10.24. A television dealer finds that the cost of holding a television in stock for a week is Rs.20, customers who can not obtain new television immediately tend to go to other dealers, and he estimates that for every customers who does not get immediate delivery he loses on an average Rs. 200. For on particular model of television the probabilities for a demand of 0, 1, 2, 3, 4 and 5 televisions in a week are 0.05, 0.10, 0.20, 0.30, 0.20 and 0.15 respectively. How many televisions per week should the dealer order? (Assume there is no time lag between ordering and delivery). [**Ans.** 4 televisions per week]

10.25. A company uses to order a new machine after a certain fixed time. It is observed that one of the parts of the machine is very expensive if it is ordered without machine. The cost of spare part when ordered with the machine is Rs.500. The cost of the downtime of the machine and the cost of arranging new part is Rs. 10,000. From the past record it is observed that spare part is required with the probabilities given below:

Demand (r)	0	1	2	3	4	5	6 or more
Probability, P(r)	0.9	0.05	0.02	0.01	0.01	0.01	0.0

Find the optimal number of spare parts, which should be ordered with the order of machine. [**Ans.** 2 spare parts]

10.26. The probability distribution of monthly sales of a certain item is as follows:

Monthly sales:	0	1	2	3	4	5	6
Probability:	0.01	0.06	0.25	0.35	0.20	0.03	0.10

The cost of carrying inventory is Rs. 30 per unit per month and the cost of unit shortage is Rs. 70 per month. Determine the optimum stock level, which minimizes the total expected cost. [**Ans.** 3]

10.27. A cycle dealer finds that the cost of holding a cycle in stock for a week is Rs.10. Customers who can not obtain new cycles immediately tend to go to other dealers, he estimates that for every customer who cannot get immediate delivery he loses an average of Rs.80. for one particular model of cycle the probabilities of demand of 0, 1, 2, 3, 4 and 5 cycles in a week are 0.05, 0.10, 0.20, 0.30, 0.20 and 0.15 respectively. How many cycles should the dealer keep in stock per week? (Assume there is no time lag between ordering and delivery).

10.28. A baking company sells cake by the kg weight. It makes a profit of Rs. 5.00 a kg on each kg sold on the day it is baked. It disposes off all cake not sold on the date it is baked at a loss of Rs.1.20 a kg. If demand is known to be rectangular between 2000 and 3000 kg, determine the optimal daily amount baked. [**Ans.** 2807 kg]

10.29. An ice-cream company sells one of its types of ice cream by weight. If the product is not sold on the day it is prepared, it can be sold at a loss of 50 paise per kg. But there is an unlimited market for one-day-old ice cream. On the other hand the company makes a profit of Rs. 3.20 on every kg of ice cream sold on the day it is prepared. Past daily orders for a distribution with $f(x) = 0.02 - 0.002x,\ 0 = x = 100$.

How many kg of ice cream should the company prepare every day?

10.30. A shop owner places orders daily for goods which will be delivered 7 days later (i.e. the re-order lead time is 7 days). On a certain day, the owner has 10 items in stock. Further more, on the six previous days, he has already placed orders, for the delivery of 2, 4, 1, 10, 11, and 5 items in that order, over each of the next 6 days. To facilitate computations, we shall assume C_1 = Rs. 0.15, C_2 = Rs. 0.95 and the distribution requirement over a 7 day period (x') is $f(x')$ equal to $0.02 - 0.0002x'$.

How many items should be ordered for the seventh day hence?

[**Ans.** 20 items]

10.31. Let the probability density of demand of a certain item during a week be

$$f(x) = 0.1 \quad 0 \le x \le 10$$
$$= 0 \quad \text{otherwise}$$

This demand is assumed to occur with a uniform pattern over the week Let the unit carrying cost of the item in inventory be Rs. 2 per week and unit shortage cost be Rs. 8 per week. How will you determine the optimal order level of the inventory? [**Ans.** 4.5]

10.32. A certain item has an annual demand of 2000 units. The cost of placing an order is Rs.400 and the annual carrying cost is Rs. 10 per unit. The cost of stock-out is estimated to average Rs.10. The demand during lead-time tends to be randomly distributed throughout the year, so that a Poisson distribution may be assumed. There are 250 working days per year and lead-time is 5 working days. Determine the optimal order quantity and reorder level.

Demand during lead time:	70	75	80	85	90	95	100
Probability:	.02	.14	.23	.24	.21	.12	.04

10.33. The daily demand of a sandwich bread at an outlet follows probability distribution as given below:

Demand:	35	36	37	38	39	40	41	42	43	44	45
Probability:	.10	.11	.20	.07	.10	.14	.12	.04	.05	.05	.02

The purchase price of the bread is Rs. 10 per packet. The selling price is Rs. 14 per packet. If the bread packets are not sold on the day of purchase, they are sold at Rs. 7 per packet to hotels for secondary use. Find the optimal order size of the sandwich bread. [**Ans.** 39 packets]

10.34. A company stocks two products. The company has limited space and cannot store more than 40 units. The demand for the distribution for the products are as follows:

Demand (Z):	0	10	20	30	40
Probability for first product :	0.10	0.20	0.35	0.25	0.10
Probability for second product:	0.05	0.20	0.40	0.20	0.15

The inventory carrying costs are Rs. 5 and Rs. 10 per unit of the ending inventories for the first and second product, respectively. The shortage costs are Rs. 20 and Rs. 50 per unit of the ending shortages for the first and second product, respectively. Find out the economic order quantities for both the products. [**Ans.** 20 units]

10.35. The demand for a particular product is as follows:

x :	0	1	2	3	4
$p(x)$:	0.01	0.20	0.39	0.20	0.20

The cost of producing one unit is Rs. 5000, I_c = Rs.500 per unit per year, Rs. 1500 per unit per year. Find optimal production level.

10.36. The probability distribution of the demand for a particular product is given as follows:

Sales :	0	1	2	3	4	5
Probability:	0.05	0.35	0.08	0.15	0.20	0.17

The cost per unit is Rs. 60 and it sells for Rs. 100. If the product is not sold, it is completely worthless. What optimum policy would you recommend?

10.37. XYZ Company wants to provide a 95 per cent service level to its customers. Using the past history of demand, the following data is available:

Daily demand follows normal distribution with average demand of 20 units and the standard deviation of 5 units. The lead-time for procurement is 4 days. The cost of placing an order is Rs. 10 and the inventory carrying cost is Rs. 1 per unit per year. There are no stock-out costs, and unfulfilled orders are supplied after the items are received. What should be the inventory policy for the company? [**Ans.** 96 units]

10.38. The demand per month for a product is distributed normally with a mean of 100 and standard deviation 25. The lead-time distribution is given below. What service level will be afforded by a reorder level of 500 units?

Lead time (months):	1	2	3	4	5
Probability:	0.10	0.20	0.40	0.20	0.10

[**Ans.** 94.5 %]

10.39. A newspaper boy buys papers for Rs. 2.60 each and sells them Rs. 3.60 each. He cannot return unsold newspaper. Daily demand has the following distribution. If each day's demand is independent of the previous day's demand, how many papers should he order each day?

No. of customers:	23	24	25	26	27	28	29	30	31	32
Probability:	.01	.03	.06	.10	.20	.25	.15	.10	.05	.05

[**Ans.** 27]

10.40. Some of the spare parts of a ship costs Rs. 1,00,000 each. These spare parts can only be ordered together with the ship. If not ordered at the time when the ship is constructed, these parts cannot be available on need. Suppose that a loss of Rs. 10,000,000 is suffered for each spare part that is needed when none is available in the stock. Further, suppose that probability that the spare part will be needed as replacement during the life-term of the class of the ship discussed are:

Spare parts required:	0	1	2	3	4	5 or more
Probability:	.9488	.0400	.0100	.0010	.0002	.0000

How many spare parts should be procured? [**Ans.** 2]

10.41. Find the optimal economic order quantity for a product having the following characteristics

Annual demand = 2400 unit

Ordering cost = Rs. 100

Cost of storage = 24% of unit cost.

Price break

Quantity	Unit cost (Rs.)
$0 \le Q < 500$	10.00
$Q \ge 5000$	9.00

[**Ans.** 472 units]

10.42. Find the optimum order quantity for a product for which the price breaks are given below:

Quantity	Unit of cost (Rs.)
$0 \le Q < 500$	15.00
$Q \ge 500$	14.00

Monthly demand for the product is 250 units, the cost of shortage is 2% of the unit cost and cost of ordering is Rs. 300. [**Ans.** 707 units]

10.43. Find the optimal order quantity for a product for which the price breaks are as follows:

Quantity	Unit cost (Rs.)
$50 \le Q_2 < 50$	Rs. 10
$50 \le Q_2 < 100$	Rs. 09
$100 \le q_3$	Rs. 08

The monthly demand for the product is 200 units, the cost of storage is 25% of the unit and ordering cost is Rs. 20 per order. [**Ans.** $Q^* = b_2 = 100$ units]

10.44. The consumption of an item is known to be fixed at 4,800 units per year. The cost of processing an order of this item is Rs. 400 and the inventory carrying charges work out to 24% per annum of the cost of the item. The cost of the item depends on the purchase lot size as per schedule given below. Deterimine the optimum ordering policy.

Quantity	Unit cost (Rs.)
up to 999	20.00
1000–1499	18.50
1500 and above	17,00

[**Ans.** $Q^* = 1000$ units]

10.45. Ten items kept in inventory by the School of Management Studies at State University are listed below. Which items should be classified as '*A*' items, '*B*' items and 'C' items? What percentage of items is in each class? What percentage of total annual value is in each class?

Items	*Annual usage*	*Value per unit (Rs.)*
1	200	40.00
2	100	360.00
3	2,000	0.20
4	400	20.00
5	6,000	0.04
6	1,200	0.80
7	120	100.00
8	2,000	0.70
9	1,000	1.00
10	80	400.00

OBJECTIVE TYPE QUESTIONS

10.1. If small orders are placed frequently (rather than placing large orders infrequently), then total inventory cost is
(*a*) increased (*b*) reduced
(*c*) either increased or reduced (*d*) minimized

10.2. If orders are placed with size determined by the EOQ, then the re-order costs component is
(*a*) equal to the holding cost component
(*b*) greater than the holding cost component
(*c*) less than the holding cost component
(*d*) either greater than or less then the holding cost component.

10.3. If EOQ is calculated, but an order is then placed which is smaller than this, will the variable cost :
(*a*) increase (*b*) decrease
(*c*) either increase or decrease (*d*) no change

10.4. If an optimal order size (Q) is calculated, but is found to be of an inappropriate size would the total cost per unit time :
(*a*) rise quickly around Q. (*b*) rise slowly around Q.
(*c*) fall quickly around Q. (*d*) fall slowly around Q.

10.5. Which costs can vary with order quantity ?
(*a*) Unit cost only (*b*) Re-order cost only.
(*c*) Holding cost only (*d*) All of these.

10.6. If we find a minimum on a total cost curve with discounted unit cost, then the optimal order size.
(*a*) at this valid minimum.
(*b*) at or to the left of this minimum.
(*c*) at or to the right of this minimum.
(*d*) anywhere.

10.7. If we find a valid minimum on a total cost curve with increasing reorder cost, then the optimal order size

(*a*) at this valid minimum.

(*b*) at or to the left of this minimum.

(*c*) at or to the right of this minimum.

(*d*) anywhere.

10.8. When compared with instantaneous replenishment, does a finite replenishment rate lead to;

(*a*) the same size batches. (*b*) larger batches.

(*c*) smaller batches

10.9. If the total investment in stock is limited, will the best order quantity for each item

(*a*) equal the economic order quantity.

(*b*) greater than the EOQ.

(*c*) less than the EOQ. (*d*) either greater or less than the EOQ.

10.10. If the unit cost rises will optimal order quantity

(*a*) increase? (*b*) decrease?

(*c*) either increase or decrease? (*d*) none of the above?

10.11. In the EOQ model

(*a*) demand is known and occurs at a constant rate.

(*b*) all demand must be satisfied

(*c*) orders arrive in a batch. (*d*) all of the above

Answers

10.1. (*c*)	10.2. (*a*)	10.3. (*d*)	10.4. (*b*)	10.5. (*d*)
10.6. (*c*)	10.7. (*b*)	10.8. (*c*)	10.9. (*c*)	10.10. (*b*)
10.11. (*d*).				

❏❏❏

Chapter 11

Simulations

11.1. INTRODUCTION

There are many problems of real life, which cannot be represented mathematically due to the complexity and probabilistic nature of problems. Simulation is used when all every thing fail. It determines the effect of alternative policies without disturbing the system. Simulation is a technique for carrying out experiments for analyzing the behaviour and evaluating the performance of real system in actual environment, on simplified system.

Simulation is a representation of reality through a use of a model or other device, which will react in the same manner as reality under a given set of conditions.

Simulation is defined as the use of a system model that has the designed characteristics of reality in order to produce the essence of actual operation.

Simulation is the technique of imitating the behaviour of some situation or system (economic, mechanical, etc.) by means of an analogous model, situation, or apparatus, either to gain information more conveniently or to train personnel.

Simulation is the process of designing a model of a real system and conducting experiments with this model for the purpose of understanding the behaviour of the operation of the system.

11.2. TYPES OF SIMULATION

To determine the behaviour of actual system in real environment, a number of experiments are performed on simulated models in laboratories or on computer. Accordingly, Simulation is mainly of two types:

(*i*) Analogue Simulation or Environmental Simulation: In this the reality is simulated in physical form. *e.g.* (*i*) Testing an aircraft model in a wind tunnel; (*ii*) Planetarium ; (*iii*) Model of traffic system with signals and crossing for joy and ride of children

(*ii*) Computer Simulation or System Simulation: For complex problems mathematical model is developed using computer programme.

11.3. APPLICATIONS OF SIMULATION

There are a number of potential areas for application of simulation. Some of the applications are described below:

1. Evaluating alternative investment opportunities and financial forecasting.
2. Testing an aircraft model in a wind tunnel.

3. It is used in banks to determine the number of tellers required to serve customers.
4. It is used to model hospitals, banks, educational institutions and urban systems.
5. Modelling, luggage handling in airports.
6. For systems that already exist, simulation can be used to test for minor design changes or to look at the control policies. For example, for bottling lines the rules used to control the flow of bottles can be investigated as well as the effects of minor breakdowns.
7. Simulation is commonly used in the area of developing new systems, particularly those that involve a high capital investment. For example, simulation can be used to test the performance of personal computer (PC) assembly lines to ascertain the throughput possible, the level of utilization of operators and any potential problems.
8. It can be used to assess the best position for work-in-progress (WIP) stores and the levels to be used.

11.4. REASONS FOR USING SIMULATION

The main reasons of using Simulation techniques are listed below:

(*i*) It is an efficient tool to solve the problems where it is expensive, difficult to conduct experiments on real system.

(*ii*) There is no sufficient time to allow the real system to operate extensively.

(*iii*) Simulation is useful in sharpening the managerial decision making skills through gaming without disturbing normal operations.

(*iv*) Simulation models can be used to conduct experiments without disrupting real system. Experimenting with real system can be very expensive.

(v) It enables the manager to provide insight into certain managerial problems where analytical solution of a problem is not possible.

(*vi*) Non-technical manager can comprehend simulation more easily.

(*vii*) It allows experimentation with a model of the real life system rather than the actual system.

(*viii*) The relationship between the variables may be nonlinear and complex to handle with.

11.5. LIMITATIONS

(*i*) It does not generate optimal solutions. It is a trial-and-error approach that may produce different solutions in repeated runs.

(*ii*) Methods generally are not efficient.

(*iii*) It is a long complicated process to develop a model and expensive also.

(*iv*) Each solution model is unique and cannot be used for other similar problems.

(*v*) It is a descriptive process rather than optimization process.

(*vi*) It is difficult to quantify the variable.

(*vii*) Simulation generates a way of evaluating solutions but it does not generate solutions by itself.

(*viii*) Simulation models cannot evaluate all the situations.

11.6. ELEMENTS AND STRUCTURE OF SIMULATION SYSTEM

1. **State variables:** Depending upon the nature of the problem, a number of variables are required to describe the physical system. These variables are known as state variables e.g. for simulation of operations in a bank, the possible state variables are the number of busy counters, the number of customers waiting in a queue or being served and the arrival time of next customers.
2. **Random variable:** The random variable is a real valued function defined over a sample space associated with the outcome of a conceptual chance experiment. Random variables are classified according to their probability density function.
3. **Event:** An event is an occurrence, which causes changes in the state variables as a result of which there is a change in the state of the system. In the example of simulation of a bank operations, the arrival or departure of a customer is an event. In Inventory system problem, the relevant events are the consumption of the material and the arrival of an order.
4. **Static and Dynamic simulation model:** A Static simulation model studies the profile of a system at a particular point of time. Static simulation model is also known as Monte Carlo Simulation. Dynamic simulation model studies a system, parameters of which changes over time. Operations of a bank from 8 A.M. to 8 P.M. is an example of Dynamic simulation.
5. **Random Number:** It is a number in a sequence of numbers whose probability of occurrence is same as that of any other number in that sequence.
6. **Pseudo Random Numbers:** Random numbers are called pseudo random numbers when they are generated by some deterministic process but have already qualified the pre-determined statistical test for randomness.
7. **Entities:** Entities are the tangible elements found in the real world, e.g. for manufacturing these could be machines or trucks. The entities may be either temporary (e.g. parts that pass through the model) or permanent (e.g. machines that remain in the model). The concepts of temporary and permanent are useful aids to understanding the overall objectives of using simulation. A customer entering in a bank is an entity. Each entity has certain properties/ attributes.
8. **Clock:** In a dynamic system the state variables are a function of time. Time in simulation is represented by a variable known as **clock**. Occurrences of events are marked on clock. The events generally occur at random. e.g. in bank it is not known when the next customer will come, and for how much time the customer will stay in bank.
9. **Simulation executive:** A key part of any simulation system is the simulation executive. The executive is responsible for controlling the time advance. A central clock is used to keep track of time. The executive will control the logical relationships between the entities and advance the clock to the new time. The process is illustrated in Figure 11.1. The simulation executive is central to providing the dynamic, time based behaviour of the model. Whilst the clock and executive are key parts of a simulation system. They are very easy to implement and are extremely simple in behaviour.
10. **Probability distribution / random number generators:** The random number generators are used to provide stochastic behaviour typical of the real world. For example, machine scrap rates will rarely be fixed but will vary between

certain ranges hence the scrap rate of a machine should be determined by a random distribution (probably a normal distribution).

The results collation and display provides the user a means of utilizing the simulation tool to provide meaningful analysis of the new or proposed system. Simulation tools will typically display tabulated raw results and possess some graphing capabilities.

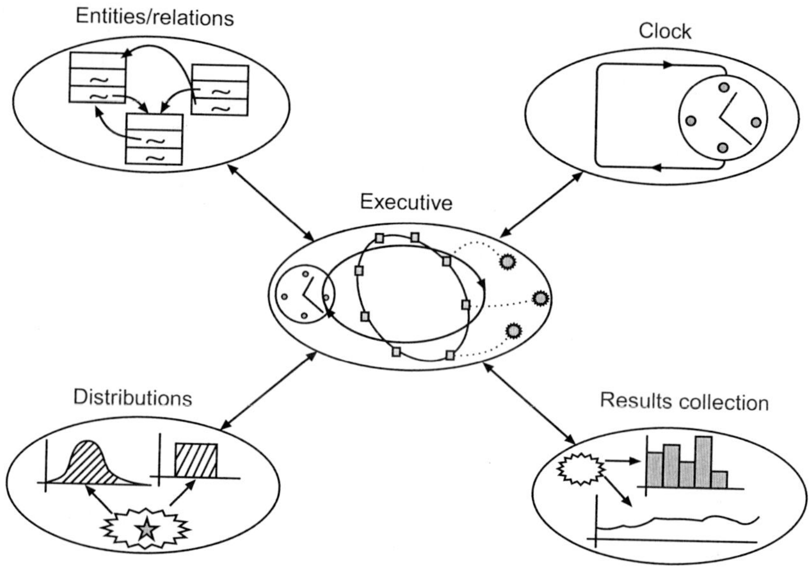

Fig.11.1. Structure of a simulation system

11.7. METHODOLOGY

The methodology of simulation consists of following steps:

(*i*) Identify and Define the problem clearly.

(*ii*) Analyze the probable costs and benefits of simulation study.

(*iii*) Construct a physical/mathematical model of the real system. The structure of model is shown in figure 11.1.

(*iv*) *Code the model:* Model is then transformed/written using suitable simulation language.

(*v*) *Validate the model:* Model is validated for representing the real system adequately. Model should behave with satisfactory accuracy consistent with the study objectives.

(*vi*) *Plan or design the experiment:* The performance of real system is checked by designing and conducting experiments on simulated model and listing various courses of actions.

(*vii*) *Analyze the results and draw conclusions:* Select the best course of action.

(*viii*) Document the features, assumptions and operating procedures and implement the findings.

11.8. SIMULATION MODELS

Simulation models can be classified into two categories:

(*i*) **Stochastic / probabilistic Simulation model:** When a system contains certain decision variables that can be represented by a probability distribution, simulation model used is called Stochastic / probabilistic simulation model. These models use random numbers to generate events. Techniques used to solve such models are known as **Monte-Carlo** techniques.

(ii) **Deterministic Simulation model:** In these models, input and output variables are not random variables. Such models are described by exact functional relationships.

Whichever software or technique is used for modeling there are a number of important considerations to be made for making a model:

- What is overall objective?
- What level of detail is actually needed?
- Has the model been checked?
- Does the model relate well to reality?
- Do the results seem credible?
- Do the results accurately reflect the model performance?

11.9. MONTE-CARLO SIMULATION

The Monte-Carlo Simulation is a simulation technique in which statistical distribution functions are created by using a sequence of random numbers. It is used to solve problems requiring decision making under uncertainty. Monte-Carlo is the code name given by Von Neuman and S.M. Ulam to the techniques of solving problems, which are quite expensive for experimental solutions and too much difficult for analytical solutions. Actually, this technique derives its name from the casinos in Monte Carlo. This method is used where solution of the model by analytical method is not possible. The use of MC methods to model physical problems allows examining more complex systems than otherwise can.

It results in a solution, which is very close to the optimal, but not necessarily the exact solution.

The simulation procedure can be summarized in following steps:

1. **Define the problem:**
   (*i*) Identify objectives of the problem.
   (*ii*) Identify main factors, which have greatest effect on objectives of the problem.
2. **Formulate an appropriate model:**
   (*i*) Identify variables and parameters of the model
   (*ii*) Form an appropriate decision rule.
   (*iii*) Collect the required information.
   (*iv*) Identify type of distribution to state the patterns of occurrence of variables.
   (*v*) Specify the manner in which time will change.
   (*vi*) Define relationship between variables and parameters.
3. Define starting conditions for simulation and specify number of simulation runs to be made.
4. Experiment with the model
   (*i*) Define a coding system to correlate factors with random numbers to be generated.
   (*ii*) Select a random number generator.
   (*iii*) Generate random numbers.
   (*iv*) Correlate generated random numbers with factors.
5. Summarize and examine results.

6. Evaluate the results and select best course of action.
7. Formulate proposals for advice to management on course of action to be adopted.

Take a simple example of a Monte Carlo simulation to illustrate the technique. The problem is to make a simulation to find the value of π . Consider a square that has one corner at the origin of a coordinate system and has sides of length 1 - it will obviously have an area of 1. Now consider inscribing a quarter of a circle inside of this with a radius of 1 its area is $\pi/4$. Use a Monte Carlo simulation to find the relative area of the quarter of circle and square and then multiply the quarter of circle's area by 4 to find π. For a point (x, y) to be inside of a quarter of circle of radius 1, its distance from the origin $\sqrt{x^2+y^2}$ will be less than or equal to 1. Generate thousands of random (x, y) positions and determine whether each of them is inside of the circle. Each time it is inside of the circle, add one to a counter. After generating a large number of points, the ratio of the number of points inside the quarter of circle to the total number of points generated(in square) will approach the ratio of the area of the quarter of circle to the area of the square. Thus the value of pi would simply be:

$$\pi \text{ (approximately)} = \frac{4\times(\text{Number of points inside quarter of circle})}{(\text{Total points generated})}$$

$$= \frac{4 \times \text{Areaof quarter of circle}}{\text{Area of square}}$$

$$= \frac{4 \times (1/4)\pi r^2}{r^2}$$

Thus we can find an approximation to π using simple math.

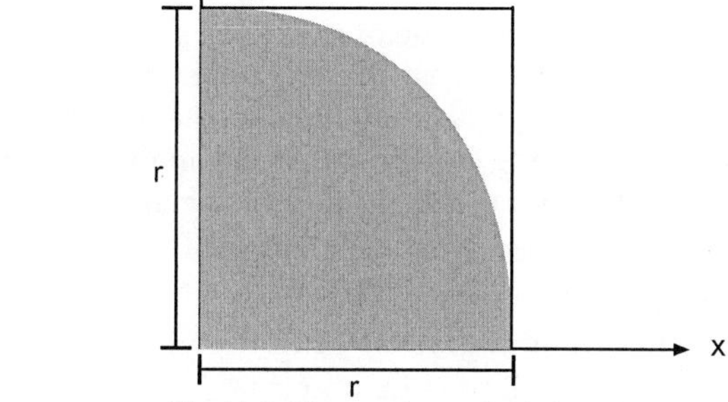

Fig.11.2. Illustration to find the value of π

11.10. GENERATION OF RANDOM NUMBERS

Monte-Carlo simulation requires the generation of a sequence of random numbers that is an integral part of the simulation model. Sequence of random numbers help in choosing random observations from probability distribution. Most common methods of generation of random numbers are- Mid-square method, Spinning arrow method, Dice rolling method and the spinning dice method. Steps to generate random numbers using Mid-square method are listed below:

(*i*) Start with an arbitrary number (seed) having **k** digits (even).

(*ii*) Square the number and extract middle **k** digits of number - next random number

(*iii*) Square the previous random number and extract middle **k** digits.

(*iv*) Continue the process.

Uniform Random Number Generators (RNGs) produce numbers with certain distribution properties. These numbers should behave similar to realizations of independent, identically distributed random variables.

We may distinguish between deterministic and non-deterministic algorithms. Deterministic RNGs are usually called pseudo-random number generators. Every RNG has its deficiencies. No RNG is appropriate for all tasks. For example, several good RNGs from the toolbox of stochastic simulation are unsuited for cryptographical applications, because they produce predictable output streams. On the other hand, cryptographic RNGs are usually (but not always) too slow for doing Monte Carlo simulations.

There are two big families of RNGs, linear generators and nonlinear ones. Cryptographers stay away from the linear algorithms because of predictability reasons, but in Monte Carlo simulations, linear RNGs are the best-known and most widely available ones. If one has to generate huge samples of random numbers as efficiently as possible, then one will use linear RNGs like the extremely fast and reliable RNGs available from Makoto Matsumoto, TT800 and the Mersenne Twister, or the (single or combined) multiple recursive generators studied and implemented by Pierre L'Ecuyer,

11.11. SIMULATION LANGUAGES

In order to ensure adequately precise results, the amount of data to be generated is very high. Therefore, manual procedures cannot be employed to simulate such real life problems. It is preferable to write simulation programme in specialized languages rather than in a general programming language. These languages are known as simulation languages. Analyst must choose a simulation language that best suits his needs. Familiarity with the system is one of the determining factors. There are two major types of simulation: continuous and discrete-event; though more modern languages can handle combinations. Most languages also have a graphical interface and at least simple statistical gathering capability for the analysis of the results. An important part of discrete-event languages is the ability to generate pseudo-random numbers and variates from different probability distributions.

Some of the important simulation languages are described below:

- **Discrete-event simulation languages: viewing the model as a sequence of random events each causing a change in state.**

  (*i*) **SIMSCRIPT:** This is high level programming language with facility specifically designed to add in the development of a simulation model. A SIMSCRIPT programme can be written in English - like statements in free format style. Such a programme is almost self documenting and can be easily explained to a non - programmer.

  SIMSCRIPT II.5 is one of the oldest computer simulation languages. Like Fortran with which it is intimately connected, programs using the language are still being used, maintained, and developed. It is used particularly for large-scale military or air-traffic control simulations though it is not limited to this type of model.

SIMSCRIPT II.5 is a powerful, free-form, English-like, general-purpose simulation programming language. It supports the application of software engineering principles, such as structured programming and modularity, which impart orderliness and manageability to simulation models.

SIMSCRIPT II.5 is available for various types of computers including PCs running Windows and NT and also for Unix workstations. The language was developed and is maintained by CACI. SIMSCRIPT and SIMSCRIPT I.5 are earlier and obsolete versions.

(*ii*) **GPSS (General Purpose Simulation System):** GPSS is a highly structured, special purpose simulation language. It is not a procedural language like SIMSCRIPT. It is a structured method for describing in detail, certain types of systems. The GPSS processor then takes this description and automatically performs a simulation. The time advancement mechanism is totally hidden from view.

GPSS is a discrete time simulation language where a simulation clock advance in uniform discrete steps. This was particularly well suited for problems such as a factory. GPSS is less flexible than simulation languages such as Simula and SIMSCRIPT II.5 that model a system with discrete events that are not constrained by a uniform time step.

(*iii*) **GASP IV:** It is a collection of subroutines written in FORTRAN language. These sub-routines and functions provide several facilities such as a time advance routine, to collect statistics, random variate generation routine and a standard report generator.

(*iv*) **Simula** is a programming language developed in the 1960s at the Norwegian Computing Centre in Oslo, primarily by Ole-Johan Dahl and Kristen Nygaard. Simula introduced the object-oriented programming paradigm and thus can be considered the first object-oriented programming language and a predecessor to Smalltalk, C++, Java, C#, and all modern class-based object-oriented languages. As its name implies, Simula was designed for doing simulations, and the needs of that domain provided the framework for many of the features of object-oriented languages today.

(*v*) **SimPy** is a process-based, object-oriented discrete-event simulation language. It is implemented in standard Python and released as Open Source under the GNU Lesser General Public License (LGPL). It provides the modeller with components for building a simulation model including Processes, for active components like customers, messages, and vehicles, and Resources, for passive components that form limited capacity congestion points like servers, checkout counters, and tunnels. It also provides Monitor objects to aid in gathering statistics. Random variates are provided by the standard Python random module.

SimPy simulates parallel processes by an efficient implementation of coroutines using Python's generators capability. It is based on ideas from Simula and SIMSCRIPT II.5.

Because it is implemented in Python, SimPy is platform-independent and can run on practically all modern operating systems.

- **Continuous simulation languages: viewing the model essentially as a set of differential equations.**
  (*i*) **DYNAMO:** This programme is suitable for simulating industrial systems that are made of chains such as factory - warehouse - distributor - retailer. The relevant variable are either levels or flows controlled by managerial decisions at specified points.
- **Other languages:**
  SPICE (Simulation Program with Integrated Circuits Emphasis) is a general purpose analog circuit simulator. It is a powerful program that is used in IC and board level design to check the integrity of circuit designs and to predict circuit behavior.

In real-world circuits, performance is affected by component value tolerances. It is difficult for designers to predict the effect of larger value tolerances. On the other hand they want to use cheaper components with large tolerances if they wish to mass produce their products. Also in radio applications, especially UHF and microwave, parasitics cannot be ignored and must be built into a generic model of the circuit being simulated. In both these cases it is usual to perform Monte Carlo simulations which are difficult or impossible to calculate by hand.

SPICE was originally developed at the Electronics Research Laboratory of the University of California, Berkeley in 1975 by Larry Nagle and Donald Pederson. Versions 1 and 2 were coded in Fortran (2G.6 in 1983 was the last) and ran on mainframe computers. Versions 3 and later are coded in C, but still use a Fortran-like syntax for circuit description. Many commercial versions of SPICE have later replaced Berkeley SPICE as the industry standard. While many are still compatible with the original Berkeley syntax, commercial vendors added proprietary extensions that limit the portability of circuit descriptions and models between different vendors. Most recent versions also include a graphical user interface for constructing circuit descriptions. For digital circuits (e.g., RAM), dedicated simulators exist that run orders of magnitude faster than the traditional Spice tools.

A new circuit simulator, based on SPICE, called ng-SPICE (for next-generation) is licensed under the GPL. LTSPICE is a free SPICE that works on Microsoft Windows and on Linux (under Wine (software) emulator).

First versions of Berkeley SPICE used Nodal analysis. However this meant that ideal voltage sources and inductors could not be included in the circuit. Later versions are using Modified nodal analysis, which does not have this drawback. Different algorithms are used to translate all circuit analysis problems into a single or multiple simpler problems of calculating an operating point of a linear circuit. Such problems can then be solved efficiently by solving a linear simultaneous equation. For example non-linear circuits are solved using a Newton-Raphson algorithm, which linearizes non-linear elements in a circuit. Transient analysis is performed using trapezoid or Gear integration algorithm.

- **Z simulation language**

11.12. COMMERCIAL VERSIONS WITH SIGNIFICANT MARKET SHARE

- HSPICE (originally from Meta Software, now owned by Synopsys)
- PSPICE
- SmartSpice (Silvaco)
- T-Spice (Tanner Research)

- Spectre (Cadence general purpose SPICE replacement, and RF simulator)
- UltraSim (Cadence fast spice tool)
- NanoSim
- NSPICE (Apache Design Solutions)
- HSIM (NASSDA)
- B2SPICE (Beige Bag)
- INTUSOFT ISpice4 Simulator
- TINA_Design_Suite DesignSoft's Tina Spice Simulator
- TINA-TI Free simulator based on DesignSoft's Tina Spice Simulator.

Table 11.1. Table of Random Numbers

78994	36244	02673	25475	84953	61793	50243	63423
04904	58485	70686	93930	34880	73059	06823	80257
46582	73570	33004	51795	86477	46736	60460	70375
29242	89792	88634	60285	07190	07795	27011	85941
68104	81339	97090	20601	78940	20228	22803	96070
17156	02182	82504	19880	93747	80910	76260	25136
50711	94789	07171	02103	99057	98775	37997	18325
39449	52409	75095	77720	39729	03205	09313	43545
75629	82729	76916	72657	58992	32756	01154	84890
01020	55151	36132	51971	32155	60735	64867	35424
08337	89989	24260	08618	66798	25889	52860	57375
76829	47229	19706	30094	69430	92399	98749	22081
39708	30641	21267	56501	95182	72442	21445	17276
89836	55817	56774	75119	06818	83043	47403	58266
25903	61370	66081	54076	67442	52964	23823	02718
71345	03422	01015	68025	19703	77313	04555	83425
61454	92263	14647	08473	34124	10740	40839	05620
80376	08909	30470	40200	46558	61742	11643	92121
45144	54373	05505	90074	24783	86299	20900	15144
12191	88527	58552	51175	11534	87218	04876	85584

Example 11.1. Star-News Shops Limited owns a large chain of news agents. The sales pattern of a particular weekly magazine is being studied.

The magazine sells for Rs. 30 per copy and costs Rs. 20 per copy from the publishers. There is a sale or return agreement whereby the publishers will take back unsold copies at cost price less a Rs. 4 per copy handling charge.

The demand pattern of a typical shop has been recorded and is shown in the table. At present, a shop orders 400 copies of the magazine a week.

Demand (copies)	*Probability*
0 < 100	0.07
100 < 200	0.12
200 < 300	0.14
300 < 400	0.19
400 < 500	0.18
500 < 600	0.09
600 < 700	0.07
700 < 800	0.07
800 < 900	0.05
900 < 1000	0.02

There has been a suggestion that an insight may be gained into the situation if the magazine's demand pattern is simulated. As a consequence, a Random Number table has been constructed, a 20 number extract from which is shown below:

00	27	74	69	32	17	98	57	71	51
03	96	15	13	56	15	83	63	32	17

(*i*) Allocate random number to the demand pattern for this shop.

(*ii*) Using a tabular formate, to simulate 20-week operations for the shop, ordering 400 copies per week, showing weekly profit or loss, returns and lost sales.

[Use random number 00 for week 1, 27 for week 2, and so on.]

(*iii*) Interpret the results of your simulation.

(*iv*) Discuss the strengths and weaknesses of simulation in this context.

Solution: For determining the demand, first consult the random numbers given in the question. For first week, random number is 00. Random number 00 is falling against demand 0 < 100. The demand is thus 50 copies (*i.e.*, $0 + \frac{1}{2} \times 100$). As the typical shop order is 400 copies, 350 copies (*i.e.*, 400–50) are returned. In second week random number is 27 which is falling between random numbers 19–32. The corresponding demand against this random number is 200 < 300, *i.e.*, 250 copies. Thus 150 copies are returned. In third week, random number is 74 which is falling between random numbers 70–78. The corresponding demand against this random number is 500< 600, *i.e.*, 550 copies. But the typical shop order is 400 copies only and therefore 150 copies will be refused which is to be teated as lost sales. The procedure may be continued for other intervals. Now

Profit per item sold = Rs. 30 – Rs. 20 = Rs. 10

Loss per item if not sold = Rs. 4

Table 11.2 Determination of Random Number Interval

Demand	*Probability*	*Cumulative Prob.*	*Random Number Allocate*
0 < 100	0.07	0.07	00 – 06
100 < 200	0.12	0.19	07 – 18
200 < 300	0.14	0.33	19 – 32
300 < 400	0.19	0.52	33 – 51
400 < 500	0.18	0.70	52 – 69
500 < 600	0.09	0.79	70 – 78
600 < 700	0.07	0.86	79 – 85
700 < 800	0.05	0.58	86 – 92
800 < 900	0.05	0.98	93 – 97
900 < 1,000	0.02	1.00	98 – 99

(*ii*) **Table 11.3 Simulation Worksheet**

Week	*Random Number*	*Demand*	*Sales*	*Returns*	*Lost Sales*	*Profit/ Loss*
1	00	50	50	350	—	Rs. – (900)
2	27	250	250	150	—	1,900
3	74	550	400	—	150	4,000
4	69	450	400	—	50	4,000
5	32	250	250	150	—	1,900
6	17	150	150	250	—	500
7	98	950	400	—	550	4,000
8	57	450	400	—	50	4,000
9	71	550	400	—	150	4,000
10	51	350	350	50	—	3,300
11	03	50	50	350	—	– (900)
12	96	850	400	—	450	4,000
13	15	150	150	250	—	500
14	13	150	150	250	—	500
15	56	450	400	—	250	4,000
16	15	150	150	250	—	500
17	83	650	400	—	250	4,000
18	62	450	400	—	50	4,000
19	32	250	250	150	—	1,900
20	17	150	150	250	—	500
Total		7,300	5,550	2,450	17,500	45,700

(*iii*) The simulation shows a total profit of Rs. 45,700 over the twenty-week period. The cost of returns in this same period was 2,450 × Rs. 4 = Rs. 9800, and the loss of profit due to lost sales was Rs. 17,500. This would increase the cost of returns, but would, at the same time, increase profit due to the shops being better able to satisfy demand. The optimum order level could be found by running the simulation several times with different order quantities.

(*iv*) ***Strength :***

— allows a variety of situation to be analysed;

— allows an optimal solution to be found ;

— a low cost way of studying the operations; and

— does not interfere with the real system.

Weakness

— assumes that the distribution is accurate;

— results depend upon the random numbers used. Consistency can be increased by doing the simulation with different random number and over a longer time period;

— demand has to be averaged ever a range of values.

Example 11.2. A company uses a high grade raw material. The consumption pattern is probabilistic and it takes two months to replenish stocks:

Consumption per month (tonnes)	1	2	3	4
Probability	0.15	0.30	0.45	0.10

The cost of placing an order is Rs. 1,000 and the cost of carrying stocks is Rs. 50 per month per tonne. The average carrying costs are calculated on the stocks held at the end of each month.

The company has two options for the purchase of raw materials as under:

Option I : Order for 5 tonnes when the closing inventory of the month plus outstanding order is less than 8 tonnes:

Option II : Order for 8 tonnes when the closing inventory of the month plus order outstanding is less than 8 tonnes.

Currently on 1st April 2004, the company has a stock of 8 tonnes of raw materials plus 6 tonnes ordered two months ago. The ordered quantity is expected to be received next month.

Using the random number given below, simulate 12 months consumption till 31.03.2005 and advise the company as to which purchase option should be accepted such that the inventory costs are minimum.

Random numbers are 88, 41, 63, 48, 74, 27, 16, 11,64, 49, 21,.

Solution:

Table 11.4 Determination of Random Numbers Interval

Demand (Tonnes)	*Probability*	*Cumulative Probability*	*Random Number Allocated*
1	0.15	0.15	00 —14
2	0.30	0.45	15 — 44
3	0.45	0.90	45 — 89
4	0.10	1.00	90 — 99

Given that stock in hand is of 8 tonnes and stock on order is 5 tonnes

Table 11.5 Simulation Worksheet (Option-1)

Random number	*Demand monthly*	*Opening stock*	*Receipt*	*Closing stock*	*Opening stock on order*	*Order quantity*	*Closing stock*
88	3	8	—	8 – 3 = 5	—	—	6
41	2	5	6	6 + 5 – 2 = 9	—	—	—
67	3	9	—	9 – 3 = 6	—	5	5
63	3	6	—	6 —3 = 3	5	—	5
48	3	3	—	3 – 3 =0	5	5	10
74	3	0	5	5– 3 =2	5	5	10
27	2	2	—	2 –2 =0	10	—	10
16	2	0	5	5 —2 =3	5	—	5
11	1	3	5	5+3 – 1 =7	—	5	5
64	3	7	—	7 – 3 =4	5	—	5
49	3	4	—	4 – 3 =1	5	5	10
21	2	1	5	5+1 – 2 =4	5	—	5

Since 5 tonnes have been ordered five times as shown in Table 11.5, total ordering cost is Rs. (5 × 1000) = Rs. 5000

Closing stock of 12 month is of 44 (= 5 + 9 + 6 + 3 + 2 + 3 + 7 + 4 + 1 + 4) tonnes. Therefore the cost of carrying stock at the rate of Rs. 50 per tonne per month is Rs. (44 × 50) = Rs. 2, 200

Total cost for 12 months = Ordering cost + Carrying cost
Rs. (5,000 + 2,200) = Rs. 7, 200.

Table 11.6 Simulation Worksheet (Option-2)

Random number	*Demand monthly*	*Opening stock*	*Receipt*	*Closing stock*	*Stock on order*	*Order quantity*	*Closing stock on order*
88	3	8	-	8 – 3 =5	-	-	6
41	2	5	6	6 + 5 –2=9	-	-	-
67	3	9	-	9 – 3 = 6	-	8	8
63	3	6	-	6 – 3 = 3	8	-	8
48	3	3	-	3 – 3 =0	8	-	8
74	3	0	8	8 + 0– 3=5	-	8	8
27	2	3	-	5 – 2 =3	8	-	8
16	2	3	-	3 – 2 = 1	8	-	8
11	1	1	8	8 + 1 – 1=8	-	-	-
64	3	8	-	8 – 3 =5	-	8	8
49	3	5	-	5 — 3 = 2	8	-	8
21	2	2	-	2 – 2 =0	8	-	8

Since 8 tonnes have been ordered three times as shows in Table 11.6. when the losing inventory of the month plus orders outstanding is less than 8 tonnes. Therefore total ordering cost is : Rs(3 × 1, 000) = Rs. 3,000.

Closing stock of 12 months, is of 47 (= 5 + 9 + 6 + 3 + 5 + 3 + 1+ 8 + 5 + 2) tonnes. Therefore cost of carrying stocks at the rate of Rs. 50 per ton per month is Rs. 47 × 50 = Rs. 2, 350.

Total cost for 12 months = Ordering cost + Carrying cost = Rs. 5, 850.

Since option 2 has lower total cost than option 1, company should choose option 2.

Example 11.3. The number of machines arriving per day at a factory repair bench has been noted over a long period of time and found to have the following distribution :

Arrivals :	0	1	2	3	4	5	6	7	8	9
Percentage:	2	7	15	20	20	16	10	6	3	1

Repair times depend on the type of fault and the number of repairs completed per day. They have been recorded and found to be represented by :

Repairs:	1	2	3	4	5	6	7	8	9	10
Percentage:	4	8	14	18	18	15	10	7	4	2

Simulate 25 days to find the maximum queue length. Use the following series of random numbers :

55 86 87 70 05 71 69 29 74 73 37 17 42 16 89 98 35 81
07 16 94 42 27 18 13 70 05 35 86 85 02 79 13 35 37 38
57 19 97 38 45 82 40 32 63 29 10 67 41 99

Solution :

Table 11.7 Determination of Random Number Intervals

Arrivals	*Prob.*	*Cumulative*	*RN Interval*	*Repairs*	*Prob.*	*Cumulative*	*RN Interval*
0	0.02	0.02	00 — 01	1	0.04	0.04	00 — 03
1	0.07	0.09	02 — 08	2	0.08	0.12	04 — 11
2	0.15	0.24	09 — 23	3	0.14	0.26	12 — 25
3	0.20	0.44	24 — 43	4	0.18	0.44	26 —43
4	0.20	0.64	44 — 63	5	0.18	0.62	44 — 61
5	0.16	0.80	64 — 79	6	0.15	0.77	62 — 76
6	0.10	0.90	80 — 89	7	0.10	0.87	77 — 86
7	0.06	0.96	90 — 95	8	0.07	0.94	87 — 93
8	0.03	0.99	96 — 98	9	0.04	0.98	94 — 97
9	0.01	1.00	99	10	0.02	1.00	98 — 99

By reading two series of random numbers from tables the arrival and repairs can be simulated over 25 days to find the queue lengths as shown in Table 11.8 below :

Table 11.8 Simulation Worksheet

Day	*Random Number (1)*	*Arrivals*	*Random Number (2)*	*Repairs*	*Queue length*
1	55	4	86	7	0
2	87	6	70	6	0
3	05	1	71	6	0
4	69	5	29	4	1
5	74	5	73	6	0
6	37	3	17	3	0
7	42	3	17	3	0
8	89	6	98	10	0
9	35	3	81	7	0
10	07	1	16	3	0
11	94	7	42	4	3
12	27	3	18	3	3
13	13	2	70	6	0
14	05	1	35	4	0
15	86	6	85	7	0
16	02	1	79	7	0
17	13	2	35	4	0
18	37	3	38	4	0
19	57	4	19	3	1
20	97	8	38	4	5
21	45	4	82	7	2
22	40	3	32	4	1
23	63	4	29	4	1
24	10	2	67	6	0
25	41	3	99	10	0

Maximum queue length = 5

Example 11.4. An Investment company wants to study the investment project based on three factors : market demand in units, price per unit minus cost per unit, and the investment required. These factors are felt to be independent of each other. In analysing a new consumer product, the corporation estimates the following probability distributions:

Annual Demand		*(Price-Cost)*		*Investment Required*	
Units	*Prob.*	*Perunit*	*Probability*	*Rs.*	*Probability*
25,000	0.05	3.00	0.10	27,50,000	0.25
30,000	0.10	5.00	0.20	30,00,000	0.50
35,000	0.20	7.00	0.40	35,00,000	0.25
40,000	0.30	9.00	0.20		
45,000	0.20	10.00	0.10		
50,000	0.10				
55,000	0.05				

Using simulation process, repeat the trial 10 times. Compute the return on investment for each trial taking these three factors into account. Approximately, what is the most likely return ?

Use the following random numbers for annual demand (price-cost) and the investment required:

30, 59, 63, 27, 64, 28, 31, 54, 64, 32, 12, 09, 94, 08, 60, 28, 23, 85, 68, 31, 16, 69, 26, 74, 61, 72, 57, 20, 18, 87.

Solution: The yearly return can be computed by using the formula :

$$\text{Return } (R) = \frac{\text{(Price – Cost)} \times \text{Number of units demanded}}{\text{Investment}}$$

Corresponding to each of the three factors, by building a cumulative probability distribution an appropriate set of random numbers is assigned is assigned to represent each of the three factors as shown in Table 11.9 :

Table 11.9 Determination of Random Number Intervals

Annual demand (000 units)	*Probability*	*Cum. Probability*	*RN Interval*
25	0.05	0.05	00 — 04
30	0.10	0.15	05 — 14
35	0.20	0.35	15 — 34
40	0.30	0.65	35 — 64
45	0.20	0.85	65 — 85
50	0.10	0.95	86 — 94
55	0.05	1.00	95 — 99

Profit units (Rs)	*Probability*	*Cum. Probability*	*RN Interval*
3.00	0.10	0.10	00—09
5.00	0.20	0.30	10—29
7.00	0.40	0.70	30—69
9.00	0.20	0.90	70—89
10.00	0.10	1.00	90—99

Investment required ('000 Rs.)	*Probability*	*Cum. Probability*	*RN Interval*
2,750	0.25	0.25	00—24
3,000	0.50	0.75	25—74
3,500	0.25	1.00	75—99

The simulation worksheet is prepared for 10 trials and the results of simulation are shown in Table 11.10 below:

Table 11.10 Simulation Worksheet

Trials	Random Number of Demand	Simulated Demand (000unit)	Random Number for Profit	Simulated Profit (000 Rs)	Random Number	Simulated Investment (0,000 Rs.)	Simulated Return %
1	30	35	12	5.00	16	275	6.36
2	59	40	09	3.00	69	300	4.00
3	63	40	94	10.00	26	275	14.55
4	27	35	08	3.00	74	300	3.50
5	64	40	60	7.00	61	300	9.33
6	28	35	28	5.00	72	300	5.83
7	31	35	23	5.00	57	300	5.83
8	54	40	85	9.00	20	275	13.10
9	64	40	68	7.00	18	275	10.18
10	32	35	31	7.00	87	350	7.00

The above table shows, the highest likely return is 14.55% which is corresponding to the annual demand of 40,00 units resulting a profit of Rs. 10 per unit and the required investment will be 27,50,000.

Example 11.5. A job has to be processed over two machines, machine M_1 and machine M_2 in that order.

The distribution of inter-arrival time of the jobs at the first machine is as follows

Time (minutes)	:	1	2	3	4
Probability	:	0.2	0.2	0.2	0.4

The processing times at the two machines are as follows :

Machine M_1		*Machine M_2*	
Time (minutes)	*Probability*	*Time (minutes)*	*Probability*
1	0.1	4	0.2
2	0.2	5	0.3
3	0.3	6	0.4
4	0.3	7	0.1
5	0.1		

On the basis of 10 simulation runs find out the average queue length before machine M_1 and the average queue length before machine M_2.

Solution: The various steps for simulation are

1. Simulate the inter-arrival time of the ith job on machine M_1 with the help of random number (i =1, 2,...)
2. Arrival time of ith job

   = Arrival time of (i–1) th job (i=2, 3,....) + Inter-arrival time of ith job.

   Arrival time of the first job

   = Inter-arrival time of the first job.
3. Simulate the processing time on machine M_1 by random numbers.

4. Departure time on machine M_1
   = Arrival time for processing on machine M_2
   = Max (Arrival time of ith job, process completion time of (i—1)th job on machine M_1) + Processing time of ith job on machine M_1
5. If arrival time of ith job< process completion time of (i–1) th, i^{th} arrival waits. If arrival time of ith job < process completion time of (i–2)th, both i^{th} and (i –1)th arrivals wait.
6. Simulate the processing time of machine M_2 with the help of random umbers.
7. Process completion by machine M_2 of ith arrival
   = Max [Arrival time of ith job, process completion time of (i—1) th job] + Process time by machine M_2.
8. If arrival time of ith < process completion time of (i–1)th, the ith arrival waits. If arrival time of ith < process completion time of (i–1) th arrivals wait, and so on.

   Using the procedure, the simulated processing data of 10 arrivals are displayed in the simulation worksheet :

Table 11.11.

Simulation run	*RN*	*Inter-arrival time*	*Arrival time on M_1*	*RN*	*Process time on M_1*	*Departure time on M_1*	*RN*	*Process time on M_2*	*Completion time on M_2*	*Queue size M_1*	*Queue size M_2*
1	8	4	4	0	1	5	3	5	10	-	-
2	7	4	8	6	4	12	0	4	16	—	-
3	8	4	12	9	5	17	0	4	21	-	-
4	9	4	16	3	3	20	0	4	25	1	1
5	4	3	19	7	4	24	0	4	29	1	1
6	4	3	22	0	1	25	2	5	34	1	1
7	0	1	23	0	1	26	4	5	39	2	2
8	6	4	27	5	3	30	5	6	45	-	2
9	8	4	31	6	4	35	1	4	49	-	2
10	7	4	35	4	3	38	2	5	54	-	3

On the basis of simulation study, the average queue length before machine M_1 is 0.5, and the average queue length before machine M_2 is 1.2

EXERCISE

11.1. What is simulation?

11.2. What are advantages and drawbacks of using simulation techniques?

11.3. Describe the elements of simulation system.

11.4. What is Monte-Carlo simulation? Why is it called so.

11.5. Write the steps ivolved in Monte-Carlo simulation.

11.6. Explain the Monte-Carlo simulation using an example. Discuss its various applications.

11.7. What are random number generators? Discuss the process of generating random numbers.

11.8. Discuss various simulation languages being used commercially.

11.9. Write various steps of methodology of simulation.

11.10. Explain why simulation is necessary technique in present world.

11.11. A company manufactures 30 items per day. The sale of these items depends upon demand which has the following distribution:

Sales (units)	*Probability*
27	0.10
28	0.15
29	0.20
30	0.35
31	0.15
32	0.05

The production cost and sale price of each unit are Rs. 40 and Rs. 50 respectively. Any unsold product is to be disposed of at a loss of Rs. 15 per unit. There is a penalty of Rs. 5 per unit if the demand is not met.

Using the following random numbers estimate total profit/loss for the company of the next 10 days: 10, 99, 65 99, 95, 01, 79, 11, 16, 20

If the company decides to produce 29 items per day, what are the advantages or disadvantages to the company ?

11.12. A manufacturer transport daily its production, whose probability distribution is given below, through trucks, having a capacity of 200 units.

Daily production :	198	199	200	201	202	202	203
Probability :	0.1	0.2	0.2	0.25	0.08	0.08	0.09

Simulate for seven days and determine average truck capacity utilisation and average number of units which remain at the factory each day.

11.13. A company trading in motor vehicles spares wish to determine the level of stock it should carry for product X. Its demand distribution is as follows:

Demand (Units/day) :	3	4	5	6	7
Probability :	0.1	0.2	0.3	0.3	0.1

Holding cost (per unit per day) 20 paise, ordering cost (per order) Rs. 5 shortage cost per unit per day Rs. 1, lead time for replenishment 3 days and beginning stock in hand 20 units.

You are required to carry out a simulation run over 10 days with the objective of evaluating the inventory rule : Order 15 units when present inventory plus any outstanding order falls below 12 units. Use random numbers 0,9, 1, 1, 5, 1, 8, 6, 3, 7, 1, 2, 9.

11.14. Given the following data of arrival pattern and service pattern at a Shoe Store. Simulate the queue behaviour for 10 arrivals. Start time is 10.00 A.M.

Time between arrival (minutes)	*Probability*	*Services Time (minutes)*	*Probability*
0.2	0.3	0.5	0.1
0.4	0.3	1.0	0.5
0.6	0.2	1.5	0.3
0.8	0.1	2.0	0.1
1.0	0.1		

Use the following Random Numbers :

For arrivals : 6, 2, 8, 6, 1, 4, 3, 5, 7, 8

For services : 4, 3, 6, 6 7, 2, 7, 0, 8, 7

11.15. The number of machines arriving per day at a factory repair bench has been noted over a long period of time and found to have the following distribution:

Arrivals	:	0	1	2	3	4	5	6	7	8	9
Percentage	:	2	7	15	20	20	16	10	6	3	1

Repair times depend on the type of fault and the number of repairs completed per day. They have been recorded and found to be represented by —

Arrivals	:	1	2	3	4	5	6	7	8	9	10
Percentage	:	4	8	14	18	18	15	10	7	4	2

Simulate 25 days to find the maximum queue length.

11.16. The following table gives the arrival pattern at a coffee counter for one minute intervals. The service is taken as 2 persons in one minute in one counter :

No. of persons arriving	:	0	1	2	3	4	5	6	7
Probability percentage	:	5	10	15	30	20	10	5	5

Using Monte-Carlo simulation techniques and the following random numbers, generate the pattern of arrivals and the queue formed when the following 20 random numbers are given : 5, 25, 16, 80, 35, 48, 67, 79, 90, 92, 9, 14, 1, 55, 20, 71, 30, 42, 60, and 85. Find the queue length if two counters are used, *i.e.*, 4 persons in one minute.

11.17. The materials manager of a company is interested in determining the reorder point for an item, the pattern of demand for which is as given below :

No. of units per day	*No. of days on which the demand occurred*
0	5
1	9
2	16
3	38
4	23
5	9

Past experience indicates that there are fluctuations in the lead time for procurement of item. The following data is available from the records for the last 30 orders:

Lead time(weeks)	*No. of time the specified lead time occurred*
1	18
2	7
3	5

The management policy is to ensure that the proportion of stockouts should not exceed 5%. Illustrate how the simulation approach can be used to determine the re-order point. (Analysis for 20 order is adequate.)

11.18. At a toll office a sample of 100 arrivals of vehicles gives the following frequency distribution of the inter arrival and service time.

Inter arrival time (min)	*Frequency*	*Service time*	*Frequency*
1.0	2	1.5	10
1.5	5	–	–
2.0	9	2	22
2.5	25	–	–
3.0	22	2.5	40
3.5	11	–	–
4.0	10	3.0	20
4.5	6	–	–
5.0	3	3.5	8
5.5	2	–	–

There is a clerk at the office. Simulate the process for 20 arrivals and estimate the average percentage vehicle waiting time and average percentage idle time of the clerk.

11.19. An airline has 20 flights leaving a base per day, each with a hostess. The airline keeps two hostesses in reserve so that they may be called in case the scheduled hostess for a flight is absent. The probability distribution for daily number of absenteeism by hostesses is as follows:

Number absent :	0	1	2	3	4	5
Probability :	0.30	0.35	0.20	0.10	0.03	0.02

Use Monte-Carlo method to estimate the utilisation of reserve hostesses and also the probability that at least one flight will be cancelled in a day because of non-availability of hostesses.

11.20. The observations of the study made for 100-day period by M/s Voltas (I) Ltd., an air-conditioner manufacturing and servicing unit, are tabulated below:

No.of air-conditioners requesting service	*Service requests*		*Servicing done*	
	Frequency	*Probability*	*Frequency*	*Probability*
0	5	0.05	8	0.08
1	10	0.10	2	0.02
2	15	0.15	30	0.30
3	20	0.20	20	0.20
4	25	0.25	15	0.15
5	15	0.15	20	0.20
6	10	0.10	5	0.05

Simulate a 10-day period arrival and service pattern. From this also out the average backlog units/day.

We assume that M/s. Voltas (I) Ltd. has only one service team and servicing is done on first come, first served basis for 24 hours.

Simulate the situation by generating random number to find out selling price, variable cost and annual sales volume. The random number for the three factors are :

00—19, 20—69, 70—99; 00—19, 20—79, 80—99; and 00—19, 20—69, 70—99 respectively.

Determine the average profit for 20 simulations (observations).

11.21. A company manufactures around 200 mopeds. Depending upon the availability of raw materials and other conditions, the daily production has been varying from 196 mopeds to 204 mopeds, whose probability distribution is as given below :

Production per day :	196	197	198	199	200	201	202	203	204
Probability :	0.05	0.09	0.12	0.14	0.20	0.15	0.11	0.08	0.06

The finished mopeds are transported in a specially designed three-storeyed lorry that can accommodate only 200 mopeds. Using the given 15 random number, *viz*, 82, 89, 78, 24, 53, 61, 18, 45, 04, 23, 50, 77, 24, 54, 10, simulate the process to find out (*i*) What will be the average number of mopeds waiting in the factory ? and (*ii*) What will be the average number of empty spaces on the lorry?

11.22. An assembly line has three work stations. The time required for each station to complete its operation is as follows :

Time (minutes)	*Probabilities*		
	Station 1	*Station 2*	*Station 3*
4	0.25	0.10	0.05
5	0.25	0.30	0.25
6	0.25	0.40	0.25
7	0.25	0.20	0.45

The times given are the only values the operation times take on. Simulate the flow of 20 items through the assembly line. What is the average time that an item takes to go through all the operations?

11.23. A railway reservation office has two counters working simultaneously to cater to the heavy rush between 8 a.m and 12 noon. The average time taken by any one reservation cleark is 10 minutes and the distribution is exponential. The passengers arrive in a Poisson process with a mean rate of 12 per hour.

Simulate the arrivals of the first 20 passengers. Through simulation estimate the average queue length and average waiting time in the queue. Compare these results with the corresponding theoretical results.

11.24. A confectioner sells confectionery items. Past data of demand per week in hundred kilograms with frequency is given below:

Demand/week (in hundred kg.)	0	5	10	15	20	25
Frequency	2	11	8	21	5	3

Generate the demand for the next 10 weeks. Also find out the average demand per week.

OBJECTIVE TYPE QUESTIONS

11.1. Once a queue model has been constructed, analysis of the model can be performed in

(*a*) Through analytical solution (*b*) Through simulation

(*c*) Either (*a*) or (*b*) (*d*) Both

(*e*) None

11.2. A quantitative analysis technique that involves building a mathematical model presents a real-world situation. The model is then experimented with to estimate the effects of various actions and decision is known as:

(*a*) Decision theory (*b*) Simulation

(*c*) Game theory (*d*) None of these

11.3. One of the disadvantages of simulation is that :

(*a*) It is very expensive and requires to develop large repetitions of data.

(*b*) Simulation solution can be cent percent accurate.

(*c*) Simulation are applicable in cases where there is an element of randomness in a system.

(*d*) All of the above.

11.4. An advantage of simulation, as opposed to optimization, is that

(*a*) Often multiple measures of goodness can be examined.

(*b*) Some appreciation for the variability of outcomes of interest can be obtained.

(*c*) More complex scenarios can be studied

(*d*) All of the above.

11.5. A random number refers to

(*a*) An observation from a set of number each of which is equally likely.

(*b*) In observation selected at random from a normal distribution.

(*c*) An observation selected at random from any distribution provided by the analyst.

(*d*) None of the above.

11.6. In a typical simulation model input provided by the analyst includes

(*a*) Value for the parameters. (*b*) Value for the decision variables.

(*c*) Value for the measure of effectiveness.

(*d*) All of the above (*e*) Both (*a*) and (*b*).

11.7. If both an analytic model and a simulation model could be used to study a problem including random events, the analytic model is often preferred because

(*a*) The simulator generally requires a number of runs just to get a good estimate of the objective value for a particular decision

(*b*) The analytic model may produce an optimal decision.

(*c*) The simulation study may require evaluating a large number of possible decisions.

(*d*) all of the above

11.8. Large complicated simulation models suffer from the following problems :
   (*a*) Average costs are not well defined.
   (*b*) It is difficult to create the appropriate random events.
   (*c*) They may be expensive to write and to use an experimental device.
   (*d*) All of the above.

11.9. In performing a simulations it is advisable to
   (*a*) Use the results of earlier decisions to suggest the next decision to try
   (*b*) Use the same number of trials for each decision.
   (*c*) Simulate all possible decisions.
   (*c*) None of the above

Answers

11.1. (*c*) 11.2. (*b*) 11.3. (*a*) 11.4. (*d*) 11.5. (*a*) 11.6. (*e*) 11.7. (*d*)
11.8. (*c*) 11.9. (*a*)

❒❒❒

Unit V

Chapter 12

Queuing Models

12.1. INTRODUCTION

Arriving units while moving towards the service facility forms a waiting line to maintain a proper discipline, due to lack of capability to serve them all at a time. This waiting line is known as **queue.** Arriving units can be persons, machines, vehicles, parts etc. The queue does not include the unit being served.

"A queuing situation develops when items arrive at a serving channel for some type of service. The service of each arrival takes some length of time and can be provided from one or more serving channel"

The process or system that provides the services to the arriving units is known as service channel or service facility. e.g. aero planes seeking to land at busy airport, ships in the harbour waiting to be unloaded, patients waiting at a doctor's clinic, persons waiting at a railway booking office, cars waiting for traffic signal, calls arriving at a telephone switchboard, customers waiting at bank teller's window and machines waiting to be repaired.

The study of waiting lines is known as Queuing Theory. Study of queues was initiated by a Danish engineer, A.K.Erlang, to eliminate bottlenecks created by telephone calls on switching board. Queuing theory is concerned with statistical behaviour of queues. e.g. finding waiting time.

Queuing theory deals with the analysis and management of waiting lines.

Queuing Theory is s set of mathematical solutions to waiting line or process type problems. In these problems, there is some process, which consumes time, units arrive in the system, and are processed through it.

12.2. QUEUING SYSTEM

- To describe a queuing system, an input process and an output process must be specified.

  Examples of input and output processes are:

Situation	*Input process*	*Output process*
Bank	Customers arrive at bank	Tellers serve the customer
Pizza parlor	Request for pizza delivery are received	Pizza parlor send out truck to deliver pizzas

The queuing system consists of arriving units/customers, waiting units, queue discipline, service facility and the units leaving the system after being served. It is shown in figure 12.1.

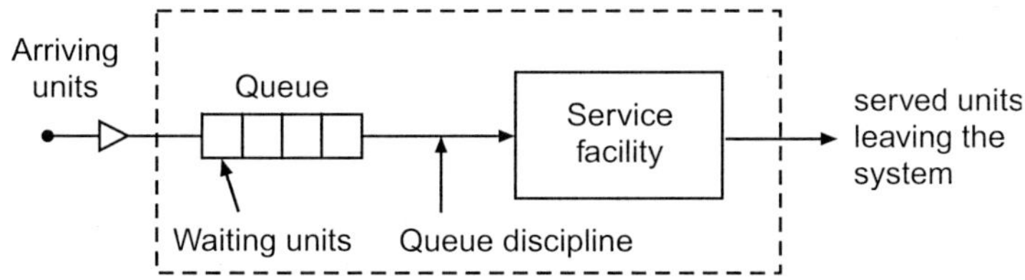

Fig. 12.1. Queuing system.

In the queuing process, units arrive at the service system and may enter the system depending upon the queue conditions. The units entering the system makes a queue for availing the service and are served by service facilities according to some rule, known as queue discipline. After being served, units leave the system.

12.3. ELEMENTS OF QUEUING SYSTEM

1. **Input Process:** It describes the pattern in which units arrive and join the system. Input source can be characterized by following factors.
   (*a*) **Size of population:** The size may be finite or infinite. If the number of potential customers requiring service is few then the size of population is finite. For example - number of machines to be repaired in a workshop. If the potential customers are very large then the size of population is infinite. For example customers arriving at railway booking counter.
   (*b*) **Arrival time distribution:** The customer arrival rate generally follows Poisson distribution, which is discrete probability distribution and inter-arrival time follow exponential distribution.
   (*c*) **Customer's behaviour:** The customer on entering the system may behave in following four ways:
      (*i*) **Balking:** If the queue appears to be very long to a customer and he/she has no time & space to wait, he/she may decide not to join the queue. This property is known as balking of customers.
      (*ii*) **Reneging:** When a waiting customer decides to leave the queue in anticipation of longer waiting time and impatience. This is known as reneging.
      (*iii*) **Jockeying:** If a customer moves from one queue to another due to its smaller size; behaviour of customer is known as jockeying. e.g. a customer in supermarket
      (*iv*) **Priorities:** In certain applications, some customers are served before others irrespective of their arrival. These customers have priorities over others.
2. **Queue Discipline:** It is a rule according to which the customers are selected for service when a queue has been formed. Most common queue disciplines are:
      (*i*) **First come, first served (FCFS) or First in, First out (FIFO)**: In this the customers are served in order of their arrival. e.g. customers at railway ticket counter or cinema ticket window are served according to this rule.
      (*ii*) **Last in, First out (LIFO) or First in, Last out (FILO)**: Here, the last arrived customer in the system is served first. e.g. in a big warehouse the items which come in last are taken out first.
      (*iii*) **Service in random order (SIRO)**: According to this rule, customers are serviced in random order irrespective of their arrival in the system.

There are various other disciplines according to which a customer is served in preference over the other. Under priority discipline, the service is of two types:

(*a*) Pre-emptive: In this system, the high priority customers are given service over the low priority customers.

(*b*) Non-pre-emptive: In this system, a customer of low priority is serviced before a customer of high priority is entertained for service.

In case of parallel channels "Fastest server rule (FSR)" is adopted. In which, the incoming customer is assigned a server of highest service rate among the free ones.

3. **Service Mechanism**: The service mechanism is concerned with service time and service facilities. Service time is the time interval from the commencement of service to the completion of service. Service time distribution can be (*i*) constant (*ii*) Poisson distribution (*iii*) Exponential distribution (iv) Erlang distribution.

   Service facilities can be of following types:

   (*i*) **Single queue - one server**

   (*ii*) **Single queue - several servers**

   (*iii*) **Several queues - one server**

   (*iv*) **Several servers**: Several servers can be arranged in parallel or in series or a complex combination of both depending upon the service mechanism. In series arrangement, a customer passes through all servers in sequence till service is completed. e.g. in a machine shop work piece passes through different machines before job is complete. While in parallel arrangement, several servers are arranged in a way that an incoming customer in a single queue can be served when one of the server is free. *e.g.* a barber shop with many chairs.

4. **Size of Queue**: Size of a queue may be finite or infinite. When only a finite number of customers can stay in the system, queue is said to be finite. For example the maximum length of queue in a petrol pump is limited by the space for waiting vehicles. When there is no limit on the size of queue, it is said to be infinite queue. For example in a sales department there is no restriction on the number of orders that can be received so a queue of orders of any size can be formed.

12.4. CHARECTERISTICS OF QUEUING SYSTEMS

Some of the operational characteristics of queuing system are as follows:

1. **Expected number of customers in the queue:** It is the average number of customers waiting in the queue. Short queues mean either good customer service or too much capacity. Similarly, long queues indicate either poor service or need to increase capacity.
2. **Expected number of customers in the system:** The number of customers in the system is the average number of customers in the system both in queue and being served. It relates to service efficiency and capacity. Large values imply congestion, customer dissatisfaction and a need for more capacity.
3. **Expected waiting time in queue:** It is the average time spent by a customer in the queue before the commencement of his service. Long lines do not reflect long waiting times if the service rate is fast.
4. **Total time in system:** It is the total elapsed time from entry into the system until exit from the system. It is equal to waiting time in queue plus service time. It may

indicate problems with customers, server efficiency or capacity. If much time is taken in the system, there may be a need to change the priority discipline, increase productivity or adjust capacity.

5. **Service facility utilization (busy period):** The utilization of the service facilities tells about the percentage of time the facilities are busy. Service utilization factor ($\rho = \lambda/\mu$) is defined as the proportion of time that a service facility spends with customers. It is also known as **traffic intensity**. Its unit is Erlang. Here, λ is the average number of customers arriving per unit time and μ is average number of customers being served per unit time.

12.5. CLASSIFICATION OF QUEUING SYSTEMS

1. **Single channel or multi-channel queues:** Queues can be classified according to the number of serving channels - whether there is a single channel or several channels. Most queuing situations in highway traffic are single channel queues.
2. **Under saturated or over saturated queues**: If the arrival rate is less than the maximum serving rate of a queuing system, it is classified as an under saturated queue. The queue length varies in length but reaches steady state. If the arrival rate is greater than the maximum service rate, the system never reaches steady state and the queue length is continually increasing.
3. **Arrival and service distributions**: The arrivals and service distributions can be one of several types ranging from uniform to random. For several reasons, most frequently both the arrival and service time distributions are treated as being random, i.e., Poisson distributed.

12.6. STATES OF SYSTEM

(*i*) **Steady state:** A system is said to be in steady state when the characteristics of the system are independent of time.

(*ii*) **Transient state:** A system is said to be in transient state when the operating characteristics of the system like input, output, average queue length etc. are dependent on time. This state occurs in early stages of operations.

(*iii*) **Explosive state:** If the arrival rate of the system is greater than its service rate, the queue length will increase with time and become infinite. This state of system is known as explosive state.

12.7. KENDALL'S NOTATION FOR REPRESENTING QUEUING MODELS

The queuing models may be completely specified in the following symbolic form: $(a/b/c):(d/e)$

Where a = Probability distribution of inter-arrival time

b = Probability distribution of inter-service time

c = Number of service channels

d = Capacity of the system (maximum number of customers allowed in the system)

e = Queue discipline

For example: $(M/E_k/1):(\infty/\text{FIFO})$ means a queuing system in which arrivals follow Poissson distribution, service time as Erlang distribution, single server, infinite capacity and the queue discipline is first in - first out.

Here, following letters are used to specify the kind of distribution:
M specifies Poisson distribution
E_k specifies Erlang or Gamma distribution
GI specifies general input distribution
G specifies General service time distribution

12.8. CLASSIFICATION OF QUEUING MODELS

Queuing models are used to:

– describe the behavior of queuing systems
– determine the level of service to provide
– evaluate alternate configurations for providing service

Queuing models are classified as follows:

(*i*) **(*M*/*M*/1) : (∞/FIFO) :** It signifies Poisson arrival rate, Poisson service rate, single server, infinite capacity and first in - first out queue discipline.

(*ii*) **(*M*/*M*/*S*) : (∞/FIFO) :** : It signifies Poisson arrival rate, Poisson service rate, **multi-server (*S* servers),** infinite capacity and first in - first out queue discipline.

(*iii*) **(*M*/*M*/1) : (N/FIFO) :** It denotes Poisson arrival rate, Poisson service rate, single server, **finite / limited capacity (N customers are permitted in the system)** and first in - first out queue discipline.

(*iv*) **(*M*/*M*/*S*) : (*N*/ FIFO) :** This model is similar to model (*ii*) except the maximum number of customers is limited to *N*, where $N > S$.

12.9. ASSUMPTIONS OF QUEUING THEORY

Following assumptions are made while analyzing the problems of Queuing theory:

(*i*) Steady state (stable) condition exist i.e. the probability that n items are in queue at any time, remains same at all time.

(*ii*) Mean service rate μ is greater than mean arrival rate λ.

(*iii*) The queue discipline is First in, First out (FIFO)

(*iv*) Both, the number of customers in the queue at any instant and the waiting time experienced by a particular customer, are random variables.

(*v*) Arrival rate follow Poisson distribution and service times follow negative exponential distribution.

12.10. NOTATIONS USED

Following notations are used in different queuing models:

n = number of units/customers in the service system

λ = mean arrival rate (average number of arrivals per unit time)

μ = mean service rate (average number of customers served per unit time)

$1/\lambda$ = mean time between arrivals

$1/\mu$ = mean service time (mean time per customer served)

$\rho = \lambda/\mu$ = Traffic intensity or utilization factor (average utilization of service facility). Its unit is Erlang.

s = number of servers

P_n = Steady state probability that there are n customers in the system

L_s = average number of customers in the system (customers in queue plus being served)
L_q = average number of customers in the queue
W_s = average waiting time spent by a customer in the system
W_q = average waiting time spent by a customer in the queue
$(W/W > 0)$ = expected waiting time of a customer who has to wait
$(L/L > 0)$ = expected number of customers in the queue when there is a queue
$P(W > 0)$ = probability of customer having to wait for service

12.11. BENEFITS OF QUEUING THEORY

1. Queuing theory provides models that determine the appropriate service and number of service channels.
2. Scheduling of aircraft at landing and take off from busy airports.
3. Routing and Scheduling of salesman and sales efforts.
4. It is used in scheduling of parts and components to assembly lines.

12.12. LIMITATIONS OF QUEUING THEORY

(*i*) Queuing problems are generally complex and probabilistic in nature and cannot be easily solved.
(*ii*) In multi-channel situations, the analysis becomes more difficult when the departure from one queue becomes the arrival of another queue.
(*iii*) Arrival rate and service time distributions assumed during analysis may not fit in actual situation.
(*iv*) If arrival rate and service time are variable, queuing models cannot analyze such problems.

12.13. MODEL 1 (M/M/1) : (∞/FIFO) OR BIRTH AND DEATH MODEL

Following conditions are assumed for analyzing single channel system:

1. Arrivals are served on first come - first served basis
2. There is no balking or reneging
3. Arrival rate dose not change with time and arrivals are independent of preceding arrivals.
4. Arrival rate follow Poisson distribution and arrivals are from very large population (infinite).
5. Service times follow negative exponential distribution.
6. Average service rate (μ) is greater than average arrival rate (λ).
7. Service times vary from one customer to the next and are independent of one another.

(*i*) Probability that service facility is idle

$$P_0 = 1 - \frac{\lambda}{\mu} = (1 - \rho)$$

(*ii*) Probability that there are '*n*' units in the system (in queue plus service)

$$P_n = \left(\frac{\lambda}{\mu}\right)^n \cdot \left(1 - \frac{\lambda}{\mu}\right) = \rho^n (1 - \rho)$$

(*iii*) Probability that queue size is greater than or equal to n

$$P(\geq n) = \left(\frac{\lambda}{\mu}\right)^n = \rho^n$$

(*iv*) Average number of units in the system (L_s)

$$L_s = \frac{\lambda/\mu}{(1-\lambda/\mu)} = \frac{\rho}{(1-\rho)}$$

(*v*) Average number of units in the queue (L_q)

$$L_q = L_s - \frac{\lambda}{\mu}$$

$$= \frac{\lambda^2}{\mu(\mu-\lambda)} = \frac{\rho^2}{(1-\rho)}$$

(*vi*) Average waiting time in the system, W_s

$$W_s = \frac{\text{Average number of units in the system}}{\text{Average arrival rate}}$$

$$= \frac{L_s}{\lambda} = \frac{1}{(\mu-\lambda)}$$

$$= W_q + \frac{1}{\mu}$$

(*vii*) Average waiting time in the queue, W_q

$$W_q = \frac{\text{Average number in queue}}{\text{Arrival rate}} = \frac{L_q}{\lambda}$$

$$= \frac{\lambda}{\mu(\mu-\lambda)}$$

(*viii*) Average length of non-empty queue

$$(L/L>0) = \frac{\mu}{(\mu-\lambda)} = \frac{1}{(1-\rho)}$$

(*ix*) Probability of a non-empty queue

$$(P/L>0) = \left(\frac{\lambda}{\mu}\right)^2$$

(*x*) Probablity that the waiting time in a queue is greater than or equal to '*t*'

$$P(\textit{waiting}\ \text{time} \geq t) = \frac{\lambda}{\mu} e^{-(\mu-\lambda)t}$$

(*xi*) Average waiting time of a customer who has to wait

$$(W/W>0) = \frac{1}{(\mu-\lambda)} = \frac{1}{\mu(1-\rho)}$$

(*xii*) Variance of queue length

$$\text{variance} = \frac{\rho}{(1-\rho)^2}$$

(*xiii*) Interrelationships

$$L_s = \lambda W_s$$
$$L_q = \lambda W_q$$
$$L_q = L_s - \frac{\lambda}{\mu}$$

Example 12.1. At a car service center, cars arrive according to Poisson process with mean 3 per hour. The time for servicing each car is exponentially distributed with mean 15 minutes per car. The facility can handle 1 car at a time. Compute: (*i*) Average queue length, (*ii*) Average number of cars in the queuing system, (*iii*) Waiting time in the system, and (*iv*) Percent utilization of service. **(UPTU: 2003-2004)**

Solution: Arrival rate, $\lambda = 3$ per hour

Service rate, $\mu = 60/15 = 4$ per hour

Traffic intensity (utilization factor), $\rho = \lambda/\mu = 3/4 = 0.75$

(*i*) Average queue length $(L_q) = \rho^2/(1-\rho) = 0.75^2/(1-0.75) = 2.25$ cars

(*ii*) Average number of cars in queuing system $(L_s) = \rho/(1-\rho) = 0.75/(1-0.75)$ $= 3$ cars

(*iii*) Waiting time in the system $(W_s) = 1/(\mu-\lambda) = 1/(4-3) = 1$ hour

(*iv*) Percent utilization of system, $\rho = \lambda/\mu = 3/4 = 0.75 = 75\ \%$

Example 12.2. A person repairing TV set finds that the time spent on the set has been exponential distribution with mean 20 minutes. If the TVs are repaired in the order in which they come and their arrival is approximately Poisson with an average rate of 15 for 8-hour day. What is the repairman's expected idle time each day? **(UPTU: 2004-2005)**

Solution: Here, μ (service rate) $= 1/20$, λ (arrival rate) $= 15/(8 \times 60) = 1/32$

Traffic intensity (fraction of time repairman is busy) $= \lambda/\mu = 20/32$

Number of hours repairman is busy in a 8 hour day $= 8 \times (20/32) = 5$ hours

Therefore, time for which repairman remains idle in a 8-hour day $= 8 - 5 = 3$ hours

Example 12.3. The arrival rate of customers at a banking counter follows Poisson distribution with a mean of 45 per hour. The service rate of the counter clerk also follows Poisson distribution with a mean of 60 per hour.

(*i*) What is the probability of having 0 customer in the system?

(*ii*) What is the probability of having 5 customers in the system?

(*iii*) What is the probability of having 10 customers in the system?

(*iv*) Find L_s, L_q, W_s, W_q

Solution: Arrival rate, $\lambda = 45$ per hour

Service rate, $\mu = 60$ per hour

Traffic intensity (fraction of time clerk is busy) $= \lambda/\mu = 45/60 = 0.75$

(*i*) Probability of having 0 customer in the system (idle system)

$$P_0 = 1 - \lambda/\mu$$
$$= 1 - 0.75 = 0.25$$

(*ii*) Probability of having 5 customers in the system

$$P_5 = (1 - \rho)\,\rho^5$$
$$= (1 - 0.75) \times .075^5$$

(*iii*) Probability of having 10 customers in the system $P_{10} = (1 - \rho)\rho^{10}$

$$= (1 - 0.75) \times 0.75^{10}$$
$$= 0.0141$$

(*iv*) $L_s = \lambda/(\mu - \lambda) = 45/(60\text{-}45) = 3$ customers

$L_q = \lambda^2/\mu\,(\mu - \lambda) = 45^2/60\,(60\text{-}45) = 2.25$ customers

$W_s = 1/(\mu - \lambda) = 1/(60 - 45) = 0.067$ hour

$W_q = \lambda/\mu\,(\mu - \lambda) = 45\,/\,60\,(60\text{-}45) = .05$ hour

Example 12.4. Vehicles pass through a toll gate at a rate of 90 per hour. The average time to pass through the gate is 36 seconds. The arrival rate and service rate follow Poisson distribution. There is a complaint that the vehicles wait for long duration. The authorities are willing to install one more gate to reduce the average time to pass through the toll gate to 30 seconds if the idle time of the toll gate is less than 10 % and the average queue length at the gate is more than 5 vehicles. Check whether the installation of the second gate is justified.

Solution: Arrival rate of vehicles at gate, $\lambda = 90$ per hour

Service rate, $\mu = 1/36$ per second $= 60 \times 60/36 = 100$ per hour

Traffic intensity (Utilization factor) $\rho = \lambda/\mu = 90/100 = 0.9$

(*i*) Number of vehicles waiting in the queue, $Lq = \lambda^2/\mu\,(\mu - \lambda)$
$= (90)^2/100\,(100 - 90) = 8.1$ vehicles

(*ii*) After installing another gate, revised time taken to pass through the gate = 30 seconds. Therefore,

Service rate, $\mu = 1/30$ per second $= 60 \times 60\,/30 = 120$ per hour

Utilization factor, $\rho = \lambda\,/\,\mu = 90/120 = 0.75$

(*iii*) Percentage idle time of the gate $= 1 - \rho = 1 - 0.75 = 0.25 = 25$ %

Since idle time of the gate is not less than 10 %, hence installation of another gate is not justified.

12.14. MODEL 2 (M/M/S) : (∞/FIFO)

Following conditions are assumed for analyzing multi-channel queuing system:

1. Arrivals are served on first come - first served basis.
2. Arrivals are from very large population (infinite).
3. A single queue is formed.
4. Arrival rate follow Poisson distribution.
5. There are s servers, each of which provides identical service.
6. Average combined service rate ($s\mu$) of all servers is greater than average arrival rate (λ).

S service stations are arranged in parallel and a single queue is formed.

(*i*) Probability that system shall be idle, P_0

$$P_o = \left[\sum_{n=0}^{s-1} \frac{(\lambda/\mu)^n}{n!} + \frac{(\lambda/\mu)^s}{s!(1-\rho)} \right]^{-1}$$

(*ii*) Probability that there will be exactly 'n' customers in the system

$$P_n = \frac{(\lambda/\mu)^n}{n!} P_o \qquad (n \le s)$$

$$P_n = \frac{(\lambda/\mu)^n}{s!\, s^{n-s}} P_o \qquad (n > s)$$

(*iii*) Average number of customers in queue

$$L_q = \frac{(\lambda/\mu)^s \rho}{s!\,(1-\rho)^2} P_o \qquad \left[\rho = \frac{\lambda}{s\mu}\right]$$

(*iv*) Average number of customers in the system

$$L_s = L_q + \frac{\lambda}{\mu}$$

(*v*) Average waiting time in queue, W_q

$$W_q = \frac{L_q}{\lambda}$$

(*vi*) Average waiting time in the system, W_s

$$W_s = W_q + \frac{1}{\mu} = \frac{L_s}{\lambda}$$

(*vii*) Number of individuals who actually wait,

$$(L/L > 0) = \frac{1}{(1-\rho)}$$

(*viii*) Average waiting time in queue for those who actually wait

$$(W/W > 0) = \frac{1}{(s\mu - \lambda)}$$

(*ix*) Probability that there will be some one waiting,

$$P(n \ge s+1) = \frac{P_s\, \rho}{(1-\rho)}$$

where, $$P_s = \frac{(\lambda/\mu)^s}{s!} P_o$$

(*x*) Probability that the customer has to wait

$$P(W > 0) = P(n \ge s) = \frac{P_s}{(1-\rho)}$$

(*xi*) Average number of idle servers

$$= s - (\text{average number of customers served})$$

(*xii*) Efficiency of model

$$= \frac{\text{Average number of customers served}}{\text{Total number of customer served}}$$

Example 12.5. At a centre store, vehicles arrive at the rate of 18 per hour and the arrival rate follows Poisson distribution. The unloading time of the vehicles follows exponential distribution and the unloading rate is 6 vehicles per hour. There are 4 unloading crews, Find:

(*i*) p_0 and p_3 (*ii*) L_q, L_s, W_q and W_s

Solution: Arrival rate, $\lambda = 18$ per hour

Unloading rate, $\mu = 6$ per hour

Number of unloading crews, $s = 4$

$$\rho = \frac{\lambda}{s\mu} = \frac{18}{4 \times 6} = \frac{3}{4} = 0.75 \; ; \; \frac{\lambda}{\mu} = \frac{18}{6} = 3$$

(*i*)
$$p_0 = \left[\sum_{n=0}^{s-1} \frac{(\lambda/\mu)^n}{n!} + \frac{(\lambda/\mu)^s}{s!(1-\rho)} \right]^{-1}$$

$$= \left[\sum_{n=0}^{3} \frac{(3)^n}{n!} + \frac{(3)^4}{4!(1-0.75)} \right]^{-1} = 0.0377$$

$$p_3 = \frac{(\lambda/\mu)^n}{n!} p_0 \text{ (For } 0 \le n \le s)$$

$$= \frac{(3)^3}{3!} \times 0.0377 = 0.1697$$

(*ii*)
$$L_q = \frac{(\lambda/\mu)^s.\rho}{s!(1-\rho)^2} p_0 = \frac{(3)^4 \times 0.75}{4! \times (1-0.75)^2} \times 0.0377$$

$$= 1.53 \approx 2 \text{ vehicles}$$

$$L_s = L_q + \frac{\lambda}{\mu} = 1.53 + 3 = 4.53 \approx 5 \text{ vehicles}$$

$$W_q = \frac{L_q}{\lambda} = \frac{1.53}{18} = 0.085 \text{ hour} = 5.1 \text{ minutes}$$

$$W_s = W_q + \frac{1}{\mu} = 0.085 + \frac{1}{6} = 0.252 \text{ hour}$$

$$= 15.12 \text{ minutes.}$$

12.15. ECONOMIC ANALYSIS OF QUEUING SYSTEM

Typically manager is interested in minimizing the total cost and reducing the service time.

TC = Expected total cost per unit time
SC = Expected service cost per unit time
WC = Expected waiting cost per unit time

The objective is then to choose the number of servers to

Minimize $TC = SC + WC$

When each server costs the same (C_s = cost of server per unit time),

$$SC = C_s s$$

When the waiting cost is proportional to the amount of waiting (C_w = waiting cost per unit time for each customer),

$$WC = C_w L \text{ or } WC = C_w L_q$$

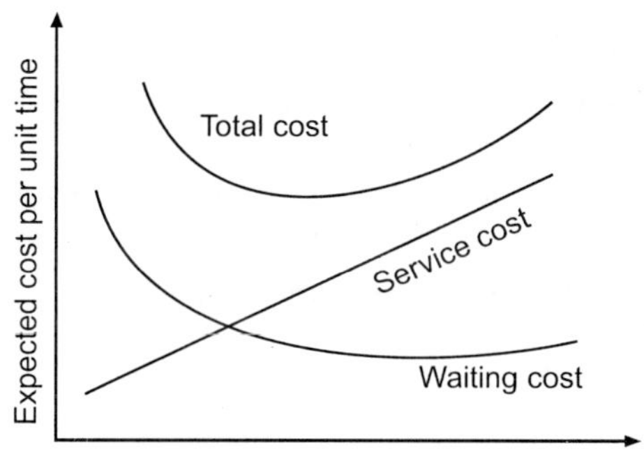

Fig. 12.2.

SOLVED EXAMPLES

Example 12.6. A TV repairman finds that the time spent on his jobs has an exponential distribution with mean 30 minutes. If he repairs sets in the order in which they come in, and if the arrival of sets is approximately Poisson with an average rate of 10 per 8-hour day, what is repairman's expected idle time each day? How many jobs are ahead of the average set just brought in?

Solution: Here, $\mu = 1/30$, $\lambda = 10/(8 \times 60) = 1/48$. Therefore, expected number of jobs are

$$L_s = \frac{\lambda/\mu}{1-\lambda/\mu} = \frac{\lambda}{\mu-\lambda}$$

$$= \frac{1/48}{1/30 - 1/48} = 12/3 \text{ jobs.}$$

Since the fraction of the time the repairman is busy (*i.e.*, traffic intensity) is equal to ρ the number of hours for which the repairman remains busy in a 8-hour day is

$$= 8. (\lambda/\mu) = 8 \times 30/48 = 5 \text{ hours.}$$

Therefore, the time for which the repairman remains idle in 8-hour day = (8 – 5) hours = 3 hours. **Ans.**

Example 12.7. At what average rate must a clerk at a supermarket work in order to ensure a probability of 0.90 that the customer will not have to wait longer than 12 minutes? It is assumed that there is only one counter to which customers arrive in a

Poission fashion at an average rate of 15 per hour. The length of service by the clerk has an exponential distribution.

Solution: Here, $\lambda = 15/60 = 1/4$ customer/minute, $\mu = ?$

Prob. [Waiting time ≥ 12] = 1 – 0.90 = 0.10.

Therefore, P (wating time ≥ 12) $= \dfrac{\lambda}{\mu} e^{-(\mu-\lambda)t}$

or $$0.1 = \frac{1/4}{\mu} e^{-(\mu - 1/4)12}$$

$$e^{(3-12\mu)} = 0.4\,\mu$$

or $$1/\mu = 2.48 \text{ minute per service. } \textbf{Ans.}$$

Example 12.8. Arrivals at a telephone booth are considered to be Poisson, with an average time of 10 minutes between one arrival and the next. The length of a phone call assumed to be distributed exponentially with mean 3 minutes. Then,

(*a*) What is the probability that a person arriving at the booth will have to wait

(*b*) What is the average length of the queues that form from time to time?

(*c*) The telephone department will install a second booth when convinced that an arrival would expect to have to wait at least three minutes for the phone. By how much must the flow of arrivals be increased in order justify a second booth?

Solution: $\lambda = 1/10$ and $\mu = 1/3$

(*a*) Probability $(W > 0) = 1 - P_0 = \lambda/\mu = \dfrac{1}{10} \times \dfrac{3}{1} = \dfrac{3}{1} = 0.3$

(*b*) $(L \mid L>0) = \mu/(\mu-\lambda) = 1/3\,(1/3 - 1/10) = 1.43$ persons

(*c*) $Wq = \lambda / \mu\,(\mu - \lambda)$

Since $Wq = 3$, $\mu = 1/3$. $\lambda = \lambda'$ for second booth, therefore

$$3 = \frac{\lambda'}{1/3(1/3-\lambda')} \text{ giving } \lambda' = 0.16.$$

Increase in arrival rate = 0.16 – 0.10 = 0.06 arrivals per minute

Example 12.9. In a railway marshalling yard, goods trains arrive at a rate of 30 trains per day. Assuming that the inter-arrival time follows an exponential distribution and the service time (the time taken to hump a train) distribution is also exponential with an average 36 minutes. Calculate the following :

(*i*) The average number of trains in the queue.

(*ii*) The probability that the queue size exceeds 10.

If the input of trains increases to an average 33 per day, what will be change in (*i*) and (*ii*)?

Solution : Here

$$\lambda = \frac{30}{60 \times 24} = \frac{1}{48} \text{ trains/minute;}$$

$$\mu = \frac{1}{36} \text{ trains/minute,}$$

and $$\rho = \frac{\lambda}{\mu} = \frac{36}{48} = 0.75$$

(*i*) $L_s = \frac{\rho}{1-\rho} = \frac{0.75}{1-0.75} = 3$ trains.

(*ii*) Prob [queue size ≥ 10] $= \rho^{10} = (0.75)^{10} = 0.06$.

When the input increases to 33 trains per day, $\lambda = 1/43$, $\mu = 1/36$.

Therefore, $\rho = \lambda/\mu = 36/43 = 0.84$.

(*i*) $Ls = 0.84/0.16 = 5$ trains

(*ii*) Prob (queue size ≥ 10) $= (0.84)^{10} = 0.2$ (approx.)

Example 12.10. A supermarket has two girls ringing up sales at the counters. If the service time for each customer is exponential with mean 4 minutes, and if people arrive in a Poisson fashion at the counter at the rate of 10 per hour.

(*a*) Calculate the probability that an arrival will have to wait for service.

(*b*) Find the expected percentage of idle time for each girl ?

(*c*) If a customer has to wait, find the expected length of his waiting time.

Solution: ***(a)*** Probability of having to wait for service is

$$P(W>0) = \frac{(\lambda/\mu)^s}{s!(1-\rho)} P_0$$

Here, $\lambda = 1/6, \mu = 1/4, s = 2, \rho = \lambda/\mu s = 1/3.$

Now, calculate

$$P_0 = \left[\sum_{n=0}^{s-1} \frac{(s\rho)^n}{n!} + \frac{(s\rho)^s}{s!(1-\rho)}\right]$$

$$= \left[\sum_{n=0}^{1} \frac{(2\times 1/3)^n}{n!} + \frac{(2.1/3)^2}{2!(1-1/3)}\right]$$

$$= \left[1 + 2/3 + \frac{4/9}{2\times\ 2/3}\right] = 1/2$$

Thus, Probability

$$P(W>0) = \frac{(4/6)^2.1/2}{2!(1-1/3)} = 1/6.$$

Example 12.11. A telephone exchange has two long distance operators. The telephone company finds that, during the peak load, long distance calls arrive in a Poisson fashion at an average rate of 15 per hour. The length of service on these calls is approximately exponentially distributed with mean length 5 minutes.

(*a*) What is the probability that a subscriber will have to wait for his long distance call during the peak hours of the day?

(*b*) If the subscribers will wait and serviced in turn, what is the expected waiting time ?

Solution : Given $s = 2$, $\lambda = 15/60 = 1/4$, $\mu = 1/5$.

Therefore,

$$P_0 = \left[\sum_{n=o}^{s-1} \frac{(s\rho)^n}{n!} + \frac{(s\rho)^s}{s!(1-\rho)}\right]^{-1} = \left[\sum_{n=0}^{1} \frac{(5/4)^n}{n!} + \frac{(5/4)^2}{2!\,(1-5/8)}\right]$$

$$= \frac{1}{1+(5/4)+(5/4)^2.1/2.8/3} = 3/13$$

(*a*) $\text{Prob.}\,(W > 0) = \dfrac{(\lambda/\mu)^s}{s!(1-\rho)} P_0 = \dfrac{(5/4)^2.3/13}{2!(1-5/8)} = 25/52 = 0.48.$

(*b*) $Wq = L_q/\lambda = \dfrac{1}{\lambda} \cdot \dfrac{\rho(s\rho)^s}{s!(1-\rho)^2} P_0.$

$$= 4.\frac{(5/8)\,(5/4)^2}{2!(1-5/8)^2}.\frac{3}{13} = \frac{125}{39} = 3.2 \text{ minutes.}$$

EXERCISE

12.1. A ticket issuing office is being manned by a single server. Customers arrive to purchase tickets according to a Poisson process with a mean rate of 30 per hour. The time required to serve a customer has an exponential distribution with a mean of 90 seconds. Find the value of P_0, L_s and W_s.

[**Ans.** $L_s = 3$ customers, $W_s = 6$ min.]

12.2. At a public telephone booth the arrivals are on the average 15 per hour. A call on the average takes 3 minutes, if there is just one phone.

(*i*) What is expected number of callers in the both at any time.

(*ii*) For what proportion of time is the booth expected to be idle. [**Ans.** 3, 1/4]

12.3. Consider a box office ticket window being manned by single server. Customers arrive to purchase tickets according to a Poisson input process with a mean rate of 30 per hour. The time required to serve a customer has an exponential distribution with a mean of 90 seconds. Determine the following :

(*i*) Fraction of the time the server is busy.

(*ii*) The average number of customers queueing for service.

(*iii*) The probability of having more than 10 customers in the system.

(*iv*) The probability that the customer has to queue for more than 3 minutes.

12.4. Patients arrive at a clinic according to Poisson distribution at the rate of 30 patients per hour. Estimated time per patient is exponential with mean rate 20 per hour. If capacity of the clinic is unlimited, find the probability that an arriving patient will not wait.

12.5. Given an average arrival rate of 20 per hour, is it better for a customer to get service at a single channel with mean service rate of 2 customers or at one of two channels in parallel with mean service rate of 11 customers for each of the two channels? Assume that both queues are of M/M/s types.

[**Ans.** It is better to get service at two channels in parallel]

12.6. Find the probability that there are no customers in the system given that :

(*a*) Number of channels in parallel =3

(*b*) Mean arrival rate = 24 /hour

(*c*) Mean service rate in each channel= 10/hour

Compare the average time that a customer is in the system for the following 2 systems:

(*i*) 5 channels in parallel with a mean service rate of 10/hour.

(*ii*) 1 channel with a mean service rate of 30/hour.

[**Ans.** $P_0 = 0.019$, (*i*) 6 minutes, (*ii*)10 minutes]

12.7. A barber shop has two barbers and three chairs for waiting customers. Assume that customers arrive in a Poisson fashion at a rate of 5 per hour and that each barber services customers according to an exponential distribution with mean of 15 minutes. Further, if a customer arrives and there are no empty chairs in the shop he will leave. Find the steady state probabilities. What is the probability that the shop is empty ? What is the expected number of customers in the shop.

[**Ans.** $P_0 = \frac{1}{n!}$ $(0.28)\,(1.2)^n, 0 \le n < 2 = (0.6)^n\,(0.28), 2 \le n \le 3, P_0 = 0.28, L_s = 3$ customers approx.]

12.8. What is Queuing theory? What are its limitations?

12.9. Define a queue with an example. What are types of queues?

12.10. What are characteristics of queuing system?

12.11. Write short note on Queuing theory. **(UPTU: 2004-2005)**

12.12. What is a "Queuing Model"? What are its objectives? Giving basic structure, explain $(M/M/1)$ model. **(UPTU: 2003-2004)**

12.13. Write important assumptions of queuing models.

12.14. Explain following terms:

(*i*) Queue discipline

(*ii*) Capacity of the system

(*iii*) Holding time

(*iv*) Balking

(*v*) Jockeying

(*vi*) Queue length

(*vii*) Traffic intensity

(*viii*) Steady and transient state

(*ix*) Utilization factor

12.15. Explain the single channel and multi channel queuing models.

12.16. What is the waiting line problem? Give three business applications of waiting line theory.

12.17. Explain the $M/M/1$ queue and write its applications.

12.18. Draw a diagram showing the physical layout of a queuing system with multi-server, multi-channel service facility.

12.19. Write short note on Birth and Death process.

12.20. Describe the different types of costs involved in a queuing system.

12.21. Explain the model $M/M/S$ in case of first come first served basis with illustrations.

12.22. Workers come to a tool store room to enquire about the special tools (required by them) for a particular job. The average time between the arrivals is 60 seconds and the arrivals are assumed to be in Poisson distribution. The average service time is 40 seconds. Determine:

(*a*) average queue length,

(*b*) average length of non-empty queue,

(*c*) average number of workers in the system including the workers being attended.

(*d*) mean waiting time of an arrival.

(*e*) average waiting time of an arrival (workers)who waits

12.23. In a bank cheques are cashed at a single 'teller' counter. Customers arrive at the counter in a Poisson manner at an average rate of 30 customers per hour. The teller takes on an average a minute and a half to cash a cheque. The service time has been shown to be exponentially distributed.

(*a*) Calculate the percentage of time the teller is busy.

(*b*) Calculate the average time a customer is expected to wait.

12.24. Telephone users arrive at a booth following a Poisson distribution with an average time of 5 minutes, one arrival and the next. The time taken for a telephone calls is on an average 3 minutes and it follows an exponential distribution. What is the probability that the booth is busy ? How many more booths should be established to reduce the waiting time to less than or equal to half of the present waiting time ?

12.25. At a computerised Rail Reservation window, the customer arrivals are considered to be Poisson with an average inter-arrival time of ten minutes. The railway clerk's time for serving the customer —(examination of slip, making reservation, collecting cash and issuing ticket) may be assumed to be distributed negative exponentially with an average of 6 minutes.

(*a*) What is the chance that the customer will straightway get a ticket without having to wait in line ?

(*b*) For what part of the time will the railway clerk be busy ?

(*c*) Find the average queue length, average number of customers in the system, average waiting of a customer and the average time a customer spends in the system.

(*d*) What is the chance that the waiting time of an arrival exceeds 45 minutes and an additional clerk is required. Determine how large the new arrival rate should be to justify a second clerk.

12.26. The arrival rate of customers at a banking counter follows Poisson distribution with a mean of 30 per hour. The service rate of the counter clerk also follows Poisson distribution with a mean of 45 per hour.

(*a*) What is the probability of having 0 customer in the system (p_0)?

(*b*) What is the probability of having 8 customers in the system (p_8)?

(*c*) What is the probability of having 12 customers in the system (p_{12})?

(*d*) Find L_s, L_q, W_s and W_q.

12.27. In a harbour, ships arrive with a mean rate of 24 per week. The harbour has 3 docks to handle unloading and loading of ships. The service rate of individual dock is 12 per week. The arrival rate and the service rate follows Poisson distribution. At any point of time, the maximum number of ships permitted in the harbour is 8. Find p_0. L_q, L_s, W_q and W_s.

OBJECTIVE TYPE QUESTIONS

12.1. A queue is formed when
  (*a*) Customers wait for services
  (*b*) Service facilities stand idle and wait for customers
  (*c*) Either *a* or *b*
  (*d*) Both (*e*) None

12.2. A model has the following characteristics
  (*a*) Input process or Arrival pattern
  (*b*) Service mechanism or service pattern
  (*c*) Queue discipline (*d*) Customer's behaviour
  (*e*) Maximum number of customers allowed in a system
  (*f*) Source of population (*g*) All of the above
  (*h*) None of the above

12.3. Customer behaviour in which he move from one queue to another in a multi server situation is
  (*a*) Balking (*b*) Reneging
  (*c*) Jockeying (*d*) Alternating

12.4. Priority queue discipline may be classified as
  (*a*) Pre-emptive and non-pre-emptive
  (*b*) Finite and infinite
  (*c*) Limited and unlimited (*d*) All of the above

12.5. Which of the following does not apply to the basic queuing model?
  (*a*) Exponentially distributed arrivals
  (*b*) Exponentially distributed service rate
  (*c*) Finite time horizon (*d*) The queue discipline is FIFO

12.6. Which of the following is not a key operating characteristic for a queuing system?
  (*a*) Utilization factor (*b*) Per cent idle time
  (*c*) Average time spent for waiting in system
  (*d*) None of the above

12.7. Which of the cost estimates and performance measures are not used for economic analysis of a queuing system?
  (*a*) Cost per server per unit time
  (*b*) Cost per unit of time for a customer waiting in the system
  (*c*) Average number of customers in the system
  (*d*) Average waiting time of customers in the system

12.8. Which symbol describes the arrival rate as Poisson distribution?
  (*a*) *D* (*b*) *M*
  (*c*) *G* (*d*) All of the above

Answers

12.1. (*c*) 12.2. (*g*) 12.3. (*c*) 12.4.(*a*) 12.5. (*c*) 12.6. (*d*) 12.7. (*d*) 12.8. (*b*)

❑❑❑

Chapter 13

Project Management (PERT and CPM)

13.1 INTRODUCTION

One of the most challenging tasks that any manager can take on is the management of a project that requires coordinating numerous activities throughout the organization. A vast detail must be considered in planning how to coordinate all these activities, in developing a realistic schedule, and then in monitoring the progress of the project.

There are two basic planning and control, operations research techniques, **PERT** (program evaluation and review technique) and **CPM** (critical path method), which are developed to help the project manager in carrying out these responsibilities. These techniques make heavy use of networks to help plan and display the coordination of all the activities.

PERT and CPM were independently developed in the late 1950s. Ever since, they have been among the most widely used OR techniques.

13.2 PROJECT AND PROJECT MANAGEMENT

A *project* is an interrelated set of activities that has a definite starting and ending point and results in the accomplishment of a unique, often major outcome. "Project management" is, therefore, the planning and control of events that, together, comprise the project. Project management aims to ensure the effective use of resources and delivery of the project objectives on time and within cost constraints.

Project management is the adept use of techniques and skills (hard and soft) in planning and controlling tasks and resources needed for the project, from both inside and outside of organisation, to achieve results.

The purpose of project management is to achieve successful project completion with the resources available. A successful project is one which:

- has been finished on time
- is within its cost budget
- performs to a technical/performance standard which satisfies the end user.

13.2.1. Project

A project is any sort of planned undertaking.

A project is a unique venture with a beginning and an end, conducted by people to meet established goals within parameters of cost, time and quality.

All of us have been involved in projects, whether, they be our personal projects or in business and industry.

Examples of typical projects are :

- Personal projects :
  - obtain an MBA
  - write a report
  - plan a wedding
  - plant a garden
  - build a house extension
- Industrial projects :
  - construct a plant
  - provide a gas supply to an industrial park
  - build a motorway
  - design a new vehicle
- Business projects :
  - develop a new course
  - develop a computer system
  - introduce a new product
  - prepare an annual report
  - set up a new office

Projects can be of any size and duration. They can be simple, like planning a party, or complex like launching space shuttle.

Generally projects are made up of :

- a defined beginning and a defined end.
- multiple activities which are performed to a plan,

Various definitions of Project are:

Therefore a project may be defined as a means of moving from a problem to a solution via a series of planned activities.

"A project is a temporary endeavour involving a connected sequence of activities and a range of resources, which is designed to achieve a specific and unique outcome and which operates within time, cost and quality constraints and which is often used to introduce change."

"A project is defined as a combination of inter related activities that must be executed in a certain order before the entire task can be completed. The activities are inter related in a logical sequence in the way that some activities cannot start until other are competed. An activity in a project is usually viewed as a job requiring time and recourses for its competition."

"A project is a temporary endeavour undertaken to create a "unique" product or service".

Typically all projects can be broken down into:

- separate *activities* (tasks/jobs) – where each activity has an associated duration or *completion time* (*i.e.* the time from the start of the activity to its finish)
- *precedence relationships* – which govern the order in which we may perform the activities, e.g. in a project concerned with building a house the activity "erect all four walls" must be finished before the activity "put roof on" can start.

and the problem is to bring all these activities together in a coherent fashion to complete the project.

13.2.2 Projects in Real-Life

- constructing houses, factories, shopping malls, athletic stadiums or arenas
- developing military weapons systems, aircraft, new ships
- launching satellite systems
- constructing oil pipelines
- developing and implementing new computer systems
- planning concert, football games, or basketball tournaments
- introducing new products into market
- building the Channel tunnel
- building the London Eye
- developing a new drug, etc.

13.2.3 Characteristics of a Project

- A project has a definite beginning and end.
- Projects consist of several activities.
- A unique in some way, one-time operational activity or effort, "non-repetitive" or "non-routine", e.g. building the very first Boeing Jumbo jet was a project - building them now is a repetitive / routine manufacturing process, not a project.
- Requires the completion of a large number of interrelated activities.
- Established to achieve specific objective.
- Resources, such as time and /or money, are limited.
- Typically has its own management structure.
- Need leadership

Two essential features are present in every project no matter how simple or complicated they are. In the first place, all projects must be *planned* out in advance if they are to be successfully executed. Secondly, the execution of the project must be *controlled* to ensure that the desired results are achieved.

13.2.4 Project Management

Project management is the combination of people, systems, and techniques required to coordinate the resources needed to complete projects within established goals.

- Project management is the application of a collection of tools and techniques to direct the use of diverse resources towards the accomplishment of a unique, complex, one time task within time, cost and quality constraints.
- Its origins lie in World War II, when the military authorities used the techniques of operational research to plan the optimum use of resources.
- One of these techniques was the use of networks to represent a system of related activities.

The most widely used and popular analytic techniques for planning scheduling and controlling of projects are Gantt Charts, Critical Path Method (CPM) and Programme Evaluation and Review Technique (PERT).

13.2.5 Project Management Process

Project management process consists of three phases: Planning, Scheduling and Controlling.

13.2.5.1 Project Planning

An important first step is to set the objectives and then define the project, breaking it down into a set of activities and related costs. It is probably too early to determine exact resource implications at this stage, but expected requirements for people, supplies and equipment should at least be estimated during the planning stage.

13.2.5.2 Project Scheduling

This phase is primarily concerned with attaching a timescale and sequence to the activities to be conducted showing the start and finish times for each activity as well as its relationship to other activities in the project. In addition, the schedule must pin-point the critical activities which require special attention if the project is to be completed on time. For the non-critical activities the schedule must show the amount of slack or float time which can be used. Materials and people needed at each stage of the project are determined and the time each is to take will be set.

A popular and easy to use technique for scheduling is the use of Gantt charts. Gantt charts, named after Henry L. Gantt, is a simple technique that can be used to attach a time scale and sequence to a project. Horizontal bars are drawn against a time scale for each project activity, the length of which represents the time taken to complete. Letters or symbols can also be added to the left of each bar to show which other activities need to be completed before that one can begin.

To construct a Gantt Chart the following steps are necessary:

1. Use the horizontal axis to represent time
2. Use the vertical axis to represent activities
3. Represent each activity by a horizontal bar of appropriate length

Take activity procedures into account by starting each activity bar to an appropriate point along the time axis after its preceding activities. Normally the start point for an activity is the earliest time that it could start after its preceding activities had finished.

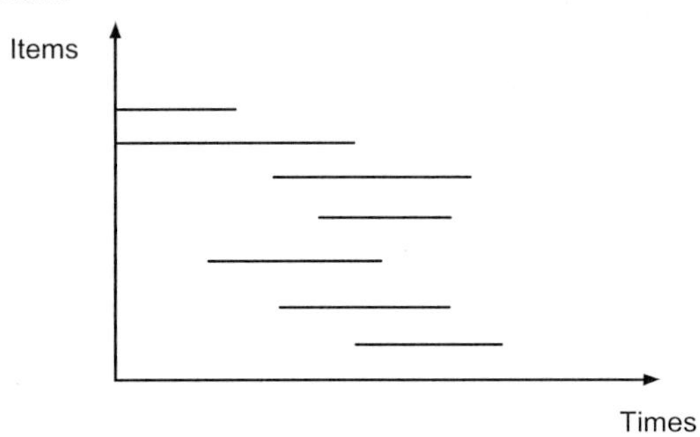

Fig. 13.1

However, where projects become complex, it becomes difficult to see relationships between activities by using a Gantt Chart. For more complex projects Network Analysis techniques are used.

13.2.5.3 Project Controlling

An approach needs to be developed for Project Implementation and Control. Control and monitoring procedures need to be put in place and appropriate information systems developed.

- Project team
  - made up of individuals from various areas and departments within a company
- Matrix organization
  - a team structure with members from functional areas, depending on skills required
- Project Manager
  - most important member of project team
- Scope statement
  - a document that provides an understanding, justification, and expected result of a project
- Statement of work
  - written description of objectives of a project
- Organizational Breakdown Structure
  - a chart that shows which organizational units are responsible for work items
- Responsibility Assignment Matrix
  - shows who is responsible for work in a project

Guidelines for Managing Projects

- Set a clear Goal
- Determine the Objectives
- Establish Checkpoints, Activities, Relationships, and Time estimates
- Create a Schedule
- Develop people individually and as a team
- Reinforce the commitment and excitement of people
- Inform everyone connected with the project
- Vitalise people by building agreements
- Empower yourself and others
- Risk approaching problems creatively

13.3 NETWORK ANALYSIS

Network analysis is the general name given to certain specific techniques which can be used for the planning, management and control of projects. It enables us to take a ***systematic quantitative structured approach*** to the problem of managing a project through to successful completion.

13.3.1 Introduction to PERT and CPM

The two most common and widely used project management techniques that can be classified under the title of Network Analysis are Program Evaluation and review Technique (PERT) and Critical Path Method (CPM). Both were developed in the 1950's to help managers schedule, monitor and control large and complex projects. CPM was first used in 1957 to assist in the development and building of chemical plants within the DuPont corporation by M.R. Walker of E.I.Du Pont de Nemours & Co. and J.E.Kelly of Remington Rand, circa. Independently developed, PERT was introduced in 1958 following research within the Special Projects Office of the US Navy. It was initially used to plan and control the Polaris missile programme which involved the coordination of thousands of contractors. The use of PERT in this case was reported to have cut eighteen months off the overall time to completion.

13.3.2 The PERT/CPM Procedure

There are six stages common to both PERT and CPM:

1. Define the project and specify all activities or tasks The Project (made up of several tasks) should have only a single start activity and a single finish activity.
2. Develop the relationships amongst activities. Decide which activities must precede and which must follow others.
3. Draw network to connect all activities. Each Activity should have unique event numbers. Dummy arrows are used where required to avoid giving the same numbering to two activities.
4. Assign time and /or costs to each activity.
5. Calculate the longest time path through the network; this is called the "critical path".
6. Use network to plan, schedule, monitor and control the project.

Finding the critical path (step 5) is a major task in controlling a project. Activities on the critical path represent tasks which, if performed behind schedule, will delay the whole project. Managers can derive flexibility by identifying the non-critical activities and re-planning, re-scheduling and reallocating resources such as manpower and finances within identified boundaries.

PERT and CPM differ slightly in their terminology and in network construction. However their objectives are the same and, furthermore, their project analysis techniques are very similar.

PERT and CPM can help to answer the following questions for projects with thousands of activities and events, both at the beginning of the project and once it is underway:

- When will the project be completed?
- What are the critical activities (i.e., the tasks which, if delayed, will effect time for overall completion)?
- Which activities are non-critical and can run late without delaying project completion time?
- What is the probability of the project being completed by a specific date?
- At any particular time, is the project on schedule?
- At any particular time, is the money spent equal to, less than or greater than the budgeted amount?
- Are there enough resources left to complete the project on time?
- If the project is to be completed in a shorter time, what is the least cost means to accomplish this and what are the cost consequences?

13.3.3 Terminology and Definitions

Network diagram represents clearly and concisely the entire project—enabling us to get a pictorial overview of what is to happen, the various activities and their relationship one to another using nodes and arrows.

A ***schedule*** allocates resources to accomplish the activities within a timeframe. The schedule sets priorities, start times and finish times.

An ***activity*** or task is the smallest unit of work effort within the project and consumes both time and resources (effort, money etc.) required to perform a part of the work which are under the control of the project manager.

It lies between two events – preceding and succeeding. The head of the arrow indicates where task ends and the tail where the task begins. An activity in the network can be represented by a number of ways: (*i*) by numbers of its head and tail events (*i.e.*, 10-20 etc.), and (*ii*) by a letter code (*i.e.*, A, B etc.). All those activities, which must be completed before the start of activity under consideration, are called its predecessor activities. All those activities, which have to follow the activity under consideration, are called its successor activities.

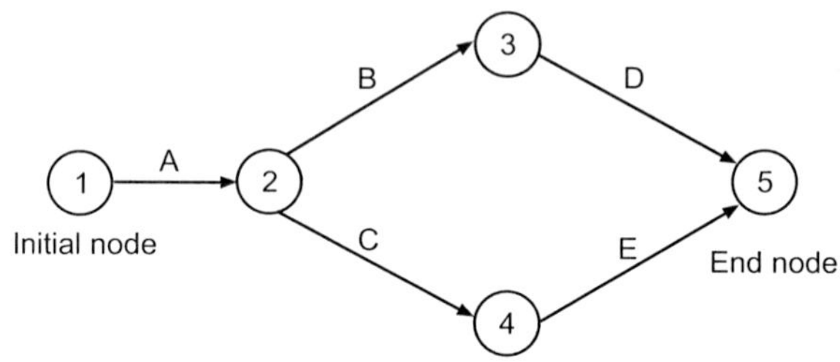

Fig. 13.2

Event : An event represents a point in time that signifies the completion of some activities and the beginning of new ones, e.g. wall built, foundations dug etc. Represented by a circle (node), where one or more activities start and / or finish. It consumes no time and resources. An event is not complete till all the activities ending into it are completed.

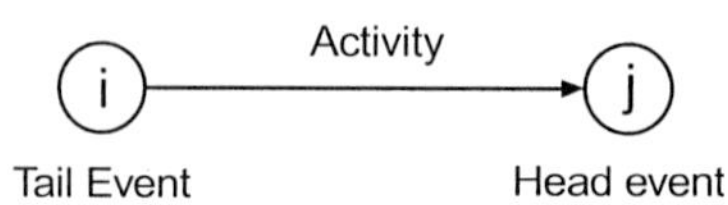

Fig. 13.3

Merge event : When more than one activity joins at event, it is known as merge event.

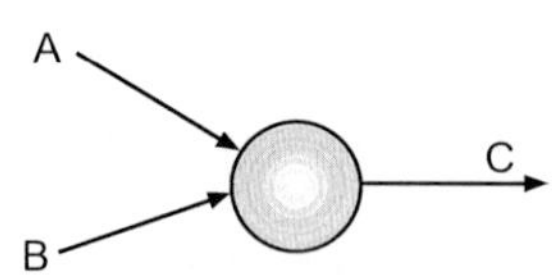

Fig. 13.4

Burst event : When more than one activity leaves an event, it is known as burst event.

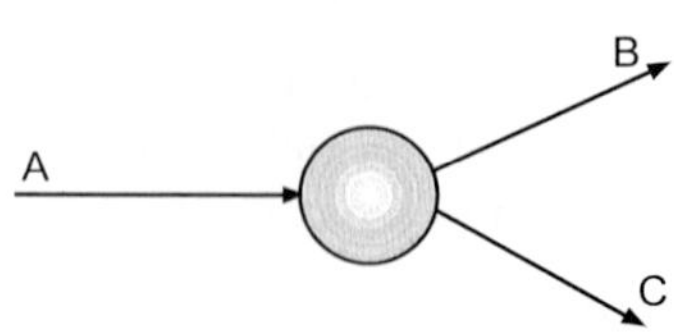

Fig. 13.5

Merge and burst event :

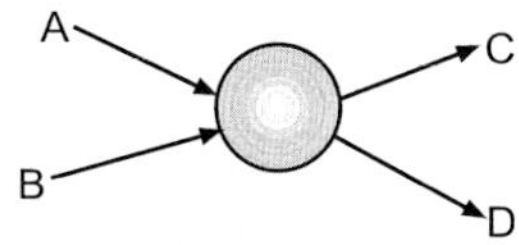

Fig. 13.6

Dummy Activity : It is an activity which does not consume time or resources. It indicates only precedence relationships. It is represented by a dashed arrow. Dummy activity is used to make activities with common starting and finishing events distinguishable.

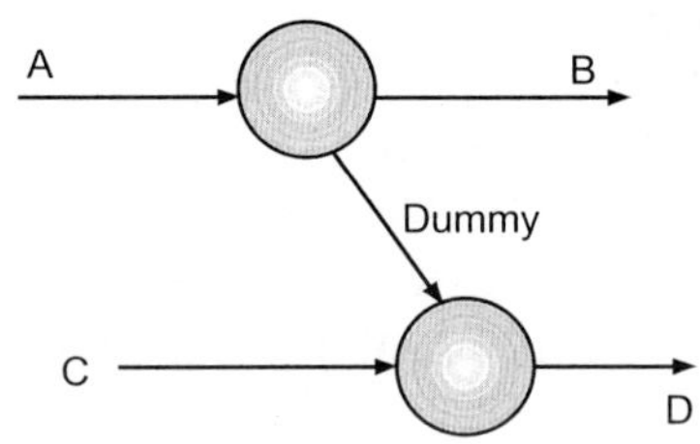

Fig. 13.7

A must finish before B can start and both A and C must finist before D can start.

Path : A connected sequence of activities leading from the starting event to the ending event.

Critical path : It gives shortest time in which the whole project can be competed.

13.3.4 Rules for Construction of Network Diagram

Rule 1. Each activity is represented by one and only one arrow in the network.

Rule 2. No two activities can be identified by the same end events

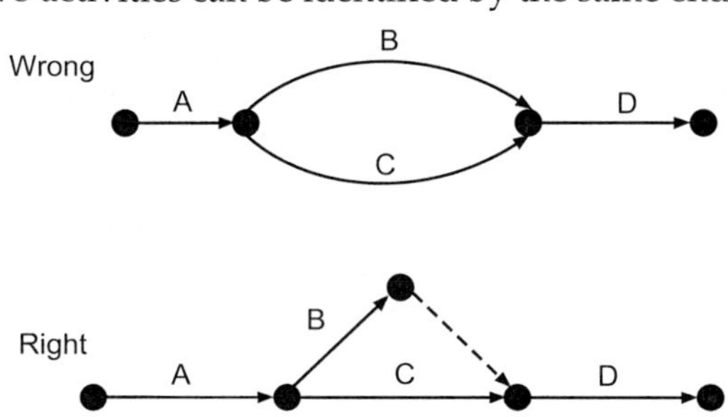

Fig. 13.8

Rule 3. Precedence relationships among all activities must always be maintained.

Rule 4. Dummy activities can be used to maintain precedence relationships only when actually required. Their use should be minimized in the network diagram.

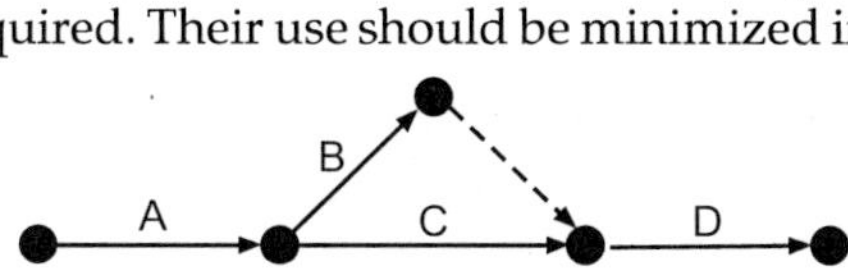

Fig. 13.9

Rule 5. Looping among the activities must be avoided.

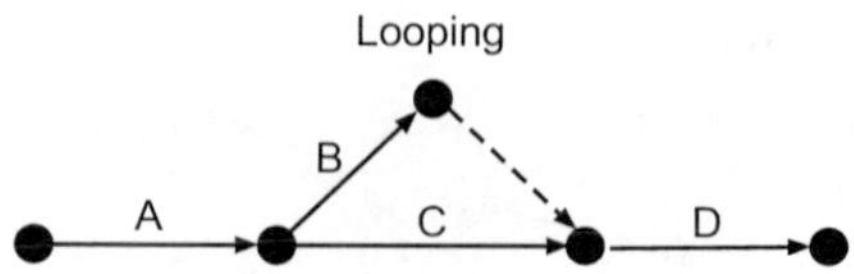

Fig. 13.10

Example 13.1 Draw a network diagram of opening a new office. Assume that following events take place in some sequence.

Event 1: Location of site started

Event 2: Location of site completed

Event 3: Building for office selected

Event 4: Cleaning of the office building started

Event 5: Interior decorator starts work

Event 6: Interior decorator finishes work

Event 7: Opening of new office advertised or announced

Event 8: List of invitees for the opening day prepared

Event 9: Invitations sent

Event 10: Office opened

Solution: Network diagram of given office problem

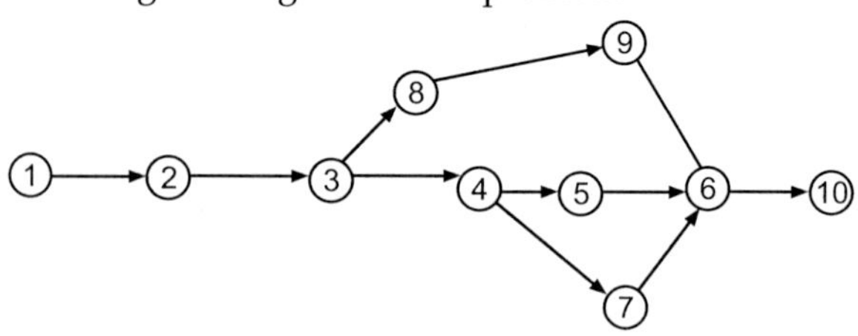

Fig. 13.11

Example 13.2 Suppose that we are going to carry out a minor redesign of a product and its associated packaging. We intend to test market this redesigned product and then revise it in the light of the test market results, finally presenting the results to the Board of the company. ***How long will it take to complete this project?***

The following list of separate activities is identified together with their associated completion times (assumed known with certainty).

Activity number		Completion time (weeks)
1.	Redesign product	6
2.	Redesign packaging	2
3.	Order and receive components for redesigned product	3
4.	Order and receive material for redesigned packaging	2
5.	Assemble products	4
6.	Make up packaging	1
7.	Package redesigned product	1
8.	Test market redesigned product	6
9.	Revise redesigned product	3
10.	Revise redesigned packaging	1
11.	Present results to the Board	1

Solution: We must prepare a list of precedence relationships indicating activities which, because of the logic of the situation, must be finished before other activities can start e.g. in the above list activity number 1 must be finished before activity number 3 can start.

For example it is plain that activity 1 must be finished before activity 9 can start but these two activities can hardly be said to have an immediate relationship (since many other activities after activity 1 need to be finished before we can start activity 9).

Activities 8 and 9 would be examples of activities that have an immediate relationship (activity 8 must be finished before activity 9 can start).

The list of immediate precedence relationships.

Activity number		Activity number	
1.	must be finished before	3	can start
2.		4	
3.		5	
4.		6	
5.		7	
6.		8	
7.		9	
8.		10	
9, 10		11	

Note here that:

- activities 1 and 2 do not appear in the right hand column of the above table, this is because there are no activities which must finish before they can start, *i.e.*, both activities 1 and 2 can start immediately.
- two activities (5 and 6) must be finished before activity 7 can start.
- it is plain from this table that non-immediate precedence relationships (e.g. "activity 1 must be finished before activity 9 can start") need not be included in the list since they can be deduced from the relationships already in the list.

Once we have completed our list of activities and our list of precedence relationships we combine them into a network diagram:

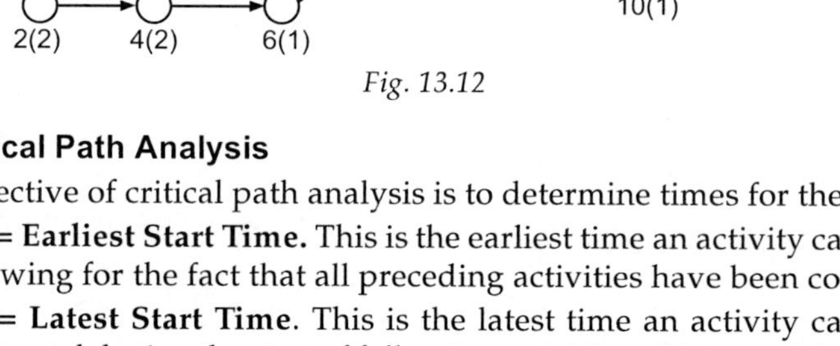

Fig. 13.12

13.3.5 Critical Path Analysis

The objective of critical path analysis is to determine times for the following:

- **ES = Earliest Start Time.** This is the earliest time an activity can be started, allowing for the fact that all preceding activities have been completed.
- **LS = Latest Start Time**. This is the latest time an activity can be started without delaying the start of following activities which would put the entire project behind schedule.

- **EF = Earliest Finish Time**. The earliest time an activity can be finished.
- **LF = Latest Finish Time**. The latest time that an activity can finish for the project to remain on schedule.
- **S = Activity Slack Time.** The amount of slippage in activity start or duration time which can be tolerated without delaying the project as a whole.

If ES and LS for any activity are known, then one can calculate values for the other three times as follows :

$$\text{EF} = \text{ES} + t \text{ (duration of activity)}$$
$$\text{LF} = \text{LS} + t$$
$$\text{S} = \text{LS} - \text{ES} \text{ or } \text{S} = \text{LF} - \text{EF}$$

Float: Float is the flexibility available in any activity which can be absorbed either by delaying that activity or by enlarging its duration. There is no difference between float and slack. FLOAT refers to an activity and SLACK refers to an event.

Total float: The total float of an activity represents the amount of time by which activity can be delayed without delaying the project completion date.

Total float = LF – EF
= LS – ES

Free float: Free float is that portion of the total float within an activity which can be manipulated without affecting the float of subsequent activities.

Free float = Total float – Head slack

Independence float: This is the amount of time activities are completed as Late as possible and all succeeding activities started as early as possible.

Independence float = Free float – Tail slack

Interfering float: Interfering float is that part of the total float which causes a reduction in the float of the successor activities.

Interfering float = LF – ES + 1 Where +1 indicates succeeding activity

When an activity has zero Total float, Free float will also be zero.

An activity in a network diagram is said to be critical, if the delay in its start will further delay the project completion time. A non critical activity is the one which allows some slack so that the start time of the activity may be advanced or delayed within certain limits without affecting the completion time of the project.

Analysis of the project normally involves:

1. *Determining the Critical Path:* The critical path is the group of activities in the project that have a zero slack time. This path of activities is critical because a delay in any activity along it would delay the project as a whole.
2. Calculating the total project completion time, T. This is done by adding the activity times of those activities on the critical path.

The steps in critical path analysis are as follows:

(*a*) Determine ES and EF values for all activities in the project: the **Forward Pass** through the network.

(*b*) Calculate LS and LF values for all activities by conducting a **Backward Pass** through the network.

(*c*) Identify the **critical path** which will be those activities with zero slack (*i.e.*: ES=LS and EF=LF).

Example 13.3 In the project network give in figure 13.13 below, find the critical path and the project duration.

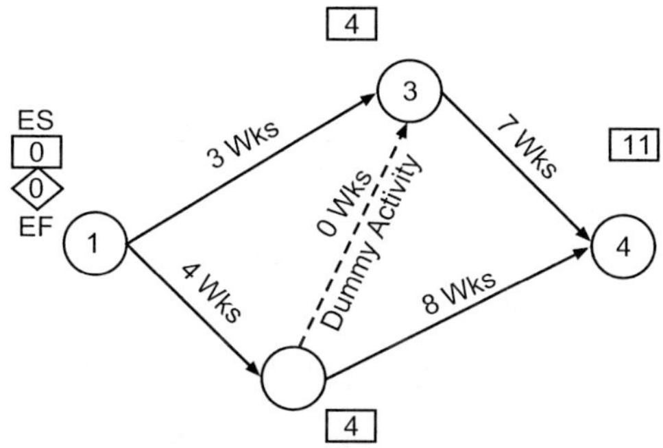

Fig. 13.13.

Solution: At Event 3, we have to evaluate two predecessor activities—Activity 1-3 and Activity 2-3, both of which are predecessor activities. Activity 1-3 gives us an Earliest Start of 3 weeks at Event 3. However, Activity 2-3 also has to be completed before Event 3 can begin. Along this route, the Earliest Start would be 4+0=4. The rule is to take the longer (bigger) of the two Earliest Starts. So the Earliest Start at event 3 is 4.

Similarly, at Event 4, we find we have to evaluate two predecessor activities—Activity 2-4 and Activity 3-4. Along Activity 2-4, the Earliest Start at Event 4 would be 10 weeks, but along Activity 3-4, the Earliest Start at Event 4 would be 11 weeks. Since 11 weeks is larger than 10 weeks, we select it as the Earliest Start at Event 4.**We have now found the longest path through the network. It will take 11 weeks along activities 1-2, 2-3 and 3-4. This is the Critical Path.**

The Backward Pass - Latest Finish Time Rule

To make the Backward Pass, we begin at the final event and work backwards to the first event.

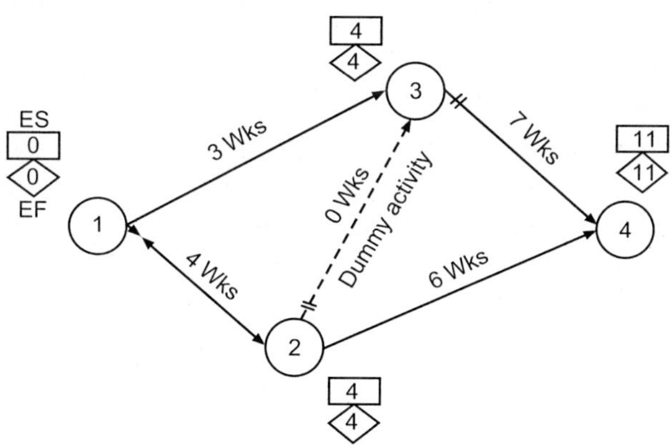

Fig. 13.14

At Event 3 there is only one activity, Activity 3-4 in the backward pass, and we find that the value is 11-7 = 4 weeks. However at Event 2 we have to evaluate 2 activities, 2-3 and 2-4. We find that the backward pass through 2-4 gives us a value of 11-6 = 5 while 2-3 gives us 4-0 = 4. We take the **smaller value** of 4 on the backward pass.

Tabulation and Analysis of Activities

We are now ready to tabulate the various events and calculate the Earliest and Latest Start and Finish times. We are also now ready to compute the SLACK or TOTAL FLOAT, which is defined as the difference between the Latest Start and Earliest Start.

Event	Duration (Weeks)	Earliest Start	Earliest Finish	Latest Start	Latest Finish	Total Float
1-2	4	0	4	0	4	0
2-3	0	4	4	4	4	0
3-4	7	4	11	4	11	0
1-3	3	0	3	1	4	1
2-4	6	4	10	5	11	1

The Earliest Start is the value in the rectangle near the tail of each activity

- The Earliest Finish is = Earliest Start + Duration
- The Latest Finish is the value in the diamond at the head of each activity
- The Latest Start is = Latest Finish - Duration

13.3.6 PERT and Activity Time Estimation - Probabilistic Approach

So far we have discussed about projects, where there is high certainty about the outcomes of activities. In other words, the cause-effect logic is well known. This is particularly the case in engineering projects.

However, in Research & Development projects, or in Social Projects which are defined as "Process Projects", where learning is an important outcome, the cause-effect relationship is not so well established. In such situations, the PERT approach is useful, because it can accommodate the variation in event completion times, based on an expert's estimates.

The major distinguishing difference between PERT and CPM is the use of three time estimates for each activity in the PERT technique, with CPM using only one time for each activity.

The three time estimates specified for each activity in PERT are:

(i) the optimistic time;

(ii) the most probable time; and

(iii) the pessimistic time.

The optimistic, most likely and pessimistic time estimates are used to calculate an expected activity completion time which, because of the skewed nature of the beta distribution, is marginally greater than the most likely time estimate. In addition, the three time estimates can be used to calculate the variance for each activity. The formulae used are as follows:

$$t_e = \frac{t_0 + 4\,t_m + t_p}{6}$$

where t_e is the expected time, t_o is the optimistic time, t_m is the most probable activity time and t_p is the pessimistic time.

It is not necessary to go into the theory behind the formula. It is enough to know that the weights are based on an approximation of the Beta distribution.

The Standard Deviation, which is a good measure of the variability of each activity, is calculated by the rather simplified formula:

$$S = \frac{t_p - t_o}{6}$$

The Variance is the square of the Standard Deviation.

$$v = \left(\frac{t_p - t_o}{6}\right)^2$$

where, v - variance of task completion time

Knowing the details of a project, its network and values for its activity times (t) and their variances (v) a complete PERT analysis can be carried out. This includes the determination of the ES, EF, LS, LF and S for each activity as well as identifying the critical path, the project completion time (T) and the variance (V) for the entire project.

Normally when using PERT, the expected times (t) are calculated first from the three values of activity time estimates, and it is these values of 't' that are then used exactly as before in CPM. The variance values are calculated for the various activity times and the variance of the total project completion time (i.e. the sum of the activity expected times of those activities on the critical path) is the sum of the variances of the activities lying on that critical path.

Once the expected completion time and variance (T and V) have been determined, the probability that a project will be completed by a specific date can be assessed. The assumption is usually made that the distribution of completion dates follows that of a normal distribution curve.

Example 13.4 In figure 13.13 below, three activities A, B and C are connected in series, (*i.e.* from a path). Their time estimates t_o-t_m-t_p are given along the activity arrow. Calculate the expected time of path 1-2-3-4.

Solution:

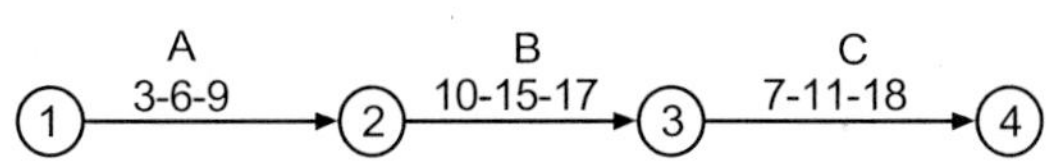

Fig. 13.15

The expected time of the path 1-2-3-4 is

$$t_E^{1-2-3-4} = t_E^{1-2} + t_E^{2-3} + t_E^{3-4}$$

$$= \frac{3 + 4 * 6 + 9}{6} + \frac{10 + 4 * 15 + 17}{6} + \frac{7 + 4 * 11 + 18}{6}$$

$$= 6 + 14.5 + 11.5 = 32$$

t_E could also be computed as

$$t_E^{1-2-3-4} = \frac{\Sigma t_o + 4\,\Sigma t_m + \Sigma t_p}{6}$$

$$= \frac{(3 + 10 + 7) + 4(6 + 15 + 11) + (9 + 17 + 18)}{6} = 32$$

The variance for the path is given by

$$V^{1-2-3-4} = \left(\frac{9-3}{6}\right)^2 + \left(\frac{17-10}{6}\right)^2 + \left(\frac{18-7}{6}\right)^2$$

$$= 1 + (1.17)^2 + (1.83)^2 = 5.72$$

As the length of the path, that is the number of activities connected in series increases, the variance of the path and hence the uncertainty of meeting the expected time increases.

Example: 13.5 In the Network of figure 13.16 below, the PERT time estimates of the activities are written along the activity arrow in the order t_o-t_m-t_p. Calculate the expected time and variance for each activity. Also calculate the expected duration and standard deviation for the following path of the Network

(*a*) 10-20-50-80-90

(*b*) 10-30-50-70-90

(*c*) 10-40-60-80-90

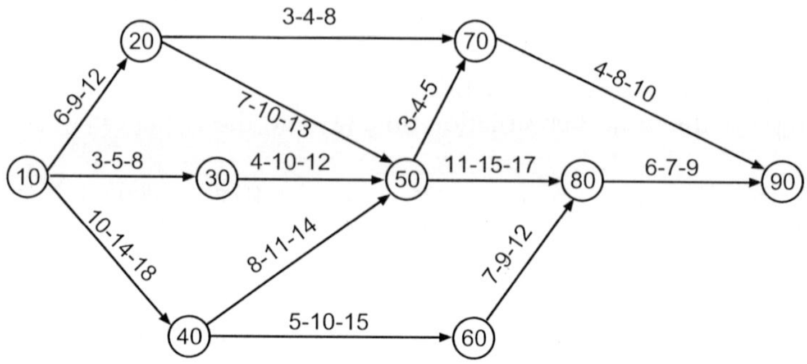

Fig. 13.16

Solution: The Computation of expected times and variance for different activities are carried in a table given below

Activity		Time Estimates			Expected Time	Variance
i	j	t_o	t_m	t_p	t_e	σ^2
10	20	6	9	12	9.00	1.00
10	30	3	5	9	5.33	1.00
10	40	10	14	18	14.00	1.78
20	50	7	10	13	10.00	1.00
20	70	3	4	8	4.5	0.69
30	50	4	10	12	9.33	1.78
40	50	8	11	14	11.00	1.00
40	60	5	10	12	10.00	2.78
50	70	3	4	5	4.00	0.11
50	80	11	15	17	14.67	1.00
60	80	4	8	10	9.17	0.69
70	90	6	7	9	7.67	1.00
80	90	6	9	12	7.17	0.25

(*a*) Expected duration and variance of the path 10 -25 - 50 - 80 - 90

Expected time $t_E = t_E^{10-20} + t_E^{20-50} + t_E^{50-80} + t_E^{80-90}$

$= 9.00 + 10.00 + 14.67 + 7.17 = 40.84$

Variance $= V^{10-20} + V^{20-50} + V^{50-80} + V^{80-90}$

$= 1.00 + 1.00 + 1.00 + 2.5 = 3.25$

Standard deviation $\sigma = \sqrt{3.25} = 1.803$

(*b*) Expected duration of path : 10-30-50-70-90

$t_E = t_E^{10-30} + t_E^{30-50} + t_E^{50-70} + t_E^{70-90}$

$= 5.33 + 9.33 + 4.00 + 7.67 = 26.33$

Variance of path : 10-30-50-70-90

$$V = 1.00 + 1.78 + 0.11 + 1.00 = 3.89$$

Standard deviation $\sigma = \sqrt{3.89} = 1.972$

(*c*) Expected duration of path 10-40-60-80-90

$$t_E = t_E^{10-40} + t_E^{40-60} + t_E^{60-80} + t_E^{80-90}$$
$$= 14.00 + 10.00 + 9.17 + 7.17 = 40.34$$

Variance, $V = 1.78 + 2.78 + 0.69 + 0.25 = 5.5$

$$\sigma = \sqrt{5.5} = 2.345$$

Example 13.6 A Social Project manager is faced with a project with the following activities:

Activity	Activity - Description	Duration
1-2	Social Work Team to live in Village	5 Weeks
1-3	Social Research Team to do survey	12 Weeks
3-4	Analyse results of survey	5 Weeks
2-4	Establish Mother & Child Health Program	14 Weeks
3-5	Establish Rural Credit Programme	15 Weeks
4-5	Carry out Immunisation of Under Fives	4 Weeks

- Draw the arrow diagram, using the helpful numbering of the activities, which suggests the following logic :
- Unless the Social Work team lives in the village, the Mother and Child Health Programme cannot be started due to ignorance and superstition of the villagers
- The Analysis of the survey can obviously be done only after the survey is complete.
- Until rural survey is done, the Rural Credit Programme cannot be started
- Unless Mother and Child Programme is established, the Immunisation of Under Fives cannot be started

  — Calculate the Earliest and Latlest Event Times

  — Tabulate and Analyse the Activities

Solution: In Social Project, the Project Manager is now not so certain that each activity will be completed on the basis of the single estimate he gave. There are many assumptions involved in each estimate, and these assumptions are illustrated in the three-time estimate he would prefer to give to each activity.

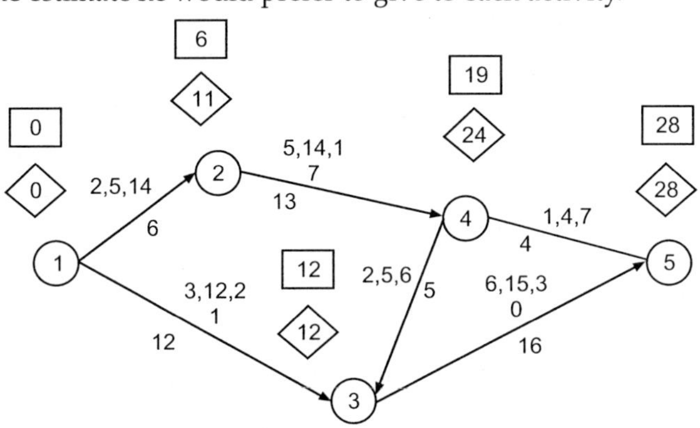

Fig. 13.17.

In Activity 1-3, the time estimates are to = 3, t_m = 12 and t_p = 21. Using PERT formula, we get:

$$t_e = \frac{3 + (4 \times 12) + 21}{6} = \frac{72}{6} = 12$$

$$s_1 = \frac{(21 - 3)}{6} = \frac{18}{6} = 3$$

We calculate the PERT event times and other details as below for each activity:

Event	t_o	t_m	t_p	t_e	ES	EF	LS	LF	TF	s.d.	Var.
1-3	3	12	21	12	0	12	0	12	0	3	9
3-5	6	15	30	16	12	28	12	28	0	4	16
1-2	2	5	14	6	0	6	5	11	5	2	4
2-4	5	14	17	13	6	19	11	24	5	2	4
3-4	2	5	8	5	12	17	19	24	7	1	1
4-5	1	4	7	4	19	23	24	28	5	1	1

Expected Length of a Project

PERT assumes that the expected length of a project (or a sequence of independent activities) is simply the sum of their separate expected lengths. Thus the summation of all the te's along the critical path gives us the length of the project.

Similarly the variance of a sum of independent activity times is equal to the sum of their individual variances.

In our example, the sum of the variance of the activity times along the critical path, VT is found to be equal to (9+16) = 25.

The square root VT gives us the standard deviation of the project length. Thus, $S_T = \sqrt{25} = 5$.

The higher the standard deviation, the greater the uncertainty that the project will be completed on the due date.

Although the t_e's are randomly distributed, the average or expected project length Te approximately follows a Normal Distribution.

A random variable drawn from a Normal Distribution has 0.68 probability of falling within one standard deviation of the distribution average. Therefore, there is a 68% chance that the actual project duration will be within one standard deviation, ST of the estimated average length of the project, t_e.

In our case, the t_e = (12+16) = 28 weeks and the S_T = 5 weeks. Assuming te to be normally distributed, we can state that there is a probability of 0.68 that the project will be completed within 28 ±5 weeks, which is to say, between 23 and 33 weeks.

Since it is known that just over 95% (.954) of the area under a Normal Distribution falls within two standard deviations, we can state that the probability that the project will be completed within 28 ± 10 is very high at 0.95.

Probability of Project Completion by Due Date

Now, although the project is estimated to be completed within 28 weeks (t_e = 28), Project Director would like to know what is the probability that the project might be completed within 25 weeks (*i.e.* Due Date or D = 25).

For this calculation, we use the formula for calculating Z, the number of standard deviations that is away from t_e.

$$Z = \frac{D - t_e}{S_t} = \frac{25 - 28}{5} = \frac{-3}{5} = -0.6$$

By looking at the following extract from a standard normal table, we see that the probability associated with a Z of -0.6 is 0.274. This means that the chance of the project being completed within 25 weeks, instead of the expected 28 weeks is about 2 out of 7. Not very encouraging.

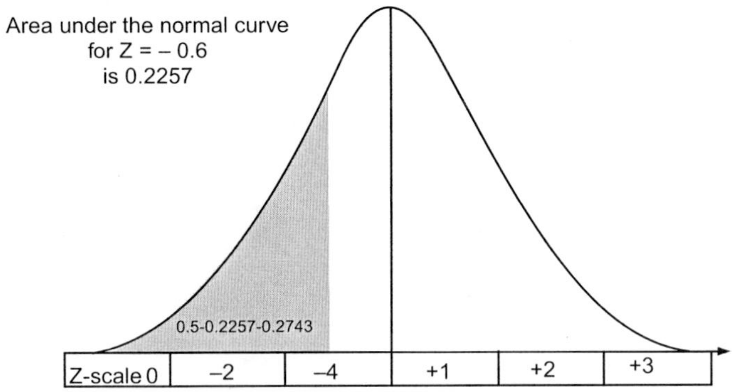

Fig. 13.18

On the other hand, the probability that the project will be completed within 33 weeks is calculated as follows:

$$Z = \frac{D - t_e}{S_t} = \frac{33 - 28}{5} = \frac{5}{5} = 1$$

The probability associated with Z = +1 is 0.84134. This is a strong probability, and indicates that the odds are 16 to 3 that the project will be completed by the due date.

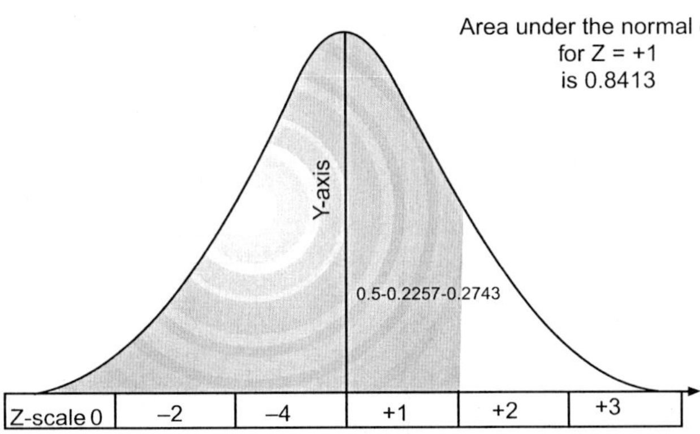

Fig. 13.19

If the probability of an event is p, the odds for its occurrence are a to b, where:

$$\frac{a}{b} = \frac{p}{1-p} = \frac{0.841434}{0.15866} \approx \frac{16}{3}$$

Example 13.7 Calculate the expected time and variance for each activity. Also calculate the expected duration and standard deviation for the critical path of the Network.

Immediate Activity	*Optimistic Predecessor.*	*Most Likely*		*Pessimistic Time (Hr.)*
		Time (Hr.)	*Time (Hr.)*	
A	—	4	6	8
B	—	1	4.5	5
C	A	3	3	3
D	A	4	5	6
E	A	0.5	1	1.5
F	B,C	3	4	5
G	B,C	1	1.5	5
H	E,F	5	6	7
I	E,F	2	5	8
J	D,H	2.5	2.7	4.5
K	G,I	3	5	7

Solution:

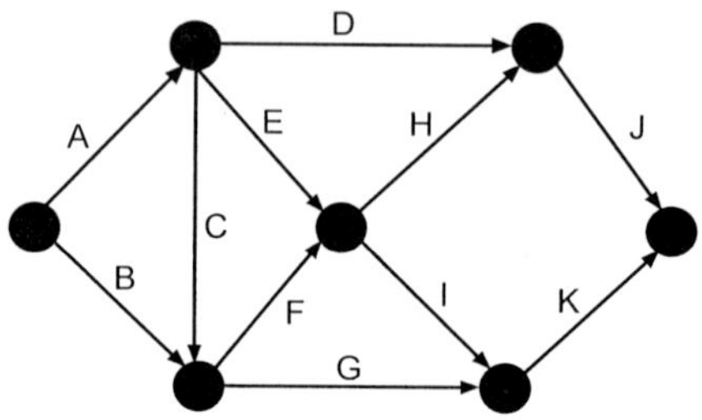

Fig. 13.20

Activity	Expected Time	Variance
A	6	4/9
B	4	4/9
C	3	0
D	5	1/9
E	1	1/36
F	4	1/9
G	2	4/9
H	6	1/9
I	5	1
J	3	1/9
K	5	4/9

Activity	ES	EF	LS	LF	Slack
A	0	6	0	6	0 *critical
B	0	4	5	9	5
C	6	9	6	9	0*
D	6	11	15	20	9
E	6	7	12	13	6
F	9	13	9	13	0*
G	9	11	16	18	7
H	13	19	14	20	1
I	13	18	13	18	0*
J	19	22	20	23	1
K	18	23	18	23	0*

$$V_{path} = VA + VC + VF + VI + VK$$
$$= \frac{4}{9} + 0 + \frac{1}{9} + 1 + \frac{4}{9} = 2$$
$$\sigma_{path} = 1.414$$
$$z = \sigma = \frac{24 - 23}{1.414} = .71$$

From the Standard Normal Distribution table:

$$P(z \leq .71) = .5 + .2612 = .7612$$

Example 13.8 For the activities of the Klone Computer project, rather than giving precise estimates of activity completion times, the company has supplied three time estimates:

Activity	Optimistic (a)	Most Likely (m)	Pessimistic (b)
A	76	86	120
B	12	15	18
C	4	5	6
D	15	18	33
E	18	21	24
F	16	26	30
G	10	13	22
H	24	28	32
I	22	27	50
J	38	43	60

Find:

1. The probability that the project will be completed within 194 days.
2. A reasonable interval estimate of the number of days to complete the project.
3. The probability that the project will be completed within 180 days.
4. The probability that the project will take longer than 210 days.
5. An upper limit for the number of days within which it can be virtually sure that the project will be completed.

Solution:

The mean, variance, and standard deviation for activity A can be found as follows:

$$\mu_A = \frac{(a + 4m + b)}{6}$$
$$= \frac{[76 + 4(86) + 120]}{6} = 90$$
$$\sigma_A = \frac{(b - a)}{6} = \frac{(120 - 76)}{6} = 7.33$$
$$\sigma_A^2 = (7.33)^2 = 53.78$$

Similar calculations for the other activities of the project give us the following table:

Activity	(μ)	(σ)	(σ^2)
A	90	7.33	53.78
B	15	1.00	1.00
C	5	.33	.11
D	20	3.00	9.00
E	21	1.00	1.00
F	25	2.33	5.44
G	14	2.00	4.00
H	28	1.33	1.78
I	30	4.67	21.78
J	45	3.67	13.44

The critical path is A-F-G-D-J.

Given this critical path, the ***expected*** completion time of the project (μ) is equal to:

$$\begin{aligned}\mu &= \mu_A + \mu_F + \mu_G + \mu_D + \mu_J \\ &= 90 + 25 + 14 + 20 + 45 = 194 \text{ days}\end{aligned}$$

The project variance, σ^2 , is equal to:

$$\begin{aligned}\sigma^2 &= \sigma_A{}^2 + \sigma_F{}^2 + \sigma_G{}^2 + \sigma_D{}^2 + \sigma_J{}^2 \\ &= 53.78 + 5.44 + 4.00 + 9.00 + 13.44 = 85.66\end{aligned}$$

And, we can calculate the standard deviation for the project by taking the square root of variance:

$$\sigma = (\sigma^2)^{1/2} = (85.66)^{1/2} = 9.255$$

If management is willing to accept the three assumptions underlying PERT,

The completion time for the project can be modeled by a normally distributed random variable having a mean of, μ = 194 days, and a standard deviation of, μ = 9.255 days.

Now, we can use this model to answer the questions posed by management.

In these analyses, we define the *random variable* X as follows:

X = completion time of the project

We can then convert the normally distributed random variable *X* into the standard normal random variable *Z* as follows:

$$Z = \frac{X - \mu}{\sigma}$$

This transformation process can be shown as follows:

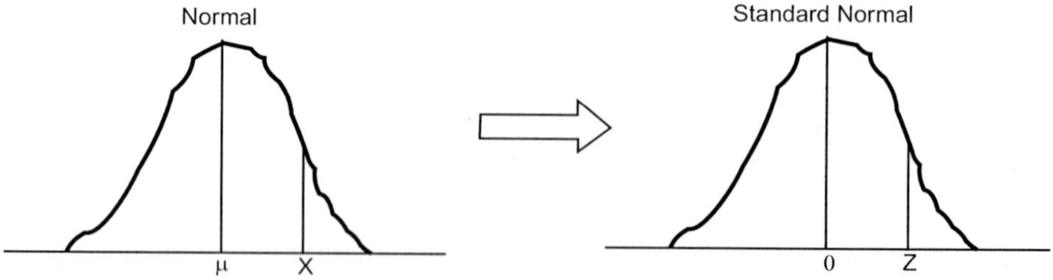

Fig. 13.21.

Here, *Z* represents the number of standard deviations X is from the mean (μ).

In other words, one standard deviation in the "normal" corresponds to exactly 1 in the "standard normal."

After this transformation, We can use a Table for the Standard Normal random variable To determine the probability of completing the project within any given time period.

Now let's answer the management's specific questions:

1. The first question was the probability that the project will be completed *within* 194 days.

This can be expressed as $P\ (X \leq 194)$.

Since 194 is the mean value ($\mu = 194$) of the distribution, we get:

$$X = \frac{X - \mu}{\sigma} = \frac{194 - 194}{9.255} = 0$$

Therefore,

$P\ (X \leq\ 94) = P\ (\ Z \leq 0) = .50$ (That is 50% probability)

We can either read this probability value from the Standard Z Table, or we can show it graphically that:

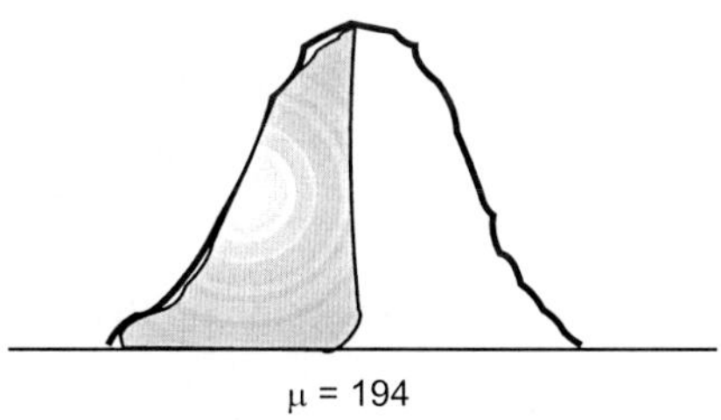

Fig. 13.22.

2. The second question looks for a reasonable interval for the number of days to complete the project.

Here, we assume that "being reasonable" means to be 95% sure. 95 percent of the total probability can be shown as follows :

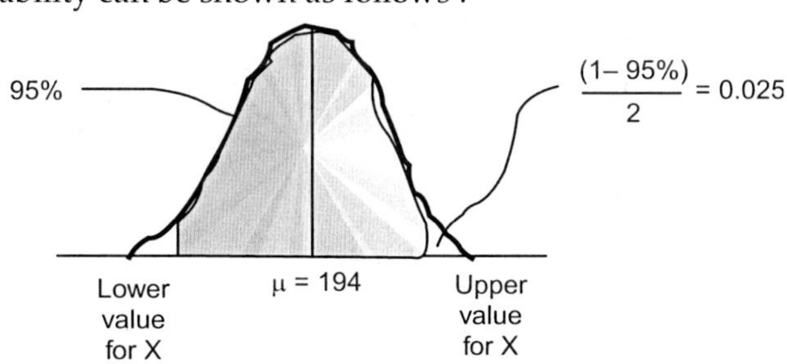

Fig. 13.23.

Since the Standard Normal Distribution Table (at Appendix A) gives the Z values that are only the right side of the shaded area, we look for the Z value corresponding to (95%)/2=0.475.

Therefore we read from the standard table that $Z_{0.475} = 1.96$.

Since, mean is μ = 194 days, and standard deviation is μ = 9.255 days,

Upper Value for X = 194 + 1.96 (9.255) = 194 + 18.14 = 212.14, and

Lower Value for X = 194 – 1.96 (9.255) = 194 – 18.14 = 175.86.

Since whole (integer valued) days are preferred, we round 18.14 to 19, which gives an interval between 175 to 213 days.

3. The probability that the project will be completed within 180 days can be expressed as $P(X \leq 180)$.

If $X = 180$

$$Z = \frac{(X - \mu)}{\sigma} = \frac{(210 - 194)}{9.255} = 1.51$$

We are interested in finding the probability shown in the following figure:

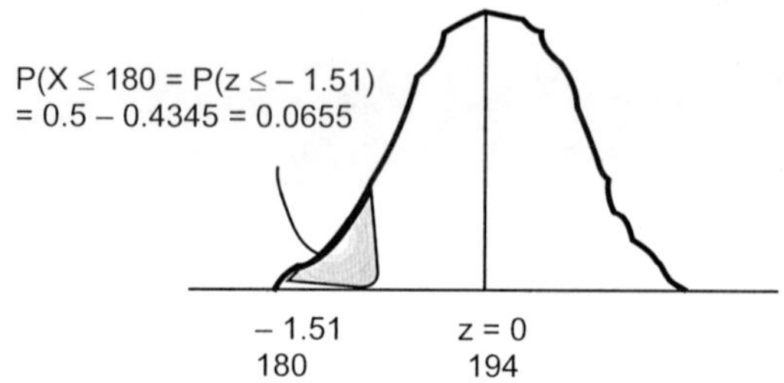

Fig. 13.24.

Then $P(X \leq 180) = P(Z \leq 1.51) = .5 - .4345 =$ **.0655**.

Here, since the Standard table values are the right side probabilities, the z value corresponding to –1.51 is found by subtracting the table value of 1.51 (that is .4345) from 0.50.

4. The probability that the project will take longer than 210 days can be expressed as $P(X > 210)$.

If $X = 210$,

$$Z = \frac{(X - \mu)}{\sigma} = \frac{(210 - 194)}{9.255} = 1.73$$

Therefore, we are interested in finding the probability shown in the following figure:

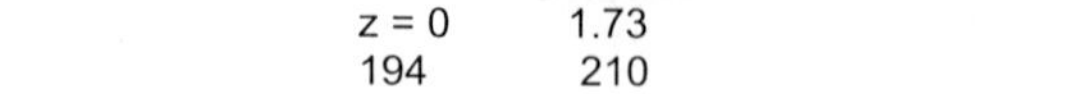

Fig. 13.25

Then $P(X > 210) = P(Z > 1.73) = .5 - .4582 =$ **.0418**.

Here, since the Standard table values are the right side probabilities, the z value corresponding to 1.73 is found by subtracting the Table value of 1.73 (that is .4582) from 0.50.

5. Being "Virtually sure" can mean different things to different people, So let us assume that management will accept a date that is 99% certain.

Therefore we will look at the z value of $\frac{(0.99)}{2}$ = 0.495 in the standard Z Table.

$Z_{0.495} = 2.58$.

Therefore, Given 99% certainty, the upper limit for X is equal to:

$X_{upper} = 194 + 2.58\ (9.255) = 194 + 23.88 = 217.88$

As a result, we can report to the management that we are virtually sure that the project will be completed within **216** days.

13.4 COST CONSIDERATION IN PROJECT: PROJECT CRASHING

- Project managers may have the option or requirement to crash the project, or accelerate the completion of the project.
- This is accomplished by reducing the length of the critical path(s).
- The length of the critical path is reduced by reducing the duration of the activities on the critical path.
- If each activity requires the expenditure of an amount of money to reduce its duration by one unit of time, then the project manager selects the least cost critical activity, reduces it by one time unit, and traces that change through the remainder of the network.
- As a result of a reduction in an activity's time, a new critical path may be created.
- When there is more than one critical path, each of the critical paths must be reduced.
- If the length of the project needs to be reduced further, the process is repeated.
- Crashing - reducing project time by expending additional resources.
- Crash time - an amount of time an activity is reduced.
- Crash cost - cost of reducing activity time.
- Goal - reduce project duration at minimum cost.

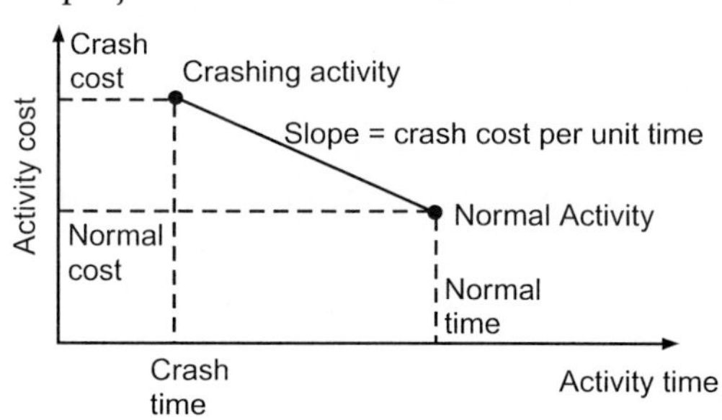

Fig. 13.26

13.5 APPLICATIONS OF PERT AND CPM

PERT and CPM have been used for a variety of projects, including the following types.

1. Construction of a new plant
2. Research and development of a new product
3. NASA space exploration projects
4. Movie productions

5. Building a ship
6. Government-sponsored projects for developing a new weapons system
7. Relocation of a major facility
8. Maintenance of a nuclear reactor
9. Installation of a management information system
10. Conducting an advertising campaign

13.6 ADVANTAGES OF PERT AND CPM

PERT/CPM has the following advantages:

1. a PERT/CPM chart explicitly defines and makes visible dependencies (precedence relationships) between the various activities,
2. PERT/CPM facilitates identification of the critical path and makes this visible,
3. PERT/CPM facilitates identification of early start, late start, and slack for each activity,
4. PERT/CPM provides for potentially reduced project duration due to better understanding of dependencies leading to improved overlapping of activities and tasks where feasible.
5. Commitments and communications
6. Detailed planning
7. Efficient monitoring and control
8. Identify potential problem areas
9. Proper use of resources
10. Re-scheduling is possible
11. Easily understood
12. Adaptive to computers
13. Decision making tools
14. Cost-time trade-off possible

13.7 DISADVANTAGES

PERT/CPM has the following disadvantages:

1. there can be potentially hundreds or thousands of activities and individual dependency relationships, which becomes difficult to handle.
2. the network charts tend to be large and unwieldy requiring several pages to print and requiring special size paper,
3. the lack of a timeframe on most PERT/CPM charts makes it harder to show status although colours can help (e.g., specific colour for completed nodes),
4. when the PERT/CPM charts become unwieldy, they are no longer used to manage the project.

13.8 DIFFERENCES BETWEEN CPM & PERT

1. CPM is activity oriented i.e. CPM network is built on the basis of activities. Also result of various calculation are considered in terms of activities of the project. On the other hand, PERT is event oriented.
2. CPM is a deterministic model i.e. it dose not take into account the uncertainties involved in the estimation of time for execution of a job or an

activity. It assumes that activity times are known and fixed, so only one time estimate is given and used for each activity. It completely ignores the probabilistic element of the problem. PERT however is a probabilistic Model. It uses three estimates of the activity time-optimistic, pessimistic and most likely with a view to take into account time uncertainty. Thus the expected duration of each activity is probabilistic, indicates that there is fifty percent probability of getting the job done within the time.

3. CPM place dual emphasis on time and cost and evaluate the trade off between project cost and project time. By deploying allows the project manager to manipulate project duration within certain limits so that project duration can be shortened to an optimal cost. On the other hand, PERT is primarily concerned with time. It helps the manager to schedule and co-ordinate various activities so that the project can be completed on schedule time.
4. CPM is used for repetitive jobs like planning and construction of a house while PERT is used for non-repetitive jobs like planning and scheduling research programmes.

13.9 LIMITATIONS OF PERT/CPM

1. Requirement of clearly defined, independent and stable activities.
2. Specified precedence relationships which are some time not possible.
3. Over emphasis on critical paths.
4. Deterministic CPM model.
5. Activity time estimates are subjective and depend on judgment.
6. PERT assumes a beta distribution for these time estimates, but the actual distribution may be different.
7. PERT consistently underestimates the expected project completion time due to alternate paths becoming critical.

EXERCISE

13.1. Write the areas of application of network techniques.

13.2. What is network analysis? When is it used?

13.3. Describe the rules for drawing the network diagram with suitable illustration.

(UPTU: 2008-09)

13.4. Write a short note on PERT / CPM networks in operations research.

13.5. Explain the meaning of crashing in network techniques.

13.6. Specify the limitation of Program Evaluation Review Technique.

(UPTU: 2008-09)

13.7. What is a critical path? Why is it so important in scheduling and controlling large projects? Can a critical path change during the course of a project? Why?

13.8. What is PERT? What information is revealed by PERT analysis?

13.9. Explain clearly the difference between an activity and an event by giving an example.

13.10. In what ways does a CPM network differ from PERT network?

13.11. What do you understand by "dummy activity"? Explain it wit the aid of an example.

13.12. Explain reasons for incorporating dummy activities in a network diagram. In what way do these differ from normal activities?

13.13. Explain terms total float, interference float and safety float.

13.14. Write short notes on independent float and free float. **(UPTU: 2008-09)**

13.15. Explain how the slacks of the critical path are determined in PERT.

13.16. Compare and contrast CPM and PERT. Under what conditions would you use PERT as opposed to CPM in project management?

13.17. Explain various basic steps in PERT/CPM techniques.

13.18. Comment on the assumptions based on which PERT/CPM analysis is done for the projects.

13.19. State the circumstances where CPM is a better technique of project management than PERT.

13.20. Write a note on the utility of CPM and PERT networks in Project management. How is PERT different from CPM?

13.21. "PERT now has proven to be an effective management tool that can be utilized by companies of all sizes in almost every industry." Discuss.

13.22. Explain following terms used in PERT

(*i*) pessimistic time (*ii*) optimistic time and (*iii*) most likely time

13.23. A certain project is composed of nine activities whose time estimates are as given below:

Activity	Duration
1-2	1
1-3	3
1-4	2
2-5	1
3-5	3
4-6	2
5-6	4
6-7	6
5-7	3

Draw the project network and find out the critical path. **(UPTU: 2008-09)**

13.24. A project comprises the following activities:

Activity	Immediate Predecessors	Activity Time (days)
A	-	12
B	-	6
C	A	13
D	A, B	12
E	C, D	11
F	D	13
G	E, F	11

(*a*) Construct a Gantt chart which will provide an overview of the planned project.

(*b*) How soon could the project be completed?

(*c*) Which activities need to be completed on time in order to ensure that the project is completed as soon as possible?

13.25. A project has the following activity details:

Activity	Immediate Predecessors	Activity Time (weeks)
A	-	3
B	-	4
C	-	3
D	C	12
E	B	5
F	A	7
G	E, F	3

Draw a Gannt chart to represent the project; determine those activities comprising the critical path; and estimate the project duration

13.26. A project has the following activities, precedence relationships, and activity durations:

Activity	Immediate Predecessors	Activity Duration (weeks)
A	-	3
B	-	4
C	-	3
D	C	12
E	B	5
F	A	7
G	E, F	3

(*a*) Draw a Gantt chart for the project.

(*b*) Construct a CPM network for the project.

(*c*) Identify those activities comprising the critical path.

(*d*) What is the project's estimated duration?

(*e*) Construct a table showing for each activity, its activity duration, earliest start time, latest start time, earliest finish time, latest finish time, and the activity slack. (**Ans:** (*c*) C, D (*d*) 15 weeks

13.27. An activity has these time estimates: optimistic time 0 = 15 weeks, most likely time $m = 20$ weeks, and pessimistic time $p = 22$ weeks.

(*a*) calculate the activity's expected time or duration t.

(*b*) calculate the activity's variance v.

(*c*) calculate the activity's standard deviation.

13.28. A project designed to refurbish a hospital operating theatre consists of the following activities, with estimated times and precedence relationships shown. Using this information draw a network diagram, determine the expected time and variance for each activity, and estimate the probability of completing the project within sixty days.

Activity	Immediate Predecessor	Optimistic Time	Most Likely Time	Pessimistic Time
A	-	5	6	7
B	-	10	13	28
C	A	1	2	15
D	B	8	9	16
E	B, C	25	36	41
F	D	6	9	18

13.29. A project has the following activities, precedence relationships, and time estimates in weeks:

Activity	Immediate Predecessor	Optimistic Time	Most Likely Time	Pessimistic Time
A	-	15	20	25
B	-	8	10	12
C	A	25	30	40
D	B	15	15	15
E	B	22	25	27
F	E	15	20	22
G	D	20	20	22

(*a*) Calculate the expected time or duration and the variance for each activity.

(*b*) Construct the network diagram

(*c*) Tabulate the values of ES,EF,LS,LF and slack for each activity

(*d*) Identify the critical path, and the project duration.

(*e*) What is the probability that the project will take longer than 57 weeks to complete?

13.30. The project detailed below has the both normal costs and "crash" costs shown. The crash time is the shortest possible activity time given that extra resources are allocated to that activity.

Activity	Immediate Predecessor	Normal Time	Normal Time Cost (Rs.)	Crash Time	Crash Time Cost (Rs.)
A	-	5	2 000	4	6,000
B	A	8	3 000	6	6,000
C	B	2	1 000	2	1,000
D	B	3	4 000	2	6,000
E	C	9	5 000	6	8,000
F	C, D	7	4 500	5	6,000
G	E, F	4	2 000	2	5,000

Assuming that the cost per day for shortening each activity is the difference between crash costs and normal costs, divided by the time saved, determine by how much each activity should be shortened so as to complete the project within twenty-six days and at the minimum extra cost.

13.31. For the following data, draw the network diagram, and then crash the activities to find the time-cost trade-off points that the company should want to consider. Start with the plan that has the longest duration.

Activity	Preceding Activity	Time (weeks) Normal Program	Crash Program	Cost (Rs. 000s) Normal Program	Crash Program
A	–	2	2	5	5
B	A	5	3	11	21
C	A	2	1	7	16
D	B, C	4	2	8	22
E	B	3	2	9	18
F	D, E	3	3	9	9

(**Ans.** The plan with the longest duration takes 14 weeks at a cost of Rs. 49,000. The one with the shortest duration takes 10 weeks at a cost of Rs. 82,000)

13.32. Draw the network diagram for the following problem and indicate a sequence of plans that the company should want to consider in making a time-cost tradeoff. The company is not interested in reducing the project duration below 29 days. Start with the plan that has the longest duration.

Activity	Preceding Activity	Time (days) Regular Program	Crash Program	Cost (Rs) Regular Program	Crash Program
A	-	10	9	5,000	5,200
B	-	14	11	3,500	3,950
C	A	8	7	4,000	4,100
D	A	7	2	2,100	3,600
E	B	5	3	2,500	3,000
F	B	10	7	2,250	3,750
G	C	9	9	5,000	5,000
H	D, E	11	9	3,850	5,250
I	G, H	5	3	2,375	3,575

(**Ans.** Beginning plan takes 35 days at a cost of Rs. 30,575. Completing the project in 29 days increases the project cost to Rs. 32,975)

13.33. The following represent activities in Sri Ram's Construction Company project.

(*a*) Draw the network to represent this situation.

(*b*) Which activities are on the critical path?

(*c*) What is the length of the critical path?

Activity	Immediate Predecessor(s)	Time (weeks)	Activity	Immediate Predecessor(s)	Time (weeks)
A	–	1	E	B	2
B	–	4	F	C, E	7
C	A	1	G	D	2
D	B	5	H	F, G	3

13.34. A small software development project at Krishna Dhir's firm has five major activities. The times are estimated and provided in the table below. Find the expected time for completing Dhir's project.

Activity	Immediate Predecessor	*a*	*m*	*b*
A	–	2	5	8
B	–	3	6	9
C	A	4	7	10
D	B	2	5	14
E	C	3	3	3

(*a*) Find the expected completion time for this project.

(*b*) What variance value would be used to find probabilities of finishing by a certain time?

OBJECTIVE TYPE QUESTIONS

13.1 Of all the paths through the network, the critical path
(*a*) has the maximum expected time
(*b*) has the minimum expected time
(*c*) has the maximum actual time
(*d*) has the minimum actual time

13.2 The PERT model assumes that
(*a*) each activity achieves its optimistic time
(*b*) the costs are uniformly distributed over the life of the activity
(*c*) that activity times are statistically independent
(*d*) none of the above

13.3 Estimating expected activity times in a PERT network
(*a*) makes use of three estimates
(*b*) puts the greatest weight on the most likely time estimate
(*c*) is motivated by a beta distribution
(*d*) all of the above
(*e*) none of the above

13.4 The marginal cost of crashing a network could change when
(*a*) the activity being crashed reaches its crash time
(*b*) the activity being crashed reaches a point where another path is also critical
(*c*) both (*a*) and (*b*)
(*d*) none of the above

Answers

13.1. (*a*) 13.2. (*b*) 13.3. (*d*) 13.4. (*c*)

Chapter 14

Replacement Problem

14.1. INTRODUCTION

The replacement problems are concerned with the situations that arise when some items such as men, machines, electric-light bulbs, etc. need replacement due to their deteriorating efficiency, failure or breakdown. Such decreased efficiency or complete breakdown may either be gradual or all of a sudden. For example, an electric-light bulb fails all of a sudden, sewer pipe line is blocked, a car becomes more and more expensive to maintain and out of date after a number of years, an employee looses his job. The replacement is carried out under following situations:

1. The old item has become in worse condition and work badly or requires expensive maintenance.
2. The old item has failed due to accident or otherwise and does not work at all, or the old item is expected to fail shortly.
3. A better or more efficient design of machine or equipment has become available in the market.

In the case of items whose efficiency go on decreasing according to their age, it requires to spend more money on account of increased operating cost, increased repair cost, increased scrap, etc. So in such cases, the replacement of an old item with new one is the only alternative to prevent such increased expenses. Thus the problem of replacement is to decide best policy to determine an age at which the replacement is most economical instead of continuing at increased cost. The need for replacement arises in many situations so that appropriate decisions may have to be taken. For example,

(i) We may decide whether to wait for complete failure of the item (which might cause some loss), or to replace earlier at the expense of higher cost of the item.

(ii) The expensive items may be considered individually to decide whether we should replace now or, if not, when it should be reconsidered for replacement.

(iii) It may be decided whether we should replace by the same type of item or by different type (latest model) of item.

The problem of replacement is encountered in the case of both men and machines. Using probability, it is possible to estimate the chance of death (or failure) at various ages.

The main objective of replacement is to direct the organization for maximizing its profit (or minimizing the cost).

14.2 FAILURE MECHANISM OF ITEMS

The term 'failure' has a wider meaning in *business* than what it has in our daily life. There are *two* kinds of failure.

1. Gradual Failure: The mechanism under this category is progressive. That is, as the life of an item increases, its efficiency deteriorates, causing:

(i) increased expenditure for operating costs,

(ii) decreased productivity of the equipments,

(iii) decrease in the value of the equipment, *i.e.*, the resale value decreases.

For example, mechanical items like *pistons, bearings, rings* etc. and *'automobile tyres'*.

2. Sudden Failure: This type of failure is applicable to those items that do not deteriorate markedly with service but which ultimately fail after some period of using. The period between installation and failure is not constant for any particular type of equipment but will follow some frequency distribution, which may be progressive, retrogressive or random in nature.

(i) *Progressive failure*: In this, probability of failure increases with the increase in the life of an item. For example, *electric light bulbs, automobile tires*, etc.

(ii) *Retrogressive failure*: Certain items have more probability of failure in the beginning of their life, and as the time passes the chances of failure become less. That is, the ability of the unit to survive in the initial period of life increases its expected life. For example, Industrial equipments like aircraft engines.

(iii) *Random failure:* Under this failure, constant probability of failure is associated with items that fail from random causes such as *physical shocks*, not related to age. In such a case, virtually all items fail before aging has any effect. For example, vacuum tubes in air-borne equipment have been shown to fail at a rate independent of the age of the tube.

The replacement situations may be placed into *five* categories:

1. Replacement of equipment that becomes worse with time, e.g. machines tools, buses in a transport organization, planes, buildings etc.
2. Group replacement of items that fail completely after certain period/amount of use, e.g., light bulbs, radio and t.v. parts, tubes, etc.
3. Replacement due to problems of mortality and staffing.
4. Replacement of an equipment that becomes out of date.
5. Miscellaneous Problems.

14.3 REPLACEMENT OF ITEMS / ASSETS THAT DETERIORATE WITH TIME

To find the optimal replacement period of items following two costs are considered:

1. maintenance and operating costs which increases with time
2. depreciation cost which diminishes with age of the equipment.

The optimal replacement of the equipment is done if (*i*) the scrap value of the equipment is zero (depreciation cost is not given), maintenance cost becomes greater than the current average cost. (*ii*) the resale value or depreciation cost, maintenance cost and the cost of the equipment is given, the replacement period is determined by minimum value of the average cost to date.

Considering a case of truck owner whose problem is to find the 'best' time at which he should replace the old truck by new one. The truck owner wants to transport goods as cheaply as possible. The associated costs are: *(i)* The running costs, and *(ii)* the capital costs of purchasing a truck.

These associated costs can be expressed as average cost per month. Now the truck owner will observe that the average monthly cost will go on decreasing, longer the replacement is postponed. However, there will come an age at which the rate of increase of running costs more than compensates the saving in average capital costs. Thus, at this age the replacement is justified.

14.3.1 Replacement Policy for Items Whose Maintenance Cost Increases With Time, and Money Value is Constant

Theorem 1. *The cost of maintenance of a machine is given, as a function increasing with time and its scrap value is constant.*

(*a*) If time is measured continuously, then the average annual cost will be minimized by replacing the machine when the average cost to date becomes equal to the current maintenance cost.

(*b*) If time is measured in discrete units, then the average annual cost will be minimized by replacing the machine when the next period's maintenance cost becomes greater than the current average cost.

(a) When time '*t*' is a continuous variable:

Let R_t = maintenance cost at time t, C = the capital cost of the item, S = the scrap value of the item.

Obviously, annual cost of the item at any time $t = R_t + C - S$.

Since the maintenance cost incurred during '*n*' years becomes $= \int_0^n R_t \, dt$, the total cost incurred on the item will become :

$$P(n) = \int_0^n R_t \, dt + C - S$$

Hence, average total cost is given by

$$F(n) = \frac{P(n)}{n} = \frac{1}{n}\int_0^n R_t \, dt + \frac{C-S}{n} \qquad \text{...(14.1)}$$

Now, we have to find such time n for which $F(n)$ is minimum. Therefore, differentiating $F(n)$ w.r.t. 'n'.

$$\frac{dF(n)}{dn} = \frac{1}{n} R_n + \left(-\frac{1}{n^2}\right)\int_0^n R_t \, d_t - \frac{C-S}{n^2} = 0, \text{ for minimum of } F(n),$$

which gives $R_n = \frac{1}{n}\int_0^n R_t \, dt + \frac{C-S}{n} = \frac{P(n)}{n}$, by virtue of equation.

Hence, maintenance cost at time n = average cost in time n.

Therefore, the equipment should be replaced when maintenance cost equals the average annual total cost.

(*b*) When time '*t*' is a discrete variable :

Since the time is considered in discrete units, the cost equation (14.1) can be written as :

$$F(n) = \frac{P(n)}{n} = \sum_{t=1}^{n} \frac{R_t}{n} - \frac{C-S}{n}$$

By using finite difference, $F(n)$ will be minimum if the following relationship is satisfied:

$$\Delta F(n-1) < 0 < \Delta F(n)$$

Now, differencing (14.4) under the summation sign by definition of first difference,

$$\Delta F(n) = F(n+1) - F(n)$$

$$= \left[\sum_{t=1}^{n+1} \frac{R_t}{n+1} - \frac{C-S}{n+1}\right] - \left[\sum_{t=1}^{n} \frac{R_t}{n} + \frac{C-S}{n}\right]$$

$$= \left(\frac{R_{n+1}}{n+1}, \sum_{t=1}^{n} \frac{R_t}{n+1}\right) - \sum_{t=1}^{n} \frac{R_t}{n} + (C-S)\left[\frac{1}{n+1} - \frac{1}{n}\right]$$

$$= \frac{R_{n+1}}{n+1} + \sum_{t=1}^{n} R_t \left(\frac{1}{n-1} - \frac{1}{n}\right) + (C-S)\left[\frac{1}{n-1} - \frac{1}{n}\right]$$

$$= \frac{R_{n+1}}{n-1} - \sum_{t=1}^{n} \frac{R_t}{n(n+1)} - \frac{C-S}{n(n+1)}$$

Since, $\Delta F(n) > 0$ for minimum $F(n)$, so

$$\frac{R_{n+1}}{n+1} > \sum_{t=1}^{n} \frac{R_t}{n(n+1)} + \frac{C-S}{n(n+1)} \quad \text{or} \quad R_{n+1} > \sum_{t=1}^{n} \frac{R_t}{n} + \frac{C-S}{n}$$

or $R_{n+1} > P(n)/n > R_n$ by virtue of equation

Similarly, it can be shown that $R_n < P(n)/n$, by virtue of $\Delta F(n-1) < 0$

Hence, $R_{n+1} > \dfrac{P(n)}{n} > R_n$

Therefore, replace the equipment at the end of n years, if the maintenance cost in the $(n+1)^{th}$ year is more then the average total cost in the nth year and the nth year's maintenance cost is less than the previous year's average total cost.

Example 14.1 The cost of a machine is Rs. 6100 and its scrap value is only Rs. 100. The maintenance costs are found from experience to be:

Year	1	2	3	4	5	6	7	8
Maintenance	100	250	400	600	900	1250	1600	2000

When should machine be replaced ?

Solution: First, find an average cost per year during the life of the machine as follows:

Total cost in first year = Maintenance cost in first year + loss in purchase price

= 100 + (6100 – 100) =Rs. 6100

∴ Average cost in first year = Rs. 6100.

Total cost up to two years = Maintenance cost up to two years + loss in purchase price

= (100 + 250) + 6000 =Rs. 6350.

∴ Average cost per year during first two years =Rs. 3175.

In a similar fashion, average cost per year during first three years

$$= \frac{6750}{3} = \text{Rs. } 2250.00,$$

average cost per year during first four years $= \frac{7350}{4}$ = Rs. 1837.50,

average cost per year during first five years $= \frac{8250}{5}$ = Rs. 1650.00,

average cost per year during first six years $= \frac{9500}{6}$ =Rs. 1583.33,

average cost per year during first seven years $= \frac{11100}{7}$ =Rs. 1585.71

These computations may be summarized in the following tabular form.

Replace at the end of year (n)	Maintenance cost (R_n)	Total maintenance cost (ΣR_n)	Difference between Price and Resale price ($C - S$)	Total Cost $P(n)$	Average cost $\frac{P(n)}{n}$
(1)	(2)	(3)	(4)	(5) = (3) + (4)	(6)
1	100	100	6000	6100	6100
2	250	350	6000	6350	3175
3	400	750	6000	6750	2250
4	600	1350	6000	7350	1837
5	900	2250	6000	8250	1650
6	1250	3500	6000	9500	1583
7	1600	5100	6000	11100	1586
8	2000	7100	6000	13100	1638

Here it is observed that the maintenance cost in the 7th year becomes greater than the average cost for 6 years *(i.e.* $R7 > P(6)/6$*)*. Hence the machine should be replaced at the end of 6th year.

Alternatively, last column of above table shows that the average cost starts increasing in the 7th year, so the machine should be replaced before the beginning of 7th year, *i.e.* at the end of 6th year.

Money Value, Present Worth Factor (PWF), And Discount Rate

Money value: Since money has a value over time, we often speak: *'money is worth 10% per year'*. This can be explained in the following two alternative ways:

(*i*) *In one way,* spending Rs. 100 today would be equivalent to spending Rs. 110 in a year's time: In other words, if we plan to spend Rs. 110 after a year from now, we could equivalently spend Rs. 100 today which would be of worth Rs. 110 next year.

(*ii*) *Alternatively,* if we borrow Rs. 100 at the rate of interest 10% per year and spend this amount today, then we have to pay Rs. 110 after one year. Thus, we conclude that Rs. 100 today will be equivalent to Rs. 110 after a year from now. Consequently, one rupee after a year from now is equivalent to $(1.1)^{-1}$ rupee today.

Discount rate (Deprecitation value): The present worth factor of unit amount to be spent after one year is given by $v = (1 + r)^{-1}$, where r is the interest rate. Then,

v is called *discount rate* (technically known as the depreciation value). Following example will explain the effect of considering money value with prescribed rate of interest.

Example 14.2 Let the value of money be assumed to be 10% per year and suppose that machine A is replaced after every 3 years whereas machine B is replaced after every six years. The yearly costs of both the machines are given as under:

Year	1	2	3	4	5	6
Machine *A*	1,000	200	400	1,000	200	400
Machine *B*	1,700	100	200	300	400	500

Determine which machine should be purchased.

Solution : Present worth factor is given by $v = \dfrac{100}{100+10} = \dfrac{10}{11}$

$\therefore$ Total discount cost (present worth) of A for 3 years

$$= \text{Rs.}\left[1{,}000 + 200 \times \frac{10}{11} + 400 \times \left(\frac{10}{11}\right)^2\right] = \text{Rs. } 1.512 \text{ (nearly)}$$

Again the total discount cost of B for six years = Rs.

$$\left[1{,}700 + 100 \times \frac{10}{11} + 200 \times \left(\frac{10}{11}\right)^2 + 300 \times \left(\frac{10}{11}\right)^3 + 400 \times \left(\frac{10}{11}\right)^4 + 500 \times \left(\frac{10}{11}\right)^5\right]$$

$= \text{Rs. } 2765$

Average yearly cost of $A = 1.512/3 = \text{Rs. } 504$

and average yearly cost of $B = 2.765/6 = \text{Rs. } 461$.

Although, this shows the apparent advantage with B, but the comparison in unfair because the periods of consideration are different.

So, if we consider 6 years period for machine A also, then the total discount of A will be

$$= 1.000 + 200 \times \frac{10}{11} + 400 \times \left(\frac{10}{11}\right)^2 + 1.000 \times \left(\frac{10}{11}\right)^3 + 200 \times \left(\frac{10}{11}\right)^4 + 400 \times \left(\frac{10}{11}\right)^5$$

$= \text{Rs. } 2647$

which is Rs. 118 less costlier than machine B over the same period.

Hence, machine A should be purchased.

14.3.2 Replacement Policy When Maintenance Cost Increases With Time and Money Value Changes With time at Constant Rate

As already explained in the preceding section, the money value can be interpreted in two different ways. Accordingly, the optimal replacement policy can be determined by the following two methods:

(*i*) The maintenance cost increases with time and the money value decreases with time at constant rate *i.e.* depreciation value is given.

(*ii*) The amount to be spent is borrowed at a given rate of interest under the condition of repaying it in pre-decided number of instalments.

Theorem 2: *The maintenance cost increases with time and the money value decreases with time at constant rate i.e. depreciation value is given. Then replacement policies will be*

(i) *Replace if the next period's cost is greater than the weighted average of previous costs.*

(ii) *Do not replace if the next period's cost is less than the weighted average of previous costs.*

Proof: *First Method:* Suppose that the item (which may be a machine or equipment etc.) is available for use over a series of time periods of equal intervals (say, one year).

Let C = purchase price of the item to be replaced

R_j = running (or maintenance) cost incurred at the beginning of year

r = rate of interest

$v = 1/(1+r)$ is the present worth of a rupee to be spent a year hence.

The process can be divided into two major steps:

Step 1: *To find the present worth of total expenditure.*

Let the item be replaced at the end of every n^{th} year. The yearwise present worth of expenditure on the item in the successive cycles of n years can be calculated as follows:

Year	1	2 ... n	$n+1$ $n+2$ ... $2n$	2_{n+1} ...
Present worth	$C+R_1$	R_{2v} ... $R_n V^{n-1}$	$(C+R_1)V^n$ R_2V^{n+1} ... R_nV^{2n+1}	$(C+R_1)V^{2n}$

Assuming that the item has no resale price at the time of replacement, the present worth of all future discounted costs associated with the policy of replacing the item at the end of every n years will be given by

$$P(n) = [C + R_1] + R_2v + ... + R_nV^{n-1} + [(C + R_1)v^n + R_2V^{n+1} + ... + R_2V^{2n-1}]$$
$$+ [(C + R_1)V^{2n} + R_2V^{2n+1} + ... + R_nV^{3n-1}] + ... \text{ and so on.}$$

Summing up the right-hand side, we get

$$P(n) = (C + R_1)(1 + v^n + v^{2n} + ...) + R_2v(1 + v^n + v^{2n} + ...) + ...$$
$$+ R_nv^{n-1}(1 + v^n + v^{2n} + ...)$$
$$= (C + R_1 + R_2v + ... + R_nv^{n-1})(1 + V_n + v^{2n} + ...)$$
$$= (C + R_1 + R_yY + ... R_vV - 1), \frac{1}{1-v^n}$$

[$\therefore$ $v < 1$, the sum of infinite G.P. is $1/(1 - v^n)$]

$$P(n) = 1] = \frac{F(n)}{1-v^n}, P(n+1) = \frac{F(n+1)}{1-v^{n+1}}$$

where, for simplicity, $F(n) = C + R_1 ... + R_nv^{n-1}$.

Step 2 : To determine replacement policy so that $P(n)$ is minimum. Since, n is measured in discrete units, we shall use the method of finite differences in order to minimize the present worth expenditure $P(n)$.

Obviously, if $P(n + 1) > P(n) > P(n - 1)$, *i.e.*, $\Delta P(n) > 0 > \Delta P(n - 1)$, then $P(n)$ will be minimum. So, by definition of first difference,

$$\Delta P(n) = P(n+1) - P(n) = \frac{F(n+1)}{1-v^{n+1}} - \frac{F(n)}{1-v^n} \text{ [from the equation]}$$

$$= \frac{F(n+1)(1-v^n) - F(n)(1-v^{n+1})}{(1-v^{n-1})(1-v^n)} \left(\frac{N^r}{D^r} \text{ form}\right)$$

For convenience, we first simplify the N^r of $\Delta P(n)$ only. That is,

$$N' = F(n+1)(1-v^n) - F(n)(1-v^{n+1})$$
$$= [F(n+1) - F(n)] + v^{n+1} F(n) - v^n F(n+1)$$
$$= R_{n+1} v^n + v^{n+1} F(n) - v^n [F(n) + v^n R_{n+1}] \qquad [\because F(n+1) = F(n) + R_{n+1} v^n]$$
$$= v^n (1-v^n) R_{n+1} - v^n (1-v) F(n)$$

$$\therefore \qquad \Delta P(n) = \frac{v^n(1-v^n) R_{n-1} - v^n (1-v) F(n)}{(1-v^{n+1})(1-v^n)}$$

$$= \frac{v^n(1-v^n)}{(1-v^{n+1})(1-v^n)} \left[\frac{1-v^n}{1-v} R_{n+1} - F(n) \right]$$

Simply setting $n-1$ for n

$$\Delta P(n-1) = \frac{v^{n-1}(1-v^n)}{(1-v^n)(1-v^{n-1})} \left[\frac{1-v^{n-1}}{1-v} R_n - F(n-1) \right]$$

After little simplification of R.H.S.

$$\Delta P(n-1) = \frac{v^{n-1}(1-v)}{(1-v^n)(1-v^{n-1})} \left[\frac{1-v^n}{1-v} R_n - F(n) \right]$$

The quantity $v^n (1-v) / (1-v^{n+1})(1-v^n)$ is always positive since $|v| < 1$. Thus, $\Delta P(n)$ has the same sign as the quantity under bracket [...], with similar explanation for $\Delta P(n-1)$ also.

Hence, the condition, $\Delta P(n-1) < 0 < \Delta P(n)$, for minimum present worth expenditure becomes,

$$= \frac{(1-v^n)}{(1-v)} R_n - F(n) < 0 < \frac{1-v^n}{1-v} R_{n+1} - F(n)$$

or
$$= \frac{1-v^n}{1-v} R_n < F(n) < \frac{1-v^n}{1-v} R_{n+1}$$

or
$$= R_n < \frac{C + R_1 + R_2 v + \ldots + R_n v^{n-1}}{1 + v + v^2 + \ldots + v^{n-1}} < R_{n+1}$$

or
$$= R_n < \frac{F(n)}{\Sigma v^{n-1}} < R_{n+1}.$$

The expression between R_n and R_{n+1} above is called the *"weighted average cost"* of previous n years with weights $1, v, v^2 \ldots v^{n-1}$, respectively.

The value of n satisfying the relationship above will be the best replacement age of the item. This proves the theorem.

How to Select the Best Equipment?

In the problem of choosing a best equipment (or item), the costs that are constant over time for each given equipment will still have to be taken into account, although these costs may differ for each equipment. Only those costs that are same for the equipments under comparison can be excluded.

Suppose two equipments E_1 and E_2 are at our choice. The data required for determining the best replacement age of each type of equipment is also given from past experience. Thus, a best selection can be done by adopting the following outlined procedure:

Step 1. First find the best replacement age for machine E_1 and E_2 both by using the relationship:

$$= R_{n+1} > \frac{F(n)}{\Sigma v^{n-1}} > R_n$$

Suppose the optimum replacement age for machines E_1 and E_2 comes out be n_1 and n_2 respectively.

Step 2. Compute the fixed annual payment (or weighted average cost) for each equipment by using the formula :

$$x = \frac{C + R_1 + R_2 + R_n V^{n-1}}{1 + v + v^2 + \ldots + v^{n-1}} = \frac{F(n)}{\Sigma v^{n-1}}$$

and substituting in this formula $n = n_1$ for equipment E_1 and $n = n_2$ for equipment E_2. Let it be X_1 and X_2 for equipment E_1 and E_2, respectively

Step 3. (*i*)If $X_1 < X_2$ then choose equipment E_1,

(*ii*) If $X_1 > X_2$, then choose equipment E_2.

(*iii*) If $X_1 = X_2$, then both equipments are equally good.

14.4 REPLACEMENT OF ASSETS / ITEMS THAT FAIL SUDDENLY

There are some practical situations where certain items fail suddenly instead of gradual deterioration e.g. radio and T.V. parts, electric-light bulb, motor belt in car or machine, etc. that result in complete breakdown of the system. Sudden failure of an item causes heavy losses to the organization. The problem can be overcome by determining the probability distribution of failures. So we find the value of 't' (the period at the end of which failure occurs) which minimizes the total cost involved for the replacement. Following two types of replacement policies are considered:

1. *Individual replacement policy*: In this, an item is replaced immediately after its failure.
2. *Group replacement policy:* In this, decision is taken as to when all the items must be replaced irrespective of the fact that items have failed or have not failed, with a provision that if any item fails before the optimal time, it may be individually replaced. This decision requires: (*i*) the rate of individual replacement during the period (*ii*) the total cost incurred for individual and group replacements during the selected interval.

The period for which the total cost incurred is minimum will be the optimal period for replacement. The time of group replacement and the total cost involved are calculated as below:

(*a*) replace the group of items at the end of tth period if the cost of individual replacement for t^{th} period is greater than the average cost per period through the end of the tth period.

(*b*) do not replace the group of items at the end of tth period if the cost of individual replacement at the end of $(t-1)^{th}$ period is less than the average cost per period through the end of the tth period.

The number of deaths per unit time becomes constant when the age distribution becomes stable (number of deaths per unit time is equal to the size of total population divided by the mean age at death).

Example 14.3: Following mortality rates have been observed for a certain type of fuses:

Week :	1	2	3	4	5
% failing by the end of the week :	5	15	35	75	100

There are 1000 fuses in use and it costs Rs. 5 to replace an individual fuse. If all fuses are replaced simultaneously it would cost Rs. 1.25 per fuse. It is proposed to replace all fuses at fixed intervals of time, whether or not they have burnt out, and to continue replacing burnt out fuses as they fail. At what intervals, the group replacement should be made? Also prove that this optimal policy is superior to the straight –forward policy of replacing each fuse only when it fails.

Solution: Let p_i be the probability that a fuse which was new when placed in position for use, fails during the ith week of its life. Following probability distribution is obtained assuming to replace burnt out bulbs as and when they fail.

$$p_1 = \text{the probability of failure in first week} = \frac{5}{100} = 0.05$$

$$p_2 = \text{the probability of failure in second week} = \frac{(15-5)}{100} = 0.10$$

$$p_3 = \text{the probability of failure in third week} = \frac{(35-15)}{100} = 0.20$$

$$p_4 = \text{the probability of failure in fourth week} = \frac{(75-35)}{100} = 0.40$$

$$p_5 = \text{the probability of failure in fifth week} = \frac{(100-75)}{100} = 0.25$$

Total = 1.0

Since the sum of all probabilities can never be greater than one, all probabilities beyond p_5 will be zero. Therefore, a fuse that has already lasted four weeks is sure to fail during the fifth week.

(*a*) fuses that fail during a week are replaced just before the end of that week, and

(*b*) actual percentage of failures during a week for a population of bulbs with the same age is the same as the expected percentage of failures during the week for that population.

If f_i is the number of replacements made at the end of the ith week and all N_0 = 1000 bulbs are new initially.

$$f_1 = N_0p_1 = 1000 \times 0.05 = 50$$

$$f_2 = N_0p_2 + f_1p_1 = 1000 \times 0.10 + 50 \times 0.05 = 102$$

$$f_3 = N_0p_3 + f_1p_2 + f_2p_1 = 1000 \times 0.20 + 50 \times 0.10 + 102 \times 0.05 = 210$$

$$f_4 = N_0p_4 + f_1p_3 + f_2p_2 + f_3p_1 = 1000 \times 0.40 + 50 \times 0.20 + 102 \times 0.10 + 210 \times 0.05 = 430$$

$$f_5 = N_0p_5 + f_1P_4 + f_2p_3 + f_3p_2 + f_4p_1$$

$$= 1000 \times 0.25 + 50 \times 0.40 + 102 \times 0.20 + 210 \times 0.10 + 430 \times 0.05 = 333$$

It is observed that the expected number of fuses failing each week increases till the 4th week and then starts decreasing. It can be shown that the value of f_i will oscillate till the system acquire a steady state in which the proportion of fuses failing each month is reciprocal of their average life. The expected life of each fuse

$= \Sigma ip_i = 1 \times 0.05 + 2 \times 0.1 + 3 \times 0.2 + 4 \times 0.4 + 5 \times 0.25 = 3.7$ weeks

Hence, average number of failures $= \dfrac{1000}{3.7} = 270$ fuses/week.

Since the replacement of all 1000 fuses simultaneously costs Rs. 1.25 per fuse and the cost of individual replacement is Rs. 5, the cost of group replacement is as given below:

End of week	Cost of individual replacement	Total cost of group replacement (Rs.)	Average cost per week (Rs.)
1.	50 × 5	1000 × 1.25 + 50 × 5 = 1500	1500
2.	102 × 5	1500 + 102 × 5 = 2010	**1005**
3.	210 × 5	2010 + 210 × 5 = 3060	1020
4.	430 × 5	3060 + 430 × 5 = 5210	1302
5.	333 × 5	5210 + 333 × 5 = 6875	1375

In 2nd week the average cost per week is minimum. Hence it is cheapest to have a group replacement after every second week. Also the average cost of individual replacement policy is Rs. 1350 where as the average cost of group replacement after every second week is Rs. 1005. Hence the optimum policy of replacing all fuses at fixed intervals of time whether or not they have burnt out and to continue replacing burnt out fuses as they fail is better than the straight forward policy of replacing each fuse when it fails.

EXERCISE

14.1. What is a replacement problem? When does it arise?

14.2. Describe various types of replacement situations.

14.3. Describe some important replacement situations and replacement policies.

14.4. Why does the problem of replacement arise? What is group replacement? When is it economical? Explain with example.

14.5. Explain the replacement procedure for items that deteriorate with time.

14.6. Explain various types of replacement decisions with suitable examples.

14.7. What are three strategies of replacement of items which follow sudden failure mechanisms? Explain each of them with example.

14.8. The cost of maintenance of a machine is given as a function increasing with time and its scrap value is constant. Show that the average annual cost will be minimized by replacing the machine when the average cost to date becomes equal to the current maintenance cost.

14.9. What is difference between group replacement and individual replacement? Under which circumstances group replacement may become profitable?

14.10. Explain how the theory of replacement is used in the replacement of items that fail completely.

14.11. Develop a model for the replacement of items whose maintenance cost increase with time and value of money remains same during the period.

14.12. Explain how the theory of replacement is used in the replacement of items whose maintenance cost varies with time. *(UPTU: 2008-09)*

14.13. A machine costs Rs. 500. Operation and maintenance costs are zero for the first year and increase by Rs. 100 every year. If money is worth 5% every year, determine the best age at which the machine should be replaced. The resale value of the machine is negligibly small. What is weighted average cost of owning and operating the machine? ***(UPTU: 2008-09)***

14.14. The maintenance cost and resale value per year of machine whose purchase price is Rs. 7000 is given below:

Year	1	2	3	4
Maintenance cost in Rs.	900	1200	1600	2100
Resale value in Rs.	4000	2000	1200	600

When should the machine be replaced? ***(UPTU: 2008-09)***

14.15. A firm is using a machine whose purchase price is Rs. 13,000. The installation charges amount to Rs. 3,600 and the machine has a scrap value of only Rs. 1,600, because the firm has a monopoly of this type of work. The maintenance cost in various years is given below:

Year :	1	2	3	4	5	6	7
Cost (Rs.)	250	750	1000	1500	2100	2900	4000

The firm wants to determine after how many years should the machine be replaced, assuming that the replacement can be done only at the year end.

14.16. An individual is planning to purchase a car. A new car will cost Rs. 1,20,000. The resale value of the car at the end of the year is 85% of the previous year value. Maintenance and operation costs during the first year are Rs. 20,000 and they increase by 15% every year. The minimum resale value of car can be Rs. 40,000. (i) When should the car be replaced to minimize average annual cost (ignoring interest)? (ii) If interest of 12% is assumed, when should the car be replaced?

14.17. A truck owner finds from his past records that the maintenance costs per year, of a truck whose purchase price is Rs. 8000 are as given below:

Year	1	2	3	4	5	6	7	8
Maintenance cost in Rs.	1000	1300	1700	2200	2900	3800	4800	6000
Resale price Rs.	4000	2000	1200	600	500	400	400	400

14.18. Let $v = 0.9$, and initial price is Rs. 5000, running cost varies as follows:

Year	1	2	3	4	5	6	7
Running cost (Rs.)	400	500	700	1000	1300	1700	2100

What would be optimal replacement interval?

14.19. The following mortality rate has been observed for an electric bulb:

Week :	1	2	3	4	5	6	7	8
Probability of: failure by end week	0.00	0.13	0.25	0.43	0.68	0.88	0.96	1.00

The cost of replacing an individual failed bulb is Rs. 3.65. It has been decided that all bulbs be replaced simultaneously at fixed intervals and also to replace individual bulbs as they fail in service. If the cost of group replacement is Rs. 1.15 per bulb, what is the optimal interval between two group replacements?

□□□

Software Programs Based on OR Applications

Program 1: Write a program to implement Bubble Sort Algorithm

```
#include<stdio.h>
#include<conio.h>
int main()
{
int arr[50],temp,i,j,n;
printf("\nEnter the size of array that you want to sort");
scanf("%d",&n);
printf("\n\tEnter The Values into ARRAY ");
for(i=0;i<n;i++)
{
printf("\n\n Enter Element no %d: ",i+1);
scanf("%d",&arr[i]);
}
for(i=1;i<n;i++)
{
for(j=0;j<n-i;j++)
{
if(arr[j] >arr[j+1])
{
temp=arr[j];
arr[j]=arr[j+1];
arr[j+1]=temp;
}
}
}
printf("\n\n-- Sorted Series --");
for(i=0;i<n;i++)
{
printf("\n \n \t %d",arr[i]);
}
getch();
return 0;
}
```

OUTPUT:

```
Enter the size of array that you want to sort :
12
            Enter The Values into ARRAY
Enter the element no 1:  90
Enter the element no 2:  78
Enter the element no 3:  56
Enter the element no 4:  65
Enter the element no 5:  70
Enter the element no 6:  23
Enter the element no 7:  88
Enter the element no 8:  92
Enter the element no 9:  56
Enter the element no 10:  43
Enter the element no 11:  33
Enter the element no 12:  31
--Sorted Series--
23
31
33
43
56
56
65
70
78
88
90
92
```

❑❑❑

Program 2: Write a program to implement Linear Search Algorithm

```
#include <iostream.h>
#include <conio.h>
#include <stdio.h>

int main()
{ int array[5],i,x;
cout<<"********************************";
cout<<"\nThis program is to implement linear search algorithm\n";
cout<<"************************************\n";
cout<<"Now enter your elements into the array ::\n";
for(int i=0;i<=4;i++)
{
cin>>array[i];
}
cout<<"Enter the number to be searched ::";
cin>>x;
for(i=0;i<=4;i++)
{
if(array[i]==x)
{
cout<<"Element found at position ::"<<i+1<<endl;
break;
}
}
if(i>4)
{cout<<"Not found\n";
}
getch();
return 0;
}
```

OUTPUT:

```
**************************************...
This program is to implement linear search algorithm
**************************************...
Now enter your elements into the array::
11
12
13
14
15
Enter the number to be searched ::15
Element found at position ::5
```

❑❑❑

Program 3: Matrix Operations:

Problem Description:

Write a C program to find out :

A. The determinant of a matrix A and Matrix B

B. Multiplication of matrix A and Matrix B

Program:

```
#include<stdio.h>
#include<conio.h>
int main()
{
int a[3][3],i,j,b[3][3],c[3][3];
long determinant1,determinant2;
printf("Enter the 9 elements of matrix A: ");
   for(i=0;i<3;i++)
      {for(j=0;j<3;j++)
         { scanf("%d",&a[i][j]);
         c[i][j]=1;
         }
      }
printf("Enter the 9 elements of matrix B: ");
   for(i=0;i<3;i++)
      {for(j=0;j<3;j++)
            {scanf("%d",&b[i][j]);
            }
      }
printf("\nThe matrix A is:\n");
for(i=0;i<3;i++)
{     printf("\n");
      for(j=0;j<3;j++)
      printf("%d\t",a[i][j]);
   }
printf("\nThe matrix B is:\n");
for(i=0;i<3;i++){
         printf("\n");
         for(j=0;j<3;j++)
         printf("%d\t",b[i][j]);
      }
   determinant1 = a[0][0]*((a[1][1]*a[2][2]) -
(a[2][1]*a[1][2])) - a[0][1]*(a[1][0]*a[2][2] - a[2][0]*a[1][2])
+ a[0][2]*(a[1][0]*a[2][1] - a[2][0]*a[1][1]);
   determinant2 = b[0][0]*((b[1][1]*a[2][2]) -
```

```
(b[2][1]*b[1][2])) -b[0][1]*(b[1][0]*b[2][2] - b[2][0]*b[1][2])
+ b[0][2]*(b[1][0]*b[2][1] - b[2][0]*b[1][1]);
   printf("\nDeterminant of 3X3 matrix A: %ld",determinant1);
   printf("\nDeterminant of 3X3 matrix B: %ld",determinant2);
   printf("\nThe resultant Matrix C obtained from multiplica-
tion of Matrix A and Matix B is:\n");
   for(i=0;i<3;i++)
      {  for(j=0;j<3;j++)
         {
            c[i][j]=a[i][j]*b[i][j];
            printf("%d",c[i][j]);
            printf("\t");
         }
         printf("\n");
      }
   getch();
      return 0;
   }
```

OUTPUT:

```
   Enter the 9 elements of matrix A:
   2
   3
   4
   5
   3
   4
   6
   7
   8
   Enter the 9 elements of matrix B:
   2
   3
   4
   3
   5
   6
   1
   2
   3
```

```
The matrix A is:
2       3       4
5       3       4
6       7       8

The matrix B is:
2       3       4
3       5       6
1       2       3

Determinant of 3X3 Matrix A is: 12
Determinant of 3X3 Matrix B is: 15
The resultant Matrix C obtained from multiplication of
Matrix A and Matix B is:
4       9       16
15      15      24
6       14      24
```

❑❑❑

Program 4: Linear Programming:Appointment Problem

Problem Description

An HR manager receives performance report of four applicants.

On the basis of the points given to each applicants for four available jobs, the manager's purpose is to appoint the four jobs so as to maximize the applicants' satisfaction. They must give points from 1 to 100 to each job (1 = very poor preference, 100 = very good).

See table below:

Performance report

	Jobs	J1	J2	J3	J4
	A1	80	70	10	30
APPLICANTS	A2	80	60	10	70
	A3	30	60	10	70
	A4	40	30	10	60

Program:

```
#include <stdio.h>
#include <math.h>
#include <conio.h>
#include <iostream.h>
#define  imax 10

   double C[imax][imax];
   int MP[imax][imax];
   int IF1,IZ,M,N,NP,NZ,ZI,ZJ;
   double A,B,XCASE,MIN;

void Data() {
   int I,J;
   cout<<"Enter the number of jobs:";
   cin>>NP;
   cout<<"\nInput Appointment costs/regrets of
application:"<<"\n";
   for (I=1; I<=NP; I++) {
      cout<<"\n Applicant"<<I<<"  : \n";
    for (J=1; J<=NP; J++) {
      cout<<"    The points for job?: "<< J<<"  : ";
      cin>>C[I][J];
      }
   }
```

```
cout<<"\n APPOINTMENTS:\n\n";
}

   void Zeroes();
   void Appoint();
   void Mark();
   void Sub_add();

void Sub_Main()  {
e10:Zeroes();
   Appoint();
   Mark();
   Sub_add();
   Appoint();
   if (IF1 != NP) goto e10;
}

void Zeroes()  {
   int I,J;
   double R;
   IZ = 0; R = 0;
   for (I=1; I<=NP; I++) {
      MIN = C[I][1];
      for (J=1; J<=NP; J++) {
         if (C[I][J] == 0) IZ = 1;
         if (C[I][J] < MIN) MIN = C[I][J];
   }
   if (IZ == 1) {
      IZ = 0; goto e10;
   }
   for (J=1; J<=NP; J++)
            C[I][J] -= MIN;
e10:;}
   for (J=1; J<=NP; J++) {
   MIN = C[1][J];
   for (I=1; I<=NP; I++) {
      if (C[I][J] == 0) IZ = 1;
         if (C[I][J] < MIN) MIN = C[I][J];
   }
   if (IZ == 1) {
         IZ = 0; goto e20;
```

```
      }
      for (I=1; I<=NP; I++)
            C[I][J] -= MIN;
   e20:;}
   }

   void Appoint() {
      int I,J,K;
      for (I=1; I<=NP; I++) {
         for (J=1; J<=NP; J++) {
                     MP[I][J] = 0; C[0][J] = 0;
         }
         C[I][0] = 0;
      }
      for (I=1; I<=NP; I++) {
         XCASE = 999999;
         for (J=1; J<=NP; J++) {
            if (C[I][J] != 0 || MP[I][J] != 0) goto e10;
                        NZ = 0;
            for (K=1; K<=NP; K++)
                     if (C[K][J] == 0) NZ++;
                  if (1.0*NZ < XCASE) {
                     XCASE = 1.0*NZ; ZI = I; ZJ = J;
                  }
    e10:;}
      MP[ZI][ZJ] = 1;
      for (K=1; K<=NP; K++)
               if (C[K][ZJ] == 0 && MP[K][ZJ] == 0)  MP[K][ZJ]
= -1;
      for (K=1; K<=NP; K++)
            if (C[ZI][K] == 0 && MP[ZI][K] == 0)  MP[ZI][K] =
-1;
    }
    IF1 = 0;
      for (I=1; I<=NP; I++)
         for (J=1; J<=NP; J++)
            if (MP[I][J] == 1) IF1++;
    }

    void Mark() {
      int I,J;
    e10: for (I=1; I<=NP; I++) {
```

```
            N = 0;
            for (J=1; J<=NP; J++)
                    if (MP[I][J] == 1)  N = 1;
                 if (N == 0 && C[I][0] == 0) {
                    C[I][0] = 1; M = 1;
            }
    }
    for (J=1; J<=NP; J++)
        for (I=1; I<=NP; I++)
              if (MP[I][J] == -1 && C[I][0] == 1 && C[0][J] ==
0) {
          C[0][J] = 1; M = 1;
                              }
        for (I=1; I<=NP; I++)
            for (I=1; I<=NP; I++)
               if (MP[I][J] == 1 && C[0][J] == 1 && C[I][0] == 0)
{
               C[I][0] = 1; M = 1;
                      }
            if (M == 1) {
        M = 0; goto e10;
        }
    }

   void Sub_add() {
       int I,J;
       MIN = 999999;
       for (I=1; I<=NP; I++)
           for (J=1; J<=NP; J++) {
           A = C[I][0]; B = C[0][J];
           if (A == 1 && B == 0 && C[I][J] < MIN)  MIN = C[I][J];
           }
       for (I=1; I<=NP; I++)
                 for (J=1; J<=NP; J++) {
              A = C[I][0]; B = C[0][J];
              if (A == 1 && B == 0)  C[I][J] -= MIN;
              if (A == 0 && B == 1)  C[I][J] += 2 * MIN;
          }
       }

  void Results()  {
      int I,J;
```

```
    for (I=1; I<=NP; I++)
       for (J=1; J<=NP; J++) {
       if (MP[I][J] != 1) goto e10;
             cout<<"  APPLICANT #=>"<<I<<" JOB"<<J<<"\n";
e10:;}
    cout<<"\n";
}

int main() {

    Data();      // INPUT COST/APPOINT
    Sub_Main();
    Results();   // PRINT RESULTS

getch();
return 0;
}
```

OUTPUT:

```
Enter the number of jobs : 4

Input Appointment costs/regrets of application:

Applicant 1:
The points for job? : 1: 80
The points for job? : 2: 70
The points for job? : 3: 10
The points for job? : 4: 30

Applicant 2:
The points for job? : 1: 80
The points for job? : 2: 60
The points for job? : 3: 10
The points for job? : 4: 70

Applicant 3:
The points for job? : 1: 30
The points for job? : 2: 60
The points for job? : 3: 10
The points for job? : 4: 70
```

```
Applicant 4:
The points for job? : 1: 40
The points for job? : 2: 30
The points for job? : 3: 10
The points for job? : 4: 60

APPOINTMENTS:
   APPLICANT #=>1  JOB4
   APPLICANT #=>2  JOB3
   APPLICANT #=>3  JOB1
   APPLICANT #=>4  JOB2
```

Program 5: Linear Programming

Problem Description:

Write a program to solve the following LP problem:

A company produces 2 types of hat. Every hat A requires twice as much labour time as the second hat B. If the company produces only hat B then it can produce a total of 500 hats a day. The market limits daily sales of hat A and B to 150 and 250 hats. The profit on hats A and B are 8 and 5 respectively. Solve graphically to get the optimal solution.

Program:

```
#include<iostream.h>
#include<conio.h>
#include<graphics.h>
#include<math.h>

int profit (int a,int b)
{
int p;
p=a*8+b*5;
return(p);
}
void main()
{
 clrscr();
int gdriver=DETECT,gmode;
initgraph(&gdriver,&gmode,"c:\\tc\\bgi");
 int a[5],p[5],pmax=0;
float t,tA,tB,n,sd1,sd2,x1,x2,x3,x4,x5,x6,x7,x8,x9,x10,y,j,
i,p1,p2,p3,p4,p5,xaxis,yaxis;
 cout<<"Enter the daily profit per hat on hat type A\n";
 cin>>tA;
 cout<<"Enter the daily profit per hat on hat type B\n";
 cin>>tB;
 cout<<"Enter the total number of hats produced in a day\n";
 cin>>n;
 cout<<"Enter the no. of hats of each type sold daily\n";
 cin>>sd1>>sd2;
 cout<<"\nProfit calculation when none of the hats are
sold:    ";
 x1=0;
 x2=0;
 p[0]=profit(x1,x2);
 cout<<p[0];
```

```
    cout<<"\nThe maximum profit when only hats of type A were
sold:  ";
    x3=sd1;
    x4=0;
    p[1]=profit(x3,x4);
    cout<<p[1];
    cout<<"\nThe maximum profit when only hats of type B were
sold:  ";
    x9=0;
    x10=sd2;
    p[2]=profit(x9,x10);
    cout<<p[2];
    cout<<"\nThe maximum profit when maximum no. of B are sold
as well as A are sold:  ";
    x8=sd2;
    x7=(500-x8)/2;
    p[3]=profit(x7,x8);

    cout<<p[3];
    cout<<"\nThe max profit when max no. of A are sold as well
as B are sold:  ";
    x5=sd1;
    x6=500-2*x5;
    p[4]=profit(x5,x6);

    cout<<p[4];

    for(j=0;j<5;j++)
    {
    if(pmax<=p[j])
    pmax=p[j];
    }

    cout<<"\nThe maximum profit is:\t"<<pmax;

    line(x1+200,x2+200,x3+200,x4+200);
    line(x3+200,x4+200,x5+200,x6+200);
    line(x5+200,x6+200,x7+200,x8+200);
    line(x7+200,x8+200,x9+200,x10+200);
    line(x1+200,x2+200,x9+200,x10+200);

    getch();
   closegraph();
    }
```

OUTPUT:

```
Enter the daily profit per hat on hat type A
8
Enter the daily profit per hat on hat type B
5
Enter the total number of hats produced in a day
500
Enter the number of hats of each type sold daily
150
250

Profit calculation when none of the hats were sold: 0
The maximum profit when only hats of type A were sold: 1200
The maximum profit when only hats of type B were sold: 1250
The maximum profit when maximum number of type B were sold
as well as A were: 2250
The maximum profit when maximum number of type A were sold
as well as B were: 2200
The maximum profit is:  2250
```

PROBLEM 6: SIMPLEX METHOD

Problem Description

A construction company build malls, cinema hall and multiplex with profit 15000$, 17000$ and 20000$.

The company must respect following conditions:

1. The number of cinema halls built each month must not exceed the number of multiplex by more than two.
2. The company can make each month up to 5, 5 and 3, respectively of kinds malls, cinema hall and multiplex
3. The company can make up to 8 malls, cinema hall and multiplex, respectively in the proportions of 3, 2 and 1.

The company having these data, wants to maximize monthly profit by knowing the number of each kind to build monthly.

Program:

```
   #include <stdio.h>
   #include <math.h>
   #include <conio.h>

   #define CMAX  10  //max. number of variables in economic
function
   #define VMAX  10  //max. number of constraints

      int NC, NV, NOPTIMAL,P1,P2,XERR;
      double TS[CMAX][VMAX];

      void Data() {
         double R1,R2;
         char R;
         int I,J;
         printf("\n LINEAR PROGRAMMING\n\n");
         printf(" MAXIMIZE (Y/N) ? "); scanf("%c", &R);
         printf("\n NUMBER OF VARIABLES OF ECONOMIC FUNCTION
? "); scanf("%d", &NV);
         printf("\n NUMBER OF CONSTRAINTS ? "); scanf("%d",
&NC);
         if (R == 'Y' || R=='y')
         R1 = 1.0;
       else
         R1 = -1.0;
      printf("\n INPUT COEFFICIENTS OF ECONOMIC FUNCTION:\n");
      for (J = 1; J<=NV; J++) {
```

```
        printf("      #%d ? ", J); scanf("%lf", &R2);
        TS[1][J+1] = R2 * R1;
    }
    printf("      Right hand side ? "); scanf("%lf", &R2);
    TS[1][1] = R2 * R1;
    for (I = 1; I<=NC; I++) {
        printf("\n CONSTRAINT #%d:\n", I);
        for (J = 1; J<=NV; J++) {
            printf("      #%d ? ", J); scanf("%lf", &R2);
            TS[I + 1][J + 1] = -R2;
        }
        printf("   Right hand side ? "); scanf("%lf",
&TS[I+1][1]);
    }
    printf("\n\n RESULTS:\n\n");
    for(J=1; J<=NV; J++)  TS[0][J+1] = J;
    for(I=NV+1; I<=NV+NC; I++)  TS[I-NV+1][0] = I;
    }

   void Pivot();
   void Formula();
   void Optimize();

   void Simplex() {
 e10: Pivot();
        Formula();
        Optimize();
        if (NOPTIMAL == 1) goto e10;
    }

    void Pivot() {

        double RAP,V,XMAX;
        int I,J;

        XMAX = 0.0;
    for(J=2; J<=NV+1; J++) {
              if (TS[1][J] > 0.0 && TS[1][J] > XMAX) {
        XMAX = TS[1][J];
        P2 = J;
    }
 }
```

```
RAP = 999999.0;
for (I=2; I<=NC+1; I++) {
   if (TS[I][P2] >= 0.0) goto e10;
   V = fabs(TS[I][1] / TS[I][P2]);
   if (V < RAP) {
      RAP = V;
      P1 = I;
   }
e10:;}
   V = TS[0][P2]; TS[0][P2] = TS[P1][0]; TS[P1][0] = V;
   }

   void Formula() {;
   //Labels: e60,e70,e100,e110;
   int I,J;

   for (I=1; I<=NC+1; I++) {
      if (I == P1) goto e70;
      for (J=1; J<=NV+1; J++) {
      if (J == P2) goto e60;
      TS[I][J] -= TS[P1][J] * TS[I][P2] / TS[P1][P2];
e60:;}
e70:;}
   TS[P1][P2] = 1.0 / TS[P1][P2];
   for (J=1; J<=NV+1; J++) {
   if (J == P2) goto e100;
   TS[P1][J] *= fabs(TS[P1][P2]);
e100:;}
   for (I=1; I<=NC+1; I++) {
   if (I == P1) goto e110;
   TS[I][P2] *= TS[P1][P2];
e110:;}
   }

   void Optimize() {
      int I,J;
      for (I=2; I<=NC+1; I++)
         if (TS[I][1] < 0.0)  XERR = 1;
   NOPTIMAL = 0;
   if (XERR == 1)  return;
   for (J=2; J<=NV+1; J++)
```

```
      if (TS[1][J] > 0.0)  NOPTIMAL = 1;
   }

   void Results() {
   //Labels: e30,e70,e100;
   int I,J;

   if (XERR == 0) goto e30;
   printf(" NO SOLUTION.\n"); goto e100;
e30:for (I=1; I<=NV; I++)
   for (J=2; J<=NC+1; J++) {
      if (TS[J][0] != 1.0*I) goto e70;
      printf("       VARIABLE #%d: %f\n", I, TS[J][1]);
e70:  ;}
   printf("\n       ECONOMIC FUNCTION: %f\n", TS[1][1]);
e100:printf("\n");
   }

int main()  {

   Data();
   Simplex();
   Results();
   getch();
   return 0;
}
```

OUTPUT:

```
MAXIMIZE ? Y

NUMBER OF VARIABLES OF ECONOMIC FUNCTION ? 3

NUMBER OF CONSTRAINTS ? 3

INPUT COEFFICIENTS OF ECONOMIC FUNCTION:
      #1 ? 15
      #2 ? 17
      #3 ? 20
      Right hand side ? 0

CONSTRAINT #1:
```

```
        #1 ? 0
        #2 ? 1
        #3 ? -1
        Right hand side ? 2

  CONSTRAINT #2:
        #1 ? 3
        #2 ? 3
        #3 ? 5
        Right hand side ? 15

  CONSTRAINT #3:
        #1 ? 3
        #2 ? 2
        #3 ? 1
        Right hand side ? 8

  RESULTS:

           VARIABLE #1: 0.333333
           VARIABLE #2: 3.000000
           VARIABLE #3: 1.000000

           ECONOMIC FUNCTION: 76.000000
```

❒❒❒

Program7: Write a program to implement Simplex Problem

Program:

```
#include <stdio.h>
#include <math.h>
#include <iostream.h>
#include <conio.h>

#define  MMAX  25
#define  NMAX  25
#define  REAL  double

typedef REAL MAT[MMAX][NMAX];

MAT  A;
int  IPOSV[MMAX], IZROV[NMAX];
int  i,j,ICASE,N,M,M1,M2,M3;
REAL R;


   void simp1(MAT,int,int *,int,int,int *,REAL *);
   void simp2(MAT,int,int,int *,int,int *,int,REAL *);
   void simp3(MAT,int,int,int,int);


void simplx(MAT a,int m,int n,int m1,int m2,int m3,int
*icase,int *izrov, int *iposv)  {
int i,ip,ir,is,k,kh,kp,m12,nl1,nl2,l1[NMAX],l2[MMAX],l3[MMAX];
REAL bmax,q1,EPS=1e-6;
if(m != m1+m2+m3) {
   printf(" Bad input constraint counts in simplx.\n");
         return;
   }
   nl1=n;
   for (k=1; k<=n; k++) {
         l1[k]=k;      /*Initialize index list of columns
admissible for exchange.
         izrov[k]=k; /*Initially make all variables right-
hand.*/
}
nl2=m;
for (i=1; i<=m; i++) {
```

```
                if (a[i+1][1] < 0.0) {
                    printf(" Bad input tableau in simplx, Con-
stants bi must be nonnegative.\n");
                        return;
        }
        l2[i]=i;
        iposv[i]=n+i;

    }
    for (i=1; i<=m2; i++) l3[i]=1;
    ir=0;
    if(m2+m3 == 0) goto e30; //The origin is a feasible start-
ing solution. Go to phase two.
    ir=1;
    for (k=1; k<=n+1; k++) /*Compute the auxiliary objective
function.*/
    {
        q1=0.0;
        for (i=m1+1; i<=m; i++) q1 += a[i+1][k];
        a[m+2][k]=-q1;
    }
    e10: simp1(a,m+1,l1,nl1,0,&kp,&bmax);
        if(bmax <= EPS && a[m+2][1] < -EPS) {
        *icase=-1;
        return;
    }
    else if (bmax <= EPS && a[m+2][1] <= EPS) {

        m12=m1+m2+1;
        if (m12 <= m)
        for (ip=m12; ip<=m; ip++)
                    if(iposv[ip] == ip+n) {
            simp1(a,ip,l1,nl1,1,&kp,&bmax);
            if(bmax > EPS) goto e1;
            }
        ir=0;
        m12=m12-1;
        if (m1+1 > m12) goto e30;
        for (i=m1+1; i<=m1+m2; i++)
            if(l3[i-m1] == 1)
                for (k=1; k<=n+1; k++)
                 a[i+1][k] *= -1.0;
        goto e30;
```

```
    }

    simp2(a,m,n,l2,nl2,&ip,kp,&q1);

    if(ip == 0) {
       *icase=-1;
       return;
    }
    e1: simp3(a,m+1,n,ip,kp);

    if(iposv[ip] >= n+m1+m2+1) {

    for (k=1; k<=nl1; k++)
       if(l1[k] == kp) goto e2;
e2: nl1=nl1-1;
    for (is=k; is<=nl1; is++)  l1[is]=l1[is+1];
}
else {
    if(iposv[ip] < n+m1+1) goto e20;
    kh=iposv[ip]-m1-n;
    if(l3[kh] == 0) goto e20;
    l3[kh]=0;

    }
    a[m+2][kp+1] += 1.0;
    for (i=1; i<=m+2; i++)  a[i][kp+1] *= -1.0;
e20: is=izrov[kp];
    izrov[kp]=iposv[ip];
    iposv[ip]=is;
    if (ir != 0) goto e10;

e30: simp1(a,0,l1,nl1,0,&kp,&bmax);
    if(bmax <= EPS) {
       *icase=0;
       return;
    }
    simp2(a,m,n,l2,nl2,&ip,kp,&q1);
    if(ip == 0) {
       *icase=1;
       return;
    }
```

```
        simp3(a,m,n,ip,kp);
        goto e20;
    }

    void simp1(MAT a,int mm,int *l1,int nl1,int iabf,int *kp,REAL
*bmax) {

      int k;
      REAL test;
      *kp=l1[1];
      *bmax=a[mm+1][*kp+1];
      if (nl1 < 2) return;
      for (k=2; k<=nl1; k++) {
       if(iabf == 0)
          test=a[mm+1][l1[k]+1]-(*bmax);
       else
          test=fabs(a[mm+1][l1[k]+1])-fabs(*bmax);
       if(test > 0.0) {
          *bmax=a[mm+1][l1[k]+1];
          *kp=l1[k];
        }
      }
      return;
    }

    void simp2(MAT a, int m,int n,int *l2,int nl2,int *ip,int
kp,REAL *q1) {
      REAL EPS=1e-6;

      int i,ii,k;
      REAL q,q0,qp;
      *ip=0;
      if(nl2 < 1) return;
      for (i=1; i<=nl2; i++)
       if (a[i+1][kp+1] < -EPS) goto e2;
       return;
    e2: *q1=-a[l2[i]+1][1]/a[l2[i]+1][kp+1];
      *ip=l2[i];
      if (i+1 > nl2) return;
      for (i=i+1; i<=nl2; i++) {
        ii=l2[i];
          if(a[ii+1][kp+1] < -EPS) {
```

```
        q=-a[ii+1][1]/a[ii+1][kp+1];
        if (q <  *q1) {
           *ip=ii;
           *q1=q;
        }
        else if (q == *q1) {
        for (k=1; k<=n; k++) {
            qp=-a[*ip+1][k+1]/a[*ip+1][kp+1];
            q0=-a[ii+1][k+1]/a[ii+1][kp+1];
            if (q0 != qp) goto e6;
          }
e6:       if (q0 < qp) *ip=ii;
        }
      }
    }
    return;
}

void simp3(MAT a,int i1,int k1,int ip,int kp) {

    int ii,kk;
    REAL piv;
    piv=1.0/a[ip+1][kp+1];
    if (i1 >= 0)
      for (ii=1; ii<=i1+1; ii++)
       if (ii-1 != ip) {
          a[ii][kp+1] *= piv;
          for (kk=1; kk<=k1+1; kk++)
            if (kk-1 != kp)
              a[ii][kk] -= a[ip+1][kk]*a[ii][kp+1];
        }
    for (kk=1; kk<=k1+1; kk++)
      if(kk-1 !=  kp) a[ip+1][kk] =-a[ip+1][kk]*piv;
    a[ip+1][kp+1]=piv;
    return;
}

int main() {
    printf("\n");
     printf(" Number of variables in E.F.: "); scanf("%d",
&N);
```

```
     printf(" Number of <= inequalities..: "); scanf("%d",
&M1);
     printf(" Number of >= inequalities..: "); scanf("%d",
&M2);
     printf(" Number of = equalities.....: "); scanf("%d",
&M3);

     M=M1+M2+M3;

     for (i=1; i<=M+2; i++)
       for (j=1; j<=N+1; j++)
         A[i][j]=0.0;

     printf(" Input Economic Function:\n");
     for (i=2; i<=N+1; i++) {
       printf(" Coefficient #%d: ", i-1);
      scanf("%lf", &A[1][i]);
     }
     printf(" Constant term : ");
     scanf("%lf", &A[1][1]);

     for (i=1; i<=M; i++) {
       printf(" Input constraint #%d: \n", i);
       for (j=2; j<=N+1; j++) {
         printf(" Coefficient #%d: ", j-1);
        scanf("%lf", &R);
        A[i+1][j] = -R;
      }
       printf(" Constant term : ");
      scanf("%lf", &A[i+1][1]);
     }

     printf("\n Input Table:\n");
     for (i=1; i<=M+1; i++) {
       for (j=1; j<=N+1; j++)
         printf("%8.2f", A[i][j]);
      printf("\n");
      getch();
     }

     simplx(A,M,N,M1,M2,M3,&ICASE,IZROV,IPOSV);
```

```
  if (ICASE==0) {
    printf("\n Maximum of E.F. = %f\n", A[1][1]);

    for (i=1; i<=N; i++) {
      for (j=1; j<=M; j++)
       if (IPOSV[j] == i)  {
          printf("  X%d = %f\n", i, A[j+1][1]);
         goto e3;
      }
        printf("  X%d = %f\n", i, 0.0);
e3:;
getch();}
  }
  else
    printf(" No solution (error code = %d).\n", ICASE);

   printf("\n");
   getch();
}
```

Program 8: Transportation Problem:

Problem Description:

Write a program to solve the following transportation problem:

Find the optimal solution of the problem given below using Northwest method:

Source/destination	D	E	F	G	Capacity
A	19	30	50	10	7
B	70	30	40	60	9
C	40	8	70	20	18
Demand	5	8	7	14	34\34

Program:

```
#include <iostream.h>
#include <conio.h>
#include <stdio.h>
#include <process.h>
#include <math.h>
#include <iomanip.h>

int main()
{
        //Data Input Level
    //clrscr();
    int      i=0,j=0,a[4][5],b[4][5],allo[3][4],
sum_d=0,sum_s=0,left,total=0;

    //Data entered in form of matrix using a[4][5] and
b[4][5] array
        for(i=0;i<3;i++)
            {

    for(j=0;j<4;j++)
    {
        cout<<"\n Enter the transportation cost for route
["<<i+1<<"]["<<j+1<<"] ::"; //Data entered row-wise
        cin>>a[i][j];
        b[i][j]=a[i][j];
        allo[i][j]=0; /*Assigning 0 to all elements in array
allo*/
    }
```

```
        cout<<endl<<"Enter the supply for city "<<i+1<<" ::";
        cin>>a[i][4];
        b[i][4]=a[i][4];
                       }

    for (i=0;i<4;i++)
    {
        cout<<"\n Enter the Demand for city "<<i+1<<" ::";
        cin>>a[3][i];
        b[3][i]=a[3][i];
    }
         //Transportation table is formed and stored

         //Data Output Level
    //clrscr();
    cout<<"\n The transporatation model built is
::"<<endl<<endl; //Statement
  cout<<setw(8)<<"Cities"<<setw(4)<<"1"<<setw(4)<<"2"<<setw(4)<<"3
  "<<setw(4)<<"4"<<setw(8)<<"Supply"<<endl; /*To display the
headings of table*/
  for(i=0;i<4;i++) //Display the data
         {
         if(i<=2)

         {
         cout<<setw(8)<<i+1;
  for(j=0;j<5;j++)
         {
         cout<<setw(4)<<a[i][j]; //Display the value
         }
         cout<<endl;
         sum_s+=a[i][4];
    }
              else if(i==3) /*Display the demand of each
column*/
              {
                    cout<<setw(8)<<"Demand";
                    for(j=0; j<4;j++)
              {
              cout<<setw(4)<<a[3][j]; //Display the value
              sum_d+=a[3][j];
              }
```

```
                    cout<<endl;
            }
        }
        cout<<endl<<endl; //Data is shownned in proper format

    if(sum_s==sum_d)  //Solve further for calculation
            {
                cout<<"The problem is balanced"<<endl;
                left=sum_s;
            }
            else //Data is not appropriate for this problem, exit
from output screen
            {
                    cout<<"the problem is not balanced";
                    getch();
                    exit(0); /*Predefined function to terminate
the program*/
            }

    // Data is appropriate.Hence, preform the calculations
        i=0;
        j=0;

        do  //run first time and check for condition given in
while loop for second time
        {
            if( b[3][j] < b[i][4]) /*Calculation for value as-
signment in each cell using North-West Cell method*/
            {
                    b[i][4]-=b[3][j];
                    allo[i][j]=b[3][j]; /*Value assigned in that
cell*/
                left-=b[3][j];
                b[3][j]=0;
                j++;
            }
            else if (b[i][4] < b[3][j]) /*Calculation for value
assignment in each cell using North-West Cell method*/
            {
                b[3][j]-=b[i][4];
                allo[i][j]=b[i][4]; //Value assigned in that cell
                left-=b[i][4];
                b[i][4]=0;
```

```
            i++;
         }
         else if (b[3][j]==b[i][4]) //Calculation for value
//assignment in each cell using North-West Cell method
         {
            b[3][j]-=b[i][4];
            allo[i][j]=b[i][4]; //Value assigned in that cell
            left-=b[i][4];
            b[i][4]=0;
            i++;
            j++;

         }


      }
      while(left>=1); //End of do-while loop

      cout<<"The solution is ::"<<endl; //Statement

      for(i=0;i<3;i++) //Loop to display the allocation in
//each cell using North-West Cell method
      {
         for(j=0;j<4;j++)
         {
               if(allo[i][j]>=1) //Display the data
               {
                  cout<<"The allocation is on route "<<i+1<<"
-> "<<j+1<<" equal to Rs. "<<allo[i][j]<<" at a cost of :
Rs."<< allo[i][j]*a[i][j]<<endl;
               }
               total+=(allo[i][j]*a[i][j]); //To calculate the
//total cost during this assignment
               }
         }
         cout<<"\n The total cost is Rs."<<total<<endl;
//Display the total cost
         getch();
         return 0;
   }
```

OUTPUT:

```
Enter the transportation cost for route [1][1] :: 19
Enter the transportation cost for route [1][2] :: 30
Enter the transportation cost for route [1][2] :: 50
Enter the transportation cost for route [1][2] :: 10
Enter the supply for city 1::7
Enter the transportation cost for route [2][1] :: 70
Enter the transportation cost for route [2][2] :: 30
Enter the transportation cost for route [2][3] :: 40
Enter the transportation cost for route [2][4] :: 60
Enter the supply for city 2 ::9
Enter the transportation cost for route [3][1] :: 40
Enter the transportation cost for route [3][2] :: 8
Enter the transportation cost for route [3][3] :: 70
Enter the transportation cost for route [3][4] :: 20
Enter the supply for city 3 ::18
Enter the Demand for city 1 ::5
Enter the Demand for city 2 ::8
Enter the Demand for city 3 ::7
Enter the Demand for city 4 ::14
```

Cities	1	2	3	4	Supply
1	19	30	50	10	7
2	70	30	40	60	9
3	40	8	70	20	18
Demand	5	8	7	14	

```
The problem is balanced
The solution is ::
The allocation is on route 1 -> 1 equal to Rs.5 at a cost of :Rs. 95
The allocation is on route 1 -> 2 equal to Rs.2 at a cost of :Rs. 60
The allocation is on route 2 -> 2 equal to Rs.6 at a cost of :Rs. 180
The allocation is on route 2 -> 3 equal to Rs.3 at a cost of :Rs. 120
The allocation is on route 3 -> 3 equal to Rs.4 at a cost of :Rs. 280
The allocation is on route 3 -> 4 equal to Rs.14 at a cost of :Rs. 280

The total cost is Rs.1015
```

□□□

Program 9: TRANSPORTATION PROBLEM:

Problem Description:

A tile manufacturer owns three plants respectively producing monthly 15000, 12000 and 16000 tons. He has to deliver the following quantities respectively to four customers in different cities: 2500 T (Mumbai), 12000 T (Delhi), 6000 T (Kolkata) and 8000 T (Chennai) at the lowest total transportation cost. The transport costs grid is the following:

	Mumbai	Delhi	Kolkata	Pune
FACTORY 1	200	700	800	very high
FACTORY 2	400	400	500	500
FACTORY 3	400	500	600	600

Program:

```
#include <stdio.h>
#include <math.h>
#include <conio.h>
#include <iostream.h>
#define  MAX  10


   double C[MAX][MAX], XM[MAX][MAX];
   double R[MAX][MAX];
   double DAO[MAX], RAD[MAX];

   int P[5][25];

   int ID,DEST,IF1,IOPTIMAL,SRC,IX,IY,LT,NR;
   double CC,CT,QT,TT,XINFCC;

int Init() {
  int I,J;
  cout<<"\n TRANSPORT PROBLEM\n\n";
  cout<<" Enter the number of sources?";
  cin>>SRC;
  cout<<" Enter the number of destinations?";
  cin>>DEST;
  cout<<"\n enter the available source quatities:\n";
  for (I=1; I<=SRC; I++) {
  cout<<"  SOURCE #"<<I <<"? ";
  cin>>DAO[I];
  }
```

```
    cout<<"\n Enter the required destination quantities:\n";
    for (I=1; I<=DEST; I++)
    {
      cout<<"  DESTINATION #"<<I<<" ? " ;
      cin>>RAD[I];
    }
    cout<<"\n Enter the transport costs matrix:\n";
    for (I=1; I<=SRC; I++)
     for (J=1; J<=DEST; J++) {
          cout<<"   FROM SOURCE #"<<I<<" TO DESTINATION
#"<<J<<"?";
        cin>>C[I][J];
      }
    cout<<"\n\n TRANSPORT MODEL\n";
    return 0;
  }

    void Corner();
    void Optimal();
    void TCost();
    void Seek_Path(int I,int J);
    void Increase(int I,int J);

  void SubMain() {
    Corner();
    Optimal();
    TCost();
  }

  void Corner() {
  //Labels: e20, e50, e60
      int I,J;
      I=1; J=1;
  e20:if (DAO[I]<=RAD[J]) goto e50;
      XM[I][J] += RAD[J]; DAO[I] -= RAD[J];
      RAD[J]=0; J++;
      goto e60;  //else
  e50:XM[I][J] += DAO[I]; RAD[J] -= DAO[I];
      DAO[I]=0; I++;
  e60:if (I<=SRC && J<=DEST) goto e20;
  }
```

```
void Optimal() {
//Labels: e10, e70, e140, e150
      int I,J;
e10: XINFCC=0.0;
      for (I=1; I<=SRC; I++)
        for (J=1; J<=DEST; J++) {
          if (XM[I][J] != 0.0) goto e70;
          Seek_Path(I,J);
          Increase(I,J);
e70:;   }
     if (XINFCC>=0.0) {
        IOPTIMAL=1;
        goto e150;
      }
      for (I=1; I<=LT; I++) {
        IX=P[3][I]; IY=P[4][I];
        if (I % 2 == 0) {
          XM[IX][IY] -= TT;
          goto e140;
        }
        XM[IX][IY] += TT;
e140:;}
e150:if (IOPTIMAL==0) goto e10;
}

void Seek_Path(int I, int J) {
//Labels: e70, e160, e260
      int I1,I2;
      for (I1=1; I1<=SRC; I1++)
        for (I2=1; I2<=DEST; I2++)
          R[I1][I2]=XM[I1][I2];
      for (I1=1; I1<=SRC; I1++) R[I1][0]=0.0;
      for (I2=1; I2<=DEST; I2++) R[0][I2]=0.0;
      R[I][J]=1.0;
e70: for (I2=1; I2<=DEST; I2++) {
        if (R[0][I2]==1.0) goto e160;
        NR=0;
        for (I1=1; I1<=SRC; I1++)
          if (R[I1][I2] != 0.0) NR++;
       if (NR!=1) goto e160;
        for (I1=1; I1<=SRC; I1++) R[I1][I2]=0.0;
```

```
          R[0][I2]=1.0; IF1=1;
   e160:;}
        for (I1=1; I1<=SRC; I1++) {
          if (R[I1][0]==1.0) goto e260;
          NR=0;
          for (I2=1; I2<=DEST; I2++)
            if (R[I1][I2] != 0.0) NR++;
          if (NR!=1) goto e260;
          for (I2=1; I2<=DEST; I2++) R[I1][I2]=0.0;
          R[I1][0]=1.0; IF1=1;
   e260:;}
        if (IF1==1) {
          IF1=0; goto e70;
        }
   }

   void Increase(int I, int J) {
   //Labels: e20,e70,e130,e170,e180,e230
        int I1,I2;
          P[1][1]=I; P[2][1]=J; IX=I; IY=J; ID=1; CC=0.0;
QT=999999.0;
   e20: ID++; IF1=0;
       for (I1=1; I1<=SRC; I1++) {
          if (R[I1][IY]==0.0 || I1==IX) goto e70;
          P[1][ID]=I1; P[2][ID]=IY; IX=I1; CC -= C[IX][IY];
          IF1=1; I1=SRC;
              if (XM[IX][IY] < QT && XM[IX][IY] > 0.0)
QT=XM[IX][IY];
   e70:;}
        if (IF1==0) goto e170;
        ID++; IF1=0;
       for (I2=1; I2<=DEST; I2++) {
          if (R[IX][I2]==0.0 || I2==IY) goto e130;
          P[1][ID]=IX; P[2][ID]=I2; IY=I2; CC += C[IX][IY];
          IF1=1; I2=DEST;
   e130:;}
        if (IF1==0) goto e170;
        if (IX!=I || IY!=J) goto e20;
        goto e180;
   e170:printf(" Degenerate solution !\n");
        return;
```

```
e180:if (CC>0.0 || CC>XINFCC) goto e230;
     TT=QT; XINFCC=CC; ID--; LT=ID;
     for (I1=1; I1<=ID; I1++) {
       P[3][I1]=P[1][I1]; P[4][I1]=P[2][I1];
     }
e230:;}

void TCost() {
    int I,J;
    CT=0.0;
    printf("\n TRANSPORTS:\n");
    for (I=1; I<=SRC; I++)
      for (J=1; J<=DEST; J++) {
        CT += XM[I][J]*C[I][J];
        if (XM[I][J]==0.0) goto e10;
         printf("     FROM SOURCE #%d TO DESTINATION #%d:
%8.2f\n", I, J, XM[I][J]);
e10:;}
    printf("\n TOTAL TRANSPORT COST: %10.1f\n\n", CT);
}

int main()  {
  Init();
  SubMain();
  getch();
  return 0;
}
```

OUTPUT:

```
NUMBER OF SOURCES ? 3

NUMBER OF DESTINATIONS ? 4

INPUT THE AVAILABLE SOURCE QUANTITIES:
   SOURCE #1 ? 15000
   SOURCE #2 ? 12000
   SOURCE #3 ? 16000

 INPUT THE REQUIRED DESTINATION QUANTITIES:
  DESTINATION #1 ? 2500
  DESTINATION #2 ? 12000
  DESTINATION #3 ? 6000
  DESTINATION #4 ? 8000
```

```
INPUT TRANSPORT COSTS MATRIX:
 FROM SOURCE #1 TO DESTINATION #1 ? 200
 FROM SOURCE #1 TO DESTINATION #2 ? 700
 FROM SOURCE #1 TO DESTINATION #3 ? 800
 FROM SOURCE #1 TO DESTINATION #4 ? 9999
 FROM SOURCE #2 TO DESTINATION #1 ? 400
 FROM SOURCE #2 TO DESTINATION #2 ? 400
 FROM SOURCE #2 TO DESTINATION #3 ? 500
 FROM SOURCE #2 TO DESTINATION #4 ? 500
 FROM SOURCE #3 TO DESTINATION #1 ? 400
 FROM SOURCE #3 TO DESTINATION #2 ? 500
 FROM SOURCE #3 TO DESTINATION #3 ? 600
 FROM SOURCE #3 TO DESTINATION #4 ? 600

TRANSPORT MODEL

TRANSPORTS:
   FROM SOURCE #1 TO DESTINATION #1:   2500.00
   FROM SOURCE #1 TO DESTINATION #2: 12000.00
   FROM SOURCE #2 TO DESTINATION #3:    500.00
   FROM SOURCE #2 TO DESTINATION #3:  5500.00
   FROM SOURCE #3 TO DESTINATION #4:  6500.00
   FROM SOURCE #3 TO DESTINATION #4:  1500.00

TOTAL TRANSPORT COST: 16200000.0
```

❑❑❑

Program 10: Assignment Problem:

Problem Description:

Write a program to solve the following Assignment problem:

The table shown below is a profit matrix, entries of which indicate the profit generated by sales person working in various zones.

Zone\Person	Salesman 1	Salesman 2	Salesman 3
1	10	17	14
2	16	8	11
3	14	15	16

Find assigned zones of sales person, so that the profit is maximum.

Program

```
#include <conio.h>
#include <math.h>
#include <stdio.h>
#include <iostream.h>

int max();                  //Function for calculating maximum
void subtract(int f);          //Function for subtraction
int assignment();                //Function for performing
assignment operations
int a[3][3],b[3][3];  //Global Array

int main()
{

int m,n,z1,z2,z3,profit,l,p;

//Array for data entry

cout<<" ASSIGNMENT PROBLEM "<<endl<<endl;
cout<<"\n Enter the profit earned by each of the sales
persons in all zones(row-wise): "<<endl<<endl;
   for(m=0;m<3;m++)
    {
     for(n=0;n<3;n++)
     {
      cout<<"a["<<m+1<<"]["<<n+1<<"] :: ";
      cin>>a[m][n];
      b[m][n]=a[m][n];
```

```
        }
      }
   cout<<endl<<endl;

   cout<<"STEP 1 :: FINDING THE MAXIMUM VALUE OF THE
MATRIX"<<endl<<endl;

   z1=max();   //Calling  function max() and storing value
returned in z1

   cout<<"The maximum of all the numbers in the matrix is:
"<<z1<<"\n";

   cout<<"\n\n";

   /*Converting Maximization array into Minimization array
and displaying it*/
   cout<<"STEP 2 :: Matrix derived by subtracting maximum from
all the elements ::"<<endl;
   for(l=0;l<3;l++)
   {
     for(p=0;p<3;p++)
     {
        a[l][p]=z1-a[l][p];   //Subtraction of maximum with
values
        cout<<a[l][p]<<" ";  //Display the values
     }
     cout<<"\n";
   }
   cout<<endl;
   cout<<"STEP 3 :: Matrix derived by subtracting smallest
number in each row from every number in that row\n";

     //Reduce the matrix using row
   subtract(0); //Function call for row 1
   subtract(1); //Function call for row 2
   subtract(2); //Function call for row3
   cout<<endl<<endl;

   cout<<"STEP 4 :: Assigning the salesman to each
zone"<<endl<<endl;

   profit=assignment(); //Function call to give profit value
   cout<<"\nThe total profit earned is: "<<profit; //Objec-
tive of program
```

```
getch();
return 0;
} //End of main block

int max()  //To calculate the maximum entered element
{
 int i,j,k,n,c,d,e,f,pmax=0;
 for(c=0;c<3;c++)
 {
   for(d=0;d<3;d++)
   {
      if(pmax<a[c][d])
          pmax=a[c][d];  //Store the value in pmax
   }
 }
return pmax;  //Sending the value of pmax to the main block
}

void subtract(int row)       /*Function subtracting small-
est number in each row from every number in that row*/
{
 int e,rmin,g;
 rmin=a[row][0];
 for(e=0;e<3;e++)
 {
   if(rmin<a[row][e])
      rmin=rmin;
   else
      rmin=a[row][e];
 }
 for(g=0;g<3;g++)  //To display the value of each row
 {
   a[row][g]=a[row][g]-rmin;
   cout<<a[row][g]<<" ";
 }
cout<<"\n";
}

int assignment() /*Function block to display the distribu-
tion and give maximum profit value*/
{
 int j,i,p=0;
```

```
cout<<"Salesperson     Sales zone      Profit\n";
 for(i=0;i<3;i++)  //Display the data
 {
   for(j=0;j<3;j++)
   {
      if(a[i][j]==0)
      {
           p+=b[i][j]; //Maximum profit value
           cout<<"S"<<i+1<<"\t\t"<<j+1<<"\t\t"<<b[i][j];
      }
   }
  cout<<"\n";
 }
 return p; //Sending the value of profit to the main block
}
```

OUTPUT:

```
ASSIGNMENT PROBLEM

Enter the profit earned by each of the sales persons in all
zones <rowwise>:
a[1][1] :: 10
a[1][2] :: 17
a[1][3] :: 14
a[2][1] :: 16
a[2][2] :: 8
a[2][3] :: 11
a[3][1] :: 14
a[3][2] :: 15
a[3][3] :: 16

STEP 1 :: FINDING THE MAXIMUM VALUE OF THE MATRIX
The maximum of all the numbers in the matrix is :: 17

STEP 2 :: Matrix derived by subtracting maximum from all
the elements ::
7  0  3
1  9  6
3  2  1

STEP 3 :: Matrix derived by subtracting smallest number in
each row from every n
umber in that row
```

```
7 0 3
0 8 5
2 1 0

STEP 4 :: Assigning the salesman to each zone
       Salesperson    Sales_zone    Profit
           s1              2          17
           s2              1          16
           s3              3          16
The total profit earned is : 49
```

❑❑❑

Program 11: Queuing Problem

Problem Description:

Write a program to solve the following queuing problem:

A barber shop customers arriving in a poisson fashion at the interval of every 20 mins. Each barber services customers according to an exponential distribution with a mean of 15 minutes. Further, if a customer arrives and there are no empty chairs in the shop he will leave.

Calculate the following:

1. Probability of not waiting at the barber shop
2. Expected no. of customers in the barber shop
3. Expected time to be spent by each customer
4. Average increase to warrant a new barber

Program:

```
#include<iostream.h>
#include<conio.h>
#include<math.h>
int main()
{

    double st,em,pm,at,po,exc,wt,incr;
    cout<<"\n Enter service time in min: ";
    cin>>st;
    cout<<"\n Enter arrival time in min: ";
    cin>>at;
    em=1/(st/60);          //Calculating exponential mean
    pm=1/(at/60);          //Calculating poission mean
    po=1-pm/em;            //Calculating probability of ot
waiting in the customer's shop
    exc=pm/(em-pm);        /*Expected no. of customers in the
barber shop*/
    wt=(pm/em)*(em-pm);        /* Expected time to be spent by
the customer*/
    incr=(wt*pow(em,2))/(1+em*wt);    /*Average increase to
warrant new barber*/
    cout<<"\nProbability of not waiting at the barber shop =
"<<po;
    cout<<"\nExpected no. of customers in the barber shop =
"<<exc;
    cout<<"\nExpected time to be spent by the customer =
"<<wt;
    cout<<"\nAverage increase to warrant new barber = "<<incr;
```

```
    getch();
    return 0;
}
```

OUTPUT :

```
Enter service time in min: 15

Enter arrival time in min: 20

Probability of not waiting at the barber shop = 0.25
Expected no. of customers in the barber shop = 3
Expected time to be spent by the customer = 0.75

Average increase to warrant new barber = 3
```

❑❑❑

Program 12: Queuing problem:

Problem Description:

Vehicles pass through a toll gate at a rate of 90 per hr. The average time to pass through the gate is 36 seconds. The arrival rate and service rate follow Poisson distribution. There is a complaint that the vehicles wait for long duration. The authorities are willing to install one more gate to reduce the average time to pass through the toll gate to 30 seconds if the idle time of the toll gate is less than 10% and the average queue length at gate is more than 5 vehicle. Check whether the installation of second gate is justified.

Program:

```
#include<iostream.h>
#include<conio.h>
#include<math.h>

void main()
{
    double lm,nrho,t1,t2,mu,nmu,rho,aql,Lq,idt,pidt,d2;

    cout<<"\nEnter the no. of vehicles passing through the
toll gate per hour :: ";
    cin>>lm;    //input value of `lambda'
    cout<<"\nEnter the avg. time to pass through the gate ::
";
    cin>>t1;    //input value of initial passing time
    cout<<"\nEnter the avg. time to pass through the gate
after installation of another gate :: ";
    cin>>t2;    //input value of final passing time
    cout<<"\nEnter the required idle time after installing
new gate :: ";
    cin>>idt;   //input value of final idle time of gate
    cout<<"\nEnter the min. queue length after new gate
installation :: ";
    cin>>aql;   //input value of final average queue length
    mu=60*60/t1;      //calculate initial 'mu ie departure
//rate of vehicles through gate' in per hour
    rho=lm/mu;        //calculate utilisation factor 'rho'
    Lq=pow(lm,2)/(mu*(mu-lm)); /*calculate no. of vehicles
waiting in queue*/
    cout<<"\nNumber of vehicles waiting in the queue, Lq =
"<<Lq;
    nmu=60*60/t2;   //calculate final 'departure rate' per hour
```

```
        nrho=lm/nmu;        //calculate final utilisation factor
        cout<<"\nUtilisation factor for new gate installation =
"<<nrho;
        pidt=(1-nrho)*100;    //calculate % idle time

        if(pidt<idt)    /*check if % idle time is within given
range*/
                cout<<"\nAs the percentage idle time after new
gate installation is less than "<<idt<<",the installation of
new gate is justified.";
        else
                cout<<"\nAs the percentage idle time after new
gate installation is greater than "<<idt<<",the installation
of new gate is not justified.";

        getch();
    }
```

OUTPUT:

```
   Enter the no.of vehicles passing through the toll gate per
hour :: 90
   Enter the average time to pass through the gate :: 36
   Enter the average time to pass through the gate after
installation of another gate :: 30
   Enter the required idle time after installing new gate :: 10
   Enter the min. Queue length after new gate installation :: 5

   Number of vehicles waiting in the queue ,lq = 8.1
   Utilisation factor for new gate installation:: 0.75
   As the percentage idle time after new gate installation is
greater than 10, the Installation of new gate is not justified.
```

❑❑❑

Program 13: Sequencing Problem:

Problem Description:

write a program to solve the following sequencing problem:

There are five jobs each of which must go through the two machines A and B in the order AB. Processing times are given below:

Processing Time					
Job	1	2	3	4	5
Machine A	5	1	9	3	10
Machine B	2	6	7	8	4

Determine a sequence for five jobs that will minimize the elapsed time T.

Program:

```
#include<iostream.h>
#include<conio.h>

int main()
{
    //clrscr();
    int a[5],b[5],c[3][5],ta[5][2],tb[5][2],f[5],i,j,k,idt=0;

    /*Taking input for the processing times for jobs on m/c
A and B simultaneously*/
    cout<<"\nEnter the processing time for jobs on m/c A & B
in Hrs. ::"<<endl;
    for(i=0;i<5;i++)
    {
        cout<<"For Job "<<i+1<<" ::"<<endl;
        cout<<"Time on Machine A :";
        cin>>a[i];
        cout<<"Time on Machine B :";
        cin>>b[i];
    }

    /*calculating the min. processing time of the two m/c,
along with the m/c with less time and for which job*/
    for(i=0;i<5;i++)
    {
```

```
        if(a[i]<b[i])          //min. m/c time in m/c A
        {
           c[0][i]=a[i];       //processing time m/c A
           c[1][i]=1;
        }
        if(b[i]<a[i])          //min. m/c time in m/c B
        {
           c[0][i]=b[i];       //processing time m/c B
           c[1][i]=2;
        }
        c[2][i]=i+1;           //job no.
     }

     //Arranging the jobs in order of least m/c ing time
     for(i=0;i<5;i++)
     {
        for(j=i+1;j<5;j++)
        {
           if(c[0][i]>c[0][j])
           {
              k=c[0][i];  //swap processing time
              c[0][i]=c[0][j];
              c[0][j]=k;

              k=c[1][i];  //swap m/c no.
              c[1][i]=c[1][j];
              c[1][j]=k;

              k=c[2][i];  //swap job no.
              c[2][i]=c[2][j];
              c[2][j]=k;
           }
           else
              continue;   /*move to next processing time with-
out swaping any values*/
        }
     }

     // Finding the optimal sequence
     for(i=0,j=0;i<5;i++)     //loop for machine A jobs
     {
        if(c[1][i]==1)    //check for m/c A
```

```
        {
            f[j]=c[2][i]-1;     //store job from left
            j++;
        }
        else
            continue;
    }

    for(i=0,j=4;i<5;i++)        //loop for m/c B
    {
        if(c[1][i]==2)          //check for m/c B
        {
            f[j]=c[2][i]-1;     //store job from right
            j--;
        }
        else
            continue;
    }

    /*Finding Values for overall processing time for each
job*/
    for(i=0;i<5;i++)            //loop for m/c A
    {
        if(i!=0)                //time in calculation
            ta[i][0]=ta[i-1][1];
        else
            ta[i][0]=0;
        ta[i][1]=ta[i][0]+a[f[i]]; //time out calculation
    }

    for(i=0;i<5;i++)            //loop for m/c B
    {
        if(i==0)                //calculation for Ist time in
        {
            tb[i][0]=ta[i][1];
        }

        else
        {
            if(ta[i][1]<tb[i-1][1])     //time in calculation
            {
                tb[i][0]=tb[i-1][1];
```

```
            }
            else
            {
                idt++;
                tb[i][0]=ta[i][1];
            }
        }
        tb[i][1]=tb[i][0]+b[f[i]];     //time out calculation
    }

    //clrscr();

    //printing the final result
    cout<<"Job Sequence is:\n ";
    for(i=0;i<5;i++)
    {
        cout<<f[i]+1<<"\t";
     }
     cout<<"\n\n";
    cout<<"          \n\t Machine A \t\t\t Machine B ";
     cout<<"\n\tTime in \tTimeout \tTime in \tTime out\n";

     for(i=0;i<5;i++)
    {

        cout<<"\t"<<ta[i][0]<<"\t\t"<<ta[i][1]<<"\t\t";
        cout<<tb[i][0]<<"\t\t"<<tb[i][1]<<"\n";
    }
    cout<<"\nThe total idle time for m/c A = 0";
    cout<<"\nThe total idle time for m/c B = "<<idt;

    getch();
}
```

OUTPUT:

```
Enter the processing time for jobs on m/c A and B in Hrs ::

For Job 1 ::
Time on Machine A  :5
Time on Machine B  :2
```

```
For Job 2 ::
Time on Machine A   :1
Time on Machine B   :6

For Job 3 ::
Time on Machine A   :9
Time on Machine B   :7

For Job 4 ::
Time on Machine A   :3
Time on Machine B   :8

For Job 5 ::
Time on Machine A   :10
Time on Machine B   :4

Job Sequence is:
2          4          3          5          1
```

Machine A	
Time in	Time out
0	1
1	4
4	13
13	23
23	28

Machine B	
Time in	Time out
1	7
7	15
15	22
23	27
28	30

```
Total idle time for m/c A = 0
Total idle time for m/c B = 2
```

❑❑❑

Program 14: Write a program to implement Gaming Theory

Program:

```
#include<stdio.h>
#include<iostream.h>
#include<stdlib.h>
#include<conio.h>
int maximum(int a,int b)
{
    if(a>b)
    return a;
    else
    return b;
}
int minimum(int c,int d)
{
    if(c<d)
    return c;
    else
    return d;

}

int main()
{
int i,j,odr;
float min1,min2,max1,max2,minc,maxr,a[6][6],b[6][6];
float s1,s2,s3,s4,value;
cout<<"\n Enter the order of matrix: \n";
cin>> odr;
cout<<"\n Enter the elements of payoff matrix:\n";
for(i=0;i< odr;i++)
{
 for(j=0;j< odr;j++)
    { cin>>a[i][j];
    }
}
cout<<"\nThe payoff matrix is: \n";
for(i=0;i< odr;i++)
{
    for(j=0;j< odr;j++)
    {
```

```
            cout<<a[i][j]<<"\t";
        }
      cout<<"\n";
    }

    //Calculating value of minc
    max1=maximum(a[0][0],a[1][0]);
    max2=maximum(a[0][1],a[1][1]);
    minc=minimum(max1,max2);

    //Calculating value of maxc
    min1=minimum(a[0][0],a[0][1]);
    min2=minimum(a[1][0],a[1][1]);
    maxr=maximum(min1,min2);

    cout<<"\nThe max of row is:"<<maxr;
    cout<<"\nThe min of col is:"<<minc;
    if(minc==maxr)
    {
    cout<<"\nThis is a pure strategy:\n";
    cout<<"\nThe value of the game is :"<<minc;
    }
    else
    {
    cout<<"\nThis is a mixed strategy\n";
    s1=(a[1][1]-a[1][0])/((a[0][0]+a[1][1])-(a[0][1]+a[1][0]));
    s2=1-s1;
    s3=(a[1][1]-a[0][1])/((a[0][0]+a[1][1])-(a[0][1]+a[1][0]));
    s4=1-s3;
    value=((a[0][0]*a[1][1])-(a[0][1]*a[1][0]))/
((a[0][0]+a[1][1])-(a[0][1]+a[1][0]));
    cout<<"The value of the game is:"<<value;
    cout<<"\nThe strategies are for A is: \n"<<s1<<"\t"<<s2;
    cout<<"\n";
    cout<<"The strategies are for B is :\n"<<s3<<"\t"<<s4;
    }
    getch();
    return 0;
    }
```

OUTPUT:

```
Enter the order of matrix: 2
Enter the element of the pay off matrix:
7
3
5
6
The payoff matrix is:
7.000000            3.000000
5.000000           6.000000
The max of row is: 5.000000
The min of col is: 6.000000
This is a mixed strategy.
The value of the game is : 5.400000
The strategies for A is:
<0.200000, 0.800000>
The strategies for B is:
<0.600000, 0.400000>
```

❑❑❑

Program 15: CPM/PERT

Problem Description:

Write a program to solve the following networking problem:

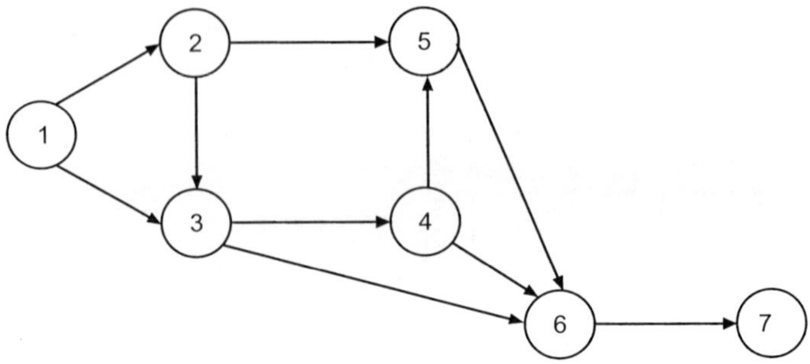

Find the critical path for the following network

Activity	Time
1-2	15
1-3	15
2-3	3
2-5	5
3-4	8
3-6	12
4-5	1
4-6	14
5-6	3
6-7	14

Program:

```
#include<iostream.h>
#include<stdio.h>
#include<conio.h>
#include<string.h>
//function for comparing three numbers and returning the larger value
int max( int a, int b, int c)
{
int mx,i;
if(a>b)
i=a;
else i=b;
if(i>c)
mx=i;
```

```
   else mx=c;

   //Returning maximum value
   return mx;
   }

   int main()
   {

     // Initialising the Variables
   int t12,t13,t23,t25,t34,t36,t45,t46,t56,t67,es[7];

   /* Taking Input for time taken to accomplish various ac-
tivities*/
      cout<<" Enter time taken from 1-2\n";
      cin>>t12;
      cout<<" Enter time taken from 1-3\n";
      cin>>t13;
      cout<<" Enter time taken from 2-3\n";
      cin>>t23;
      cout<<" Enter time taken from 2-5\n";
      cin>>t25;
      cout<<" Enter time taken from 3-4\n";
      cin>>t34;
      cout<<" Enter time taken from 3-6\n";
      cin>>t36;
      cout<<" Enter time taken from 4-5\n";
      cin>>t45;
      cout<<" Enter time taken from 4-6\n";
      cin>>t46;
      cout<<" Enter time taken from 5-6\n";
      cin>>t56;
      cout<<" Enter time taken from 6-7\n";
      cin>>t67;

   // Calculating the critical Path

   es[0]=0;
   es[1]=es[0]+t12;            /*Calculating time for ac-
tivity 1-2*/
   es[2]=max(t13,es[1]+t23,0);   /*Checking if time for 1-3
is greater than 1-2*/
```

```
    es[3]=es[2]+t34;                //Calculating time till 3-4
    es[4]=max(es[1]+t25,es[3]+t45,0);
    es[5]=max(es[2]+t36,es[4]+t56,es[3]+t46);
    es[6]=es[5]+t67;

    cout<<"\nThe resultant critical path is:\n1-2-3-4-5-6-7\n
The time taken is: "<<es[6];
    getch();
    }
```

OUTPUT:

```
    Enter time taken from 1-2
    15
    Enter time taken from 1-3
    15
    Enter time taken from 2-3
    3
    Enter time taken from 2-5
    5
    Enter time taken from 3-4
    8
    Enter time taken from 3-6
    12
    Enter time taken from 4-5
    1
    Enter time taken from 4-6
    14
    Enter time taken from 5-6
    3
    Enter time taken from 6-7
    14

    The resultant critical path is :
    1-2-3-4-5-6-7
    The time taken is:
    54
```

Appendix

TABLE 1
NORMAL DISTRIBUTION

An entry in the table is the proportion under the entire curve which is between $z = 0$ and a positive value of z. Area for negative values of z are obtained by symmetry.

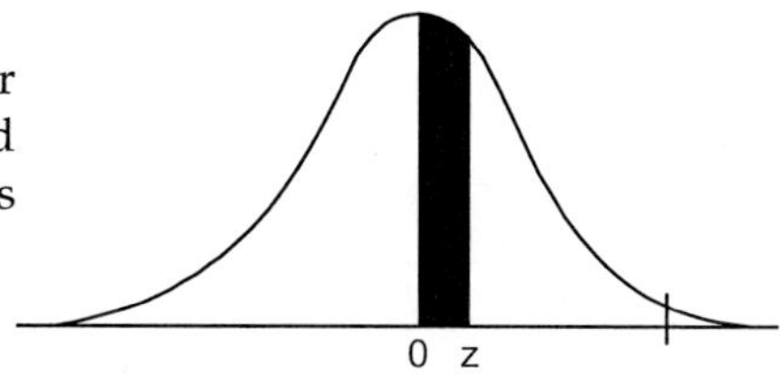

TABLE 1

z	.00	.01	.02	.03	.04	.05	.06	.07	.08	.09
0.0	.0000	.0040	.0080	.0120	.0160	.0199	.0239	.0279	.0319	.0359
0.1	.0398	.0438	.0478	.0517	.0557	.0596	.0636	.0675	.0714	.0753
0.2	.0793	.0832	.0871	.0910	.0948	.0987	.1026	.1064	.1103	.1141
0.3	.1179	.1217	.1255	.1293	.1331	.1368	.1406	.1443	.1480	.1517
0.4	.1554	.1591	.1628	.1664	.1700	.1736	.1772	.1808	.1844	.1879
0.5	.1915	.1950	.1985	.2019	.2054	.2088	.2123	.2157	.2190	.2224
0.6	.2257	.2291	.2324	.2357	.2389	.2422	.2454	.2486	.2517	.2549
0.7	.2580	.2611	.2642	.2673	.2703	.2734	.2764	.2794	.2823	.2852
0.8	.2881	.2910	.2939	.2967	.2995	.3023	.3051	.3078	.3106	.3133
0.9	.3159	.3186	.3212	.3238	.3264	.3289	.3315	.3340	.3365	.3389
1.0	.3413	.3438	.3461	.3485	.3508	.3531	.3554	.3577	.3599	.3621
1.1	.3643	.3665	.3686	.3708	.3729	.3749	.3770	.3790	.3810	.3830
1.2	.3849	.3869	.3888	.3907	.3925	.3944	.3962	.3980	.3997	.4015
1.3	.4032	.4049	.4066	.4082	.4099	.4115	.4131	.4147	.4162	.4177
1.4	.4192	.4207	.4222	.4236	.4251	.4265	.4279	.4292	.4306	.4319
1.5	.4332	.4345	.4357	.4370	.4382	.4394	.4406	.4418	.4429	.4441
1.6	.4452	.4463	.4474	.4484	.4495	.4505	.4515	.4524	.4535	.4545
1.7	.4554	.4564	.4573	.4582	.4591	.4599	.4608	.4616	.4625	.4633
1.8	.4641	.4649	.4656	.4664	.4671	.4678	.4686	.4693	.4699	.4706
1.9	.4713	.4719	.4726	.4732	.4738	.4744	.4750	.4756	.4761	.4767
2.0	.4772	.4778	.4783	.4788	.4793	.4798	.4803	.4808	.4812	.4817
2.1	.4821	.4826	.4830	.4834	.4838	.4842	.4846	.4850	.4854	.4857
2.2	.4861	.4864	.4868	.4871	.4875	.4878	.4881	.4884	.4887	.4890
2.3	.4893	.4896	.4898	.4901	.4904	.4906	.4909	.4911	.4913	.4916
2.4	.4918	.4920	.4922	.4925	.4927	.4929	.4931	.4932	.4934	.4936
2.5	.4938	.4940	.4941	.4943	.4945	.4946	.4948	.4949	.4951	.4952
2.6	.4953	.4955	.4956	.4957	.4959	.4960	.4961	.4962	.4963	.4964
2.7	.4965	.4966	.4967	.4968	.4969	.4970	.4971	.4972	.4973	.4974
2.8	.4974	.4975	.4976	.4977	.4977	.4978	.4979	.4979	.4980	.4981
2.9	.4981	.4982	.4982	.4983	.4984	.4984	.4985	.4985	.4986	.4986
3.0	.4987	.4987	.4987	.4988	.4988	.4989	.4989	.4989	.4990	.4990

TABLE 2
RANDOM NUMBERS

03991	10461	93716	16894	98953	72231	39528	72484	82474	25593
38555	95554	32886	59780	09958	18065	81616	18711	53342	44276
17546	73704	92052	46215	15917	06253	07586	16120	82641	22820
32643	52861	95819	06831	19640	99413	90767	04235	13574	17200
69572	68777	39510	35905	82244	35159	40188	28193	29593	88627
24122	66591	27699	06494	03152	19121	34414	82157	86887	55087
61196	30231	92962	61773	22109	78508	63439	75363	44989	16822
30532	21704	10274	12202	94205	20380	67049	09070	93399	45547
03788	97599	75867	20717	82037	10268	79495	04146	52162	90286
48228	63379	85783	47619	87481	37220	91704	30552	04737	21031
88618	19161	41290	67312	71857	15957	48545	32247	18619	13674
71299	23853	05870	01119	92784	26340	75122	11724	74627	73707
27954	58909	82444	99005	04921	73701	92904	13141	32392	19763
80863	00514	20247	81759	45197	25332	69902	63742	78464	22501
33564	60780	48460	85558	15191	18782	94972	11598	62095	36787
90899	75754	60833	25983	01291	41349	19152	00023	12302	80783
78038	70267	43529	06318	38384	74761	36024	00867	76378	41605
55896	66485	88722	56736	66164	49431	94458	74284	05041	49807
87539	08823	94813	31900	54155	83436	54158	34243	46978	35482
16818	60311	74457	90561	72848	11834	75051	93029	47665	64382
34677	58300	74910	64345	19325	81549	60365	94653	35075	33949
45305	07521	61318	31855	14413	70951	83799	42402	56623	34442
59747	67277	76503	34513	39663	77544	32960	07405	36409	83232
16520	69676	11654	99893	02181	68161	19322	53845	57620	52606
68652	27376	92852	55866	88448	03584	11220	94747	07399	37408
79375	95220	01159	63267	10622	48391	31751	57260	68980	05339
33521	26665	55823	47641	86225	31704	88492	99382	14454	04504
59589	49067	66821	41575	49767	04037	30934	47744	07481	83828
20554	91409	96277	48257	50816	97616	22888	48893	27499	98748
59404	72059	43947	51680	43852	59693	78212	16993	35902	91386
42614	29297	01918	28316	25163	01889	70014	15021	68971	11403
34994	41374	70071	14736	65251	07629	37239	33295	18477	65622
99385	41600	11133	07586	36815	43625	18637	37509	14707	93997
66497	68646	78138	66559	64397	11692	05327	82162	83745	22567
48509	23929	27482	45476	04515	25624	95096	67946	16930	33361
15470	48355	88651	22596	83761	60873	43253	84145	20368	07126
20094	98977	74843	93413	14387	06345	80854	09279	41196	37480
73788	06533	25897	20405	51321	92246	80088	77074	66919	31678
60530	45128	74022	84617	72472	00008	80890	18002	35352	54131
44372	15486	65741	14014	05466	55306	93128	18464	79982	68416

TABLE 2
(Continued)

18611	19241	66083	24653	84609	58232	41849	84547	46850	52326
58319	15997	08355	60860.	29735	47762	46352	33049	69248	93460
61199	67940	55121	29281	59076	07936	11087	86294	14013	31792
18627	90872	00911	08936	76335	93779	52701	08337	56303	87315
00441	58997	14060	40619	29549	69616	57275	36898	81304	48585
32624	68691	14845	46672	61958	77100	20857	73156	70284	24326
65961	73488	41839	55382	17267	70943	15633	84924	90415	93614
20288	34060	39685	23309	10061	68829	92694	48297	39904	02115
59362	95938	74416	53166	35208	33374	77613	19019	88152	00080
99782	93478	53152	67433	35663	52972	38688	32486	45134	63445
27767	43585	85301	88977	29490	69714	94015	64874	32444	48277
13025	14338	54066	15243	47724	66733	74108	88222	88570	74015
80217	36292	98525	24335	24432	24896	62880	87873	95160	59221
10875	62004	90391	61105	57411	06368	11748	12102	80580	41867
54127	57326	26629	19087	24472	88779	17944	05600	60478	03343
60311	42824	37301	42678	45990	43242	66067	42792	95043	52680
49729	71484	92003	98086	76668	73209	54244	91030	45547	70818
78626	51594	16453	94614	39014	t 97066	30945	57589	31732	57260
66692	13986	99837	00582	81232	'44987	69170	37403	86995	90307
44071	28091	07362	97703	76447	42537	08345	88975	35841	85771
59820	96163	78851	16499	87064	13075	73035	41207	74699	09310
25704	91035	26313	77463	55387	72681	47431	43905	31048	56699
22304	90314	78438	66276	18396	73538	43277	58874	11466	16082
17710	59621	15292	76139	59526	52113	53856	30743	08670	84741
25852	58905	55018	56374	35824	71708	30540	27886	61732	75454
46780	56487	75211	10271	36633	68424	17374	52003	70707	70214
59849	96169	87195	46092	26787	60939	59202	11973	02902	33250
47670	07654	30342	40277	11049	72049	83012	09832	25571	77628
94304	71803	73465	09819	58469	35220	09504	96412	90193	79568
08105	59987	21437	36786	49226	77837	98524	97831	65704	09514
64281	61826	18555	64937	64654	25843	41145	42820'.'	14924	39650
66847	70495	32350	02985	01755	14750	48968	3860^	70312	05682
72461	33230	21529	53424	72877	17334	39283	0414§	90850	64618
21032	91050	13058	16218	06554	07850	73950	79552	24781	89683
95360	67011	06651	16136	57216	39618	49856	99326	40902	05069
49712	97380	10404	55452	09971	59481	37006	21186	72682	07385
58275	61764	97586	54716	61459	21647	87417	17198	21443	41808
89514	11788	68224	23417	46376	25366	94746	49580	01176	28838
15472	50669	48139	36732	26825	05511	12459	91314	80582	71944
12120	86124	51247	44302	87112	21476	14713	71181	13177	55292

TABLE 2
(Continued)

95294	00556	70481	06905	21785	41101	49386	54480	23604	23554
66986	34099	74474	20740	47458	64809	06312	88940	15995	69321
80620	51790	11436	38072	40405	68032	60942	00307	11897	92674
55411	85667	77535	99892	71209	92061	92329	98932	78284	46347
95083	06783	28102	57816	85561	29671	77936	63574	31384	51924
90726	57166	98884	08583	95889	57067	38101	77756	11657	13897
68984	83620	89747	98882	92613	89719	39641	69457	91339	22502
36421	16489	18059	51061	67667	60631	84054	40455	99396	63680
92638	40333	67054	16067	24700	71594	47468	03577	57649	63266
21036	82803	77501	97427	76479	68562	43321	31370	28977	23896
13173	33365	41468	85149	49554	17994	91178	10174	29420	90438
86716	38746	94559	37559	49678	53119	98189	81851	29651	84215
92581	02262	41615	70360	64114	58660	96717	54244	10701	41393
12470	56500	50273	93113	41794	86861	39448	93136	25722	08564
01016	00857	41396	80504	90670	08289	58137	17820	22751	36518
34030	60726	25807	24260	71529	78920	47648	13885	70669	93506
30259	46345	06170	97965	88302	98041	11947	56203	19324	20504
73959	76145	60808	54444	74412	81105	69181	96845	38525	11600
46874	37088	80940	44893	10408	36222	14004	23153	69249	05747
60883	52109	19516	90120	46759	71643	62342	07589	08999	05985

Tutorial Question Bank

Tutorial Sheet No. 1

1. Solve

$$\text{Max } Z = 5A + 8B$$

$$\text{Subjected to, } 3A + 2B \geq 3$$
$$A + 4B \geq 4$$
$$A + B \leq 5$$
$$A, B \geq 0$$

(Z = 40)

2. In relation to linear programming, explain the implications of the following assumption of the model.
   (*i*) Linearity of the objective function
   (*ii*) Continuous variables
   (*iii*) Certainty.
3. Define the term
   (*i*) Basic variable
   (*ii*) Basic Solution
   (*iii*) Basic feasible solution
   (*iv*) Degeneracy.
4. Write short notes on
   (*i*) Inconsistency and redundancy of constraint
   (*ii*) Cycling in linear programming problems.
5. Explain the use of artificial variable in linear programming.
6. Define a basic solution to a given system of m simultaneous linear equations in n unknowns.
7. In the course of simplex table calculations, describe how you will detect a degenerate, an unbounded and a non existing feasible solution.
8. How would you resolve the following complications in LPP?
   (*i*) Minimization
   (*ii*) Equalities in constraints
   (*iii*) Tie for the leaving basic variable.
9. Short notes on
   (*i*) Alternative optima in LPP
   (*ii*) Generalized LPP.
10. Solve, Max $Z = 3A + 2B$

$$\text{Subject to, } 2A + B \leq 2$$
$$3A + 4B \geq 12$$
$$A, B \geq 0$$

(No solution)

Tutorial Sheet No. 2

1. Solve by Big M method

   Minimize $Z = 4A + B$

   Subject to, $3A + B = 3$

   $4A + 3B \geq 6$

   $A + 2B \leq 4$

   $A, B \geq 0$ [**Ans.** $A = 2/5$, $B = 9/5$, $Z = 17/5$]

2. Solve the problem 1 by 2-Phase method.
3. Use 2 phase Simplex Method to:

   Maximise $Z = 5A + 2B$

   Subject to $2A + B \leq 1$

   $A + 4B \geq 6$

   $A, B = 0$ [**Ans.** No Feasible Solution]

4. What is the function of Minimum ratio rule in Simplex Method?
5. What is sensitivity analysis and why do we perform it?
6. Solve the following LP problem.

   Maximize $Z = 3A + 2B - 5C$

   Subject to $A + B \leq 2$

   $2A + B + 6C \leq 6$

   $A - B + 3C = 0$

   $A, B, C \geq 0$ [**Ans.** $A = 1$, $B = 2$, $C = 0$, $Z = 5$]

7. Solve the above LP problem if the RHS of primal is changed from (2, 6, 0) to (2,10,5). Find the new optimal solution. [**Ans.** $A = 2$, $B = 0$, $C = 1$, $Z = 1$]
8. Solve the following LPP using Dual- simplex method:

   Minimize $Z = A + 2B + 3C$

   Subject to $2A - B + C \geq 4$

   $A + B + 2C \leq 8$

   $A - C \geq 2$

   $A, B, C \geq 0$ [**Ans.** $A = 3$, $B = 2$, $Z = 7$]

9. Solve the following LP problem

   Maximize $Z = 20A + 80B$

   Subject to $4A + 6B \leq 90$

   $8A + 6B \leq 100$

   $A, B \geq 0$ [**Ans.** $A = 0$, $B = 15$, $Z = 1200$]

10. In the above problem if the following new constraint is added to the model, find the solution of the problem.

    New Constraint is $5A + 4B \leq 80$ [**Ans.** No Change]

□□□

Tutorial Sheet No. 3

1. Find the initial basic feasible solution of the following transportation problem by Vogel's approximation method.

		Warehouses				
		W1	W2	W3	W4	CAPACITY
	F1	10	30	50	10	7
Factory	F2	70	30	40	60	9
	F3	40	8	70	20	18
Requirement		5	8	7	14	34

2. A company has four salesmen who are to be assigned to four different sales territories. The monthly sales increases estimated for each salesman in different territories (in lac rupees) are shown in the following table

		Sales territory				
		1	2	3	4	5
	A	75	80	85	70	90
Salesman	B	91	71	82	75	85
	C	78	90	85	80	80
	D	65	75	88	85	90

Suggest optimal assignment for the salesman who is to be assigned to four different sales territories. Which sales territories will remain un-assigned? What will be the maximum sales increase every month?

3. In above question, if for certain reasons D can't be assigned to territory 3,will the optimal assignment schedule be different? If so, show the new assignment schedule.

4. Find the shortest path from 1 to 12 through the network given

[**Ans.** 1-3-6-10-12, Cost 12]

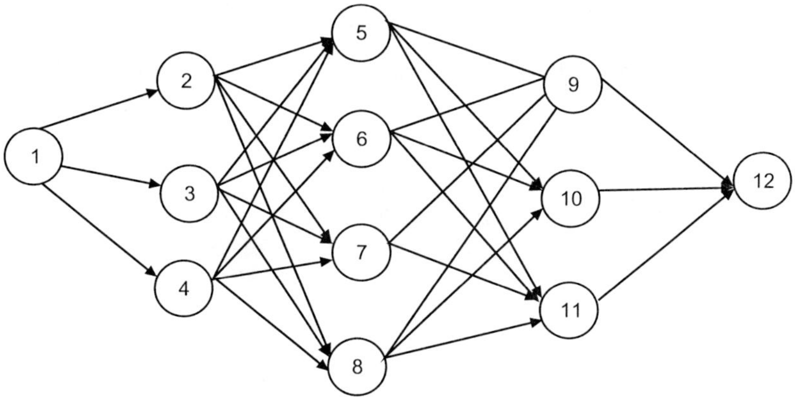

4. In problem 4 find the longest path connecting 1 and 12.

[**Ans.** 1-3-8-9-12, Cost 33]

5. Show that the assignment model is a special case of transportation model.
6. If the matrix element represents the unit transportation time solve the following transportation problem

		To				Available
		D1	D2	D3	D4	
From	O1	10	0	20	11	15
	O2	1	7	9	20	25
	O3	12	14	16	18	5
	Requirement	12	8	15	10	45

[**Ans.** 12]

7. Use least cost method to find initial basic feasible solution and then find the maximum transportation cost for the following problem [**Ans.** 180]

	D_1		D_2		D_3		D_4	
O_1	1	(20)	2		1	(10)	4	
O_2	3		3	(20)	2	(20)	1	(10)
O_3	4		2	(20)	5		9	

9. An enterprise has three factories at locations A, B and C which supplies three shops located at D, E and F. Monthly factory capacities are 10, 80 and 15 units respectively. Monthly shop requirements are 75, 20 and 50 units respectively. Unit shipping costs (in Rs) are given below

Factory	Shops		
	D	E	F
A	5	1	7
B	6	4	6
C	3	2	5

The penalty costs are not satisfying demand at the shops D, E and F are Rs. 5, Rs. 3 and Rs. 2 respectively. Determine the optimal distribution.

(**Ans** (*A*, *E*) = 10 units, (*B*, *D*) = 60 units, (*B*, *E*) = 10 units, (*B*, *F*) =10 units, (*C*, *D*) =15 units, Penalty Rs. 80)

10. Consider a transportation problem with m=3 and n=4 where

$C_{11} = 2$	$C_{12} = 3$	$C_{13} = 11$	$C_{14}=7$
$C_{21} = 1$	$C_{22} = 0$	$C_{23} = 6$	$C_{24}=1$
$C_{31} = 5$	$C_{32} = 8$	$C_{33} = 15$	$C_{34}=9$

Suppose $S_1 = 6$, $S_2 = 1$ and $S_3 = 10$ whereas $D_1 = 7$, $D_2 = 5$ and $D_4 = 2$. Apply the transportation simplex method to find out an optimal solution.

[**Ans.** Minimum cost = Rs. 100]

Tutorial Sheet No. 4

1. State the Transportation problem in general terms.
2. Explain the problem of degeneracy. How does one overcome it?
3. Differentiate between Transportation and Assignment Problems.
4. How transportation problem is solved when demand and supply are not equal.
5. Using VAM, solve the following transportation problem for maximum profit:

	A	B	C	D	SUPPLY
	12	18	06	25	200
	8	7	10	18	500
	14	03	11	20	300
DEMAND	180	320	100	400	

[**Ans.** 15300]

6. Find the optimal transportation cost for the following:

	A	B	C	D	SUPPLY
	03	07	09	10	120
	04	05	07	12	170
	09	12	05	06	210
DEMAND	290	150	70	90	

[**Ans.** Rs. 2430/-]

7. A student has to take examination in three courses X, Y, Z. He has three days available for studies. He feels that it would be best to devote the whole day to study the same course so that he may study the course for one day, two days or three days or not at all. His estimate of grades he may get by studying are as follows:

Study Days/ courses	X	Y	Z
0	1	2	1
1	2	2	2
2	2	4	4
3	4	5	4

How should he plan to study so that he maximizes the sum of his grades?

[**Ans.** Maximum return = 5, X =1, Y = 0, Z = 2]

8. A company has plants A, B, C that have capacities to produce 300 kg, 200 kg, 500 kg respectively of a particular chemical per day. The production cost

per kg in these plants are Rs. 70/-, Rs. 60/- and Rs. 66/- respectively. Four bulk consumers order for the product on the following basis:

		kg required /day	Price offered (Rs/kg)
Consumer	I	400	100
	II	250	100
	III	350	102
	IV	150	103

Shipping costs (Rs/kg) from plants to consumers are given in the table below

		I	II	III	IV
From	A	3	5	4	6
	B	8	11	9	12
	C	4	6	2	8

Work out an optimal schedule for the above situation. Under what conditions would you change the schedule? [**Ans.** Total profit Rs 30700]

9. Consider the following transportation problem

Factory	Godowns						Stock available
	1	2	3	4	5	6	
A	7	5	7	7	5	3	60
B	9	11	6	11	-	5	20
C	11	10	6	2	2	8	90
D	9	10	9	6	9	12	50
DEMAND	60	20	40	20	40	40	

It is not possible to transport any quantity from factory B to godown 5. Determine

1. Initial solution by VAM
2. Optimal basic solution
3. Is the optimum solution unique? If not, find the alternative optimum solution. [**Ans.** degeneracy, Rs. 1120, no, min. cost Rs. 1120/-]

10. Salesman has to reach city no 10 starting from city no 1 by a motorcar. Though its starting and destination points are fixed, he has considerable choice as to which city to travel through en route. The cost Cij (in Rs.) for the standard policy on the motorcar run from city i to city j is given below

	1	2	3		5	6	7		8	9		10
1	2	4	3	2	7	4	6	5	1	4	8	3
				3	3	2	4	6	6	3	9	4
				4	4	1	2	7	3	3		

Find the safest route of traveling so that the total traveling cost is minimum.

[**Ans.** min. cost = Rs. 11)(1 » 4 » 5 » 8 »10), (1» 4 » 6 » 9 » 10)]

Tutorial Sheet No. 5

1. What are the steps in decision making analysis?
2. Explain various quantitative methods, which are useful for decision making under uncertainty.
3. With the given payoff tables, determine the optimal strategy under
   (*i*) Maximin criteria
   (*ii*) Minimax criteria
   (*iii*) Hurwicz criteria with $\alpha = 0.6$
   (*iv*) Laplace criteria [**Ans.** *E.A*]

Events	Strategies			
	A	B	C	D
E1	3	4	4	5
E2	6	2	1	2
E3	1	8	8	3

4. Explain 2 person zero sum game by giving suitable example.
5. A manager has a choice between
   (*i*) A risky contact promising Rs. 7 lakh with probability 0.6 to Rs. 4 lakh with prob. 0.4.
   (*ii*) A diversified portfolio consisting of 2 contracts with independent outcomes each promising Rs 3.5 lakh with pro. 0.6 and Rs. 2 lakh with prob 0.4. Construct a decision tree for it using EMV criterion.
   [**Ans.** EMV Rs. 5.8 lakh]
6. Give an example of a good decision you made not resulted in bad outcome. Also give an example of a good decision you made that had a good outcome. Why was each decision good or bad?
7. Game theory provides a systematical quantitative approach for analyzing competitive simulation in which the competitors make use of logical processes to techniques in order to determine an optimal strategy for winning. Comment.
8. Consider the game for following payoff matrix

	Player B	
	B1	B2
Player B	2	b
	– 2	λ

   (*i*) Show that the game is strictly determinable whatever λ may be.
   (*ii*) Determine the value of game [**Ans.** 2]
9. Solve the following game using Dominance property

		Player B		
		I	II	III
	I	1	7	2
Player A	II	6	2	7
	III	5	2	6

[**Ans.** *A* (2/5, 3/5, 0), *B* (1/2, 1/2, 0), $V = 4$]

10. A producer of boats has estimated the following distribution of demand for a particular kind of boat.

No. demanded	0	1	2	3	4	5	6
Probability	0.14	0.27	0.27	0.18	0.09	0.09	0.01

Each boat costs him Rs. 7000 and he sells them for Rs. 10,000 each. Boats left unsold at the end of the season must be disposed off for Rs. 6000 each. How many boats should be in stock so as to minimize his expected profit.

[**Ans.** Max. EMV 4080 stock 3 boats]

□□□

Tutorial Sheet No. 6

1. A flower merchant purchase roses at Rs. 10/dozen and sell them for Rs. 30. Unsold flowers are donated to the temple. The daily demand for rose has the following probability distribution

Demand(dozen)	7	8	9	10
Probability	0.1	0.2	0.3	0.4

   How many dozen of roses should be purchased in order to minimize the profit. [**Ans.** 9 dozen]
2. In prob.1, also apply EOL criterion.
3. Explain minimax criterion as applied to theory of games.
4. Let (U_{ij}) be the payoff matrix for a 2 person zero sum game. If V' denotes the maximin value and V″ the minimax value of the game, then prove that $V' \geq V''$.
5. A decision problem has been expressed in the following payoff table.

Action	Events		
	I	II	III
A	10	20	26
B	– 30	30	60
C	40	30	20

   Determine which action the executive has to choose if he adopts
   (*i*) Maximin criterion [**Ans.** (20, C)]
6. Solve the above problem by
   Minimax criterion [**Ans.** (60, *B*)]
7. Solve the problem 5 using Hurwicz criterion with $\alpha = 0.7$ [**Ans.** (34, C)]
8. Solve the problem 4 using Laplace criterion. [**Ans.** (30, C)]
9. Solve the problem 4 using Minimax criterion [**Ans.** (34, *A*)]
10. Find the value of game. Also indicate whether they are fair or strictly determinable.

B				
A	1	9	6	0
	2	3	8	– 1
	– 5	– 2	10	– 3
	7	4	– 2	– 5

[**Ans.** $V = 0(A1, B4)$, fair]

Tutorial Sheet No. 7

1. Define reorder point.
2. How is reorder point related to lead time demand?
3. Explain the difference between a periodic review system and a continuous review system.
4. Out of periodic review and continuous review system, which one is better and why?
5. What is the unit for holding cost parameter.
6. Is it possible to have a negative inventory level? If so, explain how?
7. Discuss the probability of inventory control when the stochastic demand is uniform, production of commodity is instantaneous and lead-time is negligible (discrete case).
8. In a private canteen, the daily demand for packet meal follows uniform distribution as presented below,

   $P(X) = 1/(450 - 230) \; 230 \leq X \leq 450$

   The cost of production per packet of meals is Rs. 8. The S.P. is Rs. 16 /packet. The surplus packets on each day are sold at Rs. 6 /pkt in a nearby public place. Find the optimal no. of packets of meals to be prepared each day.

   [**Ans.** 406]
9. A fish stall is planning for its optimal purchases of a costly variety of fish. The daily demand of fish follows normal; distribution with a mean of 800 kg and standard deviation of 75 kg. The purchase price of the fish is 1.50/- per kg. The S.P. is 200/-/kg. If the fish is not sold on the day of purchase, it is sold to a dry fish manufacturing firm at 110/- per kg. Find the optimal daily purchase quantity of fish. [**Ans.** 810.5 kg]
10. Describe the basic characteristic of an inventory system.

❑❑❑

Tutorial Sheet No. 8

1. Explain the nature of the probabilistic model in inventory control.
2. Consider a shop that produces 3 items. The items are produced in lots. The demand rate for each item is constant and can be assumed to be deterministic. No back orders are to be allowed. The data for the items are given below

Items	1	2	3
Holding cost (Rs)	20	20	20
Set up cost (Rs)	50	40	60
Unit cost (Rs)	6	7	5
Demand rate/year	10000	12000	7500

Determine approximately the economic order quantities when the total value of average inventory levels of these items is Rs 1000.

[**Ans.** 114, 110.5, 109 units]

3. A baking company sells one of its types of cakes by weight. It makes a profit of Rs 3.20 per pound on every pound of cake sold on the day it is baked. It disposes off all the cakes not sold on the day they are baked at the rate of 50 paise per pound. If demand is known to be triangular with probability density function,

$f(R) = 0.02 - 0.0002R \quad 0 \le X \le 100$

Find the optimum amount of cake the company should bake daily.

[**Ans.** 63.3 pounds)

4. A baking company sells one of its types of biscuits by weight. It makes a profit of Rs. 9.50 per pound on every kg of biscuits sold on the day it is baked. It disposes off all the biscuits not sold on the day they are baked at the rate of Rs. 1.50 per kg. If demand is known to be rectangular between 300 kg and 400 kg, find the optimum amount of biscuits the company should bake daily. [**Ans.** 386.4 kg]
5. Derive the expression for optimum order level for the case; demand, set up cost zero, stock levels discrete and lead-time zero.
6. A person wants to decide the constituents of a diet which will fulfill his daily requirement of proteins, fats & carbohydrates at the minimum cost. The choice is to be made from 4 different types of foods. The yield per unit of these foods are given as

Food type	Yield per unit Protein	Fat	Carbohydrate	Cost per unit
1	3	2	6	45
2	4	2	4	40
3	8	7	7	85
4	6	5	4	65
Minimum requirement	800	200	700	

Formulate linear programming model.

7. Solve the following game by reducing them to 2 x 2 game by graphical method

A				
3	0	6	– 1	7
– 1	5	– 2	2	1

8. Solve the following 3 x 3 game by linear programming

	Player B		
Player A	1	– 1	– 1
	– 1	– 1	3
	– 1	2	– 1

9. A company gives a lease on a certain property. It may sell it for Rs. 75,000 or may dull the said property for oil. Various possible results are as under along with the probabilities of happening and rupees consequence

Possible result	Probabilities	Rupees consequences
Dry well	.10	– 100000
Gas well only	.40	45000
Oil and gas combination	.30	98000
Oil well	.20	199000

(*a*) Draw a discussion tree for the problem & determine whether the company should drill or sell.

10. The number of machines arriving per day at a factory repair bench has been noted over a range period of time and joined to have the following distribution

Arrivals:	0	1	2	3	4	5	6	7	8	9
Percentage	2	7	15	20	20	16	10		6	31

Repair times depend on the type of fault. The number of repairs completed per day have been recorded & found to be represented by.

Arrival :	1	2	3	4	5	6	7	8	9	10
Percentage:	4	8	14	18	18	15	10	7	4	2

Simulate 25 days to find the maximum queue length using the following series of random numbers.

55	86	87	70	05	71	69	29	74	73	37	17	42	16	89	98	35	81
07	16	94	42	27	18	13	70	05	35	86	85	02	79	13	35	17	38
57	19	97	38	45	82	40	32	63	29	10	67	41	99				

❑❑❑

Tutorial Sheet No. 9

1. List five disadvantages in using simulation analysis.
2. Explain why a simulation experiment must usually be executed over long periods of simulated time.
3. What is a random number generator?
4. Why is this term (random no. generator) a misnomer?
5. What are the desirable properties of random number generators?
6. What is the difference between model verification and model validation.
7. Describe the general probability of M/M/K queuing system and also deduce an explicit expression for the steady state probability of the length of the queue in an M/M/1 system.
8. Four counters are being run on the frontier of a country to check the passport and necessary papers of the tourists. The tourists choose the counter at random. If the arrival at the frontier is poisson at the rate λ and the service time is exponential with parameter $\lambda/2$, what is the steady state average queue at each counter? (**Ans.** 4/23)
9. In a railway yard, goods train arrives at a rate of 30 trains per day. Assuming that the inter-arrival time follows an exponential distribution and the service time is also exponential with an average 36 min, calculate

   (*a*) The mean queue size (line length)

   (*b*) The probability that the queue size exceeds 10.

   If the input of trains increases to an average 33 per day, what will be the change in (*a*) and (*b*). [**Ans.** 3 trains, 0.06, 4.8 or 5 trains, 0.2 approx]
10. List 5 reasons why one would want to use a micro computer based simulation analysis.

❑❑❑

Tutorial Sheet No.10

1. List 5 advantages of using a simulation language.
2. Problems arriving at the computer center in a poisson function at an average rate of 5/day. The rules of the computing centre are that any man waiting to get his problem solved must aid the man whose problem is solved. If the time for solving one problem has an exponential distribution with mean time of $1/3^{rd}$ day and if the average solving time is inversely proportional to the no. of people working on the problem, approximate the expected time in the center for a person entering. [**Ans.** 1/3 day or 8 days]
3. In the production shop of a company, the breakdown of the machines is found to be poisson with an average rate of 3 machines/hour. Breakdown time at one machine costs Rs. 40/hr to the company. There are two choices before the company to hiring the repairman. One of the repairman is slow but cheap and other fast but expensive. The slow cheap repairman demands Rs. 20/hr and will repair the broken down machines exponentially at the rate of 4/hr. The fast –expensive repairman demands Rs. 30/hr and will repair machines exponentially at an average rate of 6/hr. Which repairman should be hired. [**Ans.** cost- 180, cost 70, fast-expensive repairman]
4. Show that for a single service station, Poisson arrivals and exponential service times, the probability that exactly n calling units are in queuing system is $(1 - \rho)\,\rho n$, $n \geq 0$. (ρ = traffic intensity). Also find the expected line length.
5. The material manager of the firm wishes to determine the expected mean demand for a particular item in stock during the re-order lead time. The information is needed to determine how far in advance to reorder, before the stock level is reduced to zero. However, both the lead-time and the demand/day for the item are random variables described by the probability distribution,

Lead time (in days)	Probability	Demand/day(units)	Probability
1	0.4	1	0.15
2	0.35	2	0.25
3	0.25	3	0.40
		4	0.20

   Manually simulate the probability for 30 re-orders to estimate the demand during lead-time.
6. In a railway yard, goods trains arrive at a rate of 30 trains per day. Assuming that the inter-arrival time follows an exponential distribution and the service time distribution is also exponential with an average 36 min., calculate the following,
   (*i*) Average no. of customers in the system
   (*ii*) The probability that the queue size exceeds 10. [**Ans.** 3, 0.06]

7. For the above problem, if the input of the trains increases to an average 33/day, what will be the change in (i) and (ii). [**Ans.** 0.2]
8. A mechanic repairs 4 machines. The mean time between service requirements is 5 hrs for each machine and forms an exponential distribution. The mean repair time is 1 hr and also follows the same distribution pattern. Machine downtime costs Rs. 25 per hr and the machine costs Rs. 55 per day. Find the expected number of operating machines. [**Ans.** 3]
9. Determine the expected downtime lost per day in the above problem. [**Ans.** Rs. 200/day]
10. In problem no. 8, would it be economical to engage 2 mechanics, each repairing only 2 machines. [**Ans.** 0-4 hr/day]

❑❑❑

Tutorial Sheet No. 11

1. Most of the businessmen view "Inventory is a necessary evil". Do you agree with this?
2. Formulate and solve a discrete stochastic model for a single product with lead-time zero. The shortage and shortage costs are independent of time. Set up cost is constant.
3. Discuss the probabilistic inventory models with instantaneous demand and no set up cost.
4. Consider an inventory model in which the holding cost of one unit in an inventory for a specified period is C and the cost of shortage per unit is B. Suppose the demand follows a known continuous probability distribution. Determine the optimum inventory level in the beginning of the period.
5. Write a note on Newspaper boy problem.
6. Define inventory and its importance.
7. What are the advantages and disadvantages of having inventories.
8. Formulate and solve continuous probabilistic reorder point for lot size model to determine optimal reorder point for a presented lot size. Lead-time is finite. Shortages are allowed and fully backlogged.
9. A newspaper boy buys paper for 60 paise each and sales them for Rs. 1.40 each. He cannot return unsold papers. Daily demand forms the following distribution.

NO. OF CUSTOMERS	23	24	25	26	27	28	29	30	31	32
PROBABILITY	0.01	0.03	0.06	0.10	0.2	0.25	0.15	0.10	0.05	0.05

If each day's demand is independent of the previous day demand, how many papers should be ordered each day. **[Ans.** 28]

10. Some of the spare parts of a ship cost Rs. 100000 each. These spare parts can only be ordered with the ship. If not ordered at the time when shipping was construed, these parts can't be available on need. Suppose that a loss of Rs. 1,00,00,000 is suffered for each spare part that is needed as replacement during the life term of the class of the ship discussed are,

Spare parts required	0	1	2	3	4	5
Prob.	0.9488	0.0400	0.0100	0.0010	0.0002	0.0

How many spare parts should be procured? **[Ans.** 3 spare parts]

❑❑❑

Model Question Papers

Model Question Paper-1

OPERATIONS RESEARCH (TME-601)

Time: 3 Hours *Total Marks:* 100

Note : (*i*) Attempt all questions.
(*ii*) All questions carry equal marks.

1. Attempt any one part: (20)

(*a*) (*i*) Define a model. List and describe various types of models used in Operations Research.

(*ii*) Solve of the following Linear Programming problem graphically.

Maximize $Z = 4X_1 + 5X_2$

subject to $4X_1 + 7X_2 \leq 30$

$5X_1 + 2X_2 \leq 15$

$6X_1 + 8X_2 \leq 40$

X_1 and $X_2 \geq 0$

(*b*) Solve the following linear programming problem by Simplex method.

Maximize $Z = x_1 + 2x_2 + 3x_3 - x_4$

subject to $x_1 + 2x_2 + 3x_3 = 15$

$2x_1 + x_2 + 5x_3 = 20$

$x_1 + 2x_2 + x_3 + x_4 = 10$

$x_1, x_2, x_3, x_4 \geq 0$

2. Attempt any two parts of the following: (10 × 2 = 20)

(*a*) Solve the following transportation problem. The figures in the matrix indicate cost of transportation.

		Destinations 1	2	3	4	*Availability*
Sources	1	21	16	25	13	11
	2	17	18	14	23	13
	3	32	27	18	41	19
Requirements		6	10	12	15	

(*b*) A college is having a degree programme for which the effective semester time available is very less and the degree course requires field work. Hence, the savings in the total number of class hours handled can be utilized for such field work. Based on past experience, the college has estimated the number of hours required by each faculty to teach each subject. The course in its present semester has 5 subjects and the college has considered 6 existing faculty members to teach these courses. The objective is to assign the best 5 teachers, out of these 6 faculty members to teach 5 different subjects such that the total number of class hours required is minimized. The data for this problem is summarized in table. Solve this assignment problem optimaly.

		Subject 1	2	3	4	5
Faculty	1	30	39	31	38	40
	2	43	37	32	35	38
	3	34	41	33	41	34
	4	39	36	43	32	36
	5	32	49	35	40	37
	6	36	42	35	44	42

(c) Define dynamic programming problem. State and explain Bellman's principle of optimally in dynamic programming. What are its advantages and disadvantages.

3. **Attempt any two parts of the following:** **(10 × 2 = 20)**

(*a*) Two firms valued at Rs. 75 lakhs and Rs. 125 lakhs are available for sale in an open bid. There are only two parties to bid. Party A has Rs. 100 lakh ans B has Rs. 130 lakhs. what should be the strategies if each party is interested in maximizing its return.

(*b*) What are saddle point and oddments in Game Theory? Define mixed strategy and a fair game.

(*c*) Dr Thomas has been thinking about starting his own independent nursing home. The problem is to decide how large the nursing home should be. The annual return bill depends on both the size of nursing home and a number of marketing factors. After a careful analysis Dr. Thomas developed the following table:

Size of nursing home	*Good market (Rs.)*	*Fair market (Rs.)*	*Poor market (Rs.)*
Small (S)	50,000	20,000	–10,000
Medium (M)	70,000	35,000	– 25,000
Large (L)	90,000	35,000	– 45,000
Very Large(VL)	2,00,000	25,000	– 1,20,000

(*i*) What are the maximax and maximin decisions?

(*ii*) What is equally likely decision?

(*iii*) What is the criterion of Realism decision? For a = 0.8

4. **Attempt any two parts of the following:** **(10 × 2 = 20)**

(*a*) A baking company sells cake by the kg weight. It makes a profit of Rs. 5.00 a kg on each kg sold on the day it is baked. It disposes off all cake not sold on the date it is baked at a loss of Rs.1.20 a kg. If demand is known to be rectangular between 2000 and 3000 kg, determine the optimal daily amount baked.

(*b*) What are the reasons for stocking items in inventory? Explain different types of costs associated with inventory system.

(c) The daily demand of sweet bread at an outlet follows a discrete distribution as given in table. The purchase price of the sweet bread is Rs. 8 per packet and the selling price is Rs. 11 per packet. If the bread packets are not sold within the day of purchase, they are sold at Rs. 4 per packet to the hotels. Find the optimal order size of the sweet bread.

Demand, (D)	25	26	27	28	29	30	31	32	33	34	35
Probability (p)	0.2	0.11	0.10	0.09	0.08	0.12	0.14	0.05	0.04	0.04	0.03

5. Attempt any two parts of the following: **(10 × 2 = 20)**

(*a*) Explain the Monte-Carlo simulation using an example. Discuss its various applications.

(*b*) Explain following terms:

(*i*) Queue discipline (*ii*) Jockeying (*iii*) Traffic intensity
(*iv*) Balking (*v*) Reneging

(c) Workers come to tool room to recieve special tools for accomplishing a particular project assigned to them. The average time between two arrivals is 60 seconds and the arrivals are assumed to be in Poisson distribution. Average service time of the tool room attendent is 40 seconds. Determine:

(*i*) Average queue length. (*ii*) Average number of workers in the system including the workers being attended. (*iii*) Average length of non - empty queue.(*iv*) Mean waiting time of an arrival (*v*) Average waiting time of an arrival (worker who waits).

□□□

Model Question Paper-2

OPERATIONS RESEARCH (TME-601)

Time: 3 Hours | *Total Marks:* 100

Note : (*i*) Attempt all questions.
(*ii*) All questions carry equal marks.

1. Attempt any four of the following **(5 × 4 = 20)**

(*a*) Explain important characteristics of the situations to which linear programming method can be successfully applied. Describe application of this technique with a suitable example.

(*b*) Define Modelling, Discuss different types of models giving suitable examples.

(*c*) Define any four of the following:

(i) Basic solution (ii) Feasible solution
(iii) Basic feasible solution (iv) Slack variable
(v) Surplus variable

(*d*) Use duality to solve

Minimize $Z_x = 3x_1 + x_2$

Subject to $x_1 + x_2 \geq 1$

$2x_1 + 3x_2 \geq 2$

$x_1, x_2 \geq 0$

(*e*) An animal feed company must produce 200 kg of a mixture consisting of ingredients x_1 and x_2. The ingredient x_1 costs Rs. 3 per kg and x_2 costs Rs 5 per kg. Not more than 80 kg of x_1 can be used and at least 60 kg of x_2 must be used. Find the minimum cost mixture using graphical method.

2. Attempt any two parts **(10 × 2 = 20)**

(*a*) A company has 4 warehouses and 6 stores. The cost of shipping one unit from warehouse *i* to store *j* is:

7	10	7	4	7	8
5	1	5	5	3	3
4	3	7	9	1	9
4	6	9	0	0	8

Requirements of stores are 4, 4, 6, 2, 4, 2 and quantities at the warehouses are 5, 6, 2, 9. Find the minimum cost solution.

(*b*) A company is engaged in manufacturing of 5 brands of a product. It has five manufacturing setups, each capable of manufacturing any of its brands, one at a time. The cost to make a brand of these setups vary according to the following table. Find the optimum assignment of products on these setups resulting in minimum cost.

	S_1	S_2	S_3	S_4	S_5
B_1	4	6	7	5	11
B_2	7	3	6	9	5
B_3	8	5	4	6	9
B_4	9	12	7	11	10
B_5	7	5	9	8	11

(*c*) Describe the concept of dynamic programming and state the principle of optimality. Write the basic features of dynamic programming.

3. Attempt any two parts **(10 × 2 = 20)**

(*a*) Solve the following game by algebraic method

$$V = \begin{bmatrix} 1 & -1 & -2 \\ -1 & 1 & 1 \\ 2 & -1 & -2 \end{bmatrix}$$

(*b*) Explain the following terms: (*i*) Decisions under certainty (*ii*) Decision under risk (*iii*) Decision under uncertainty.

(*c*) The following table gives the pay off (in Rs) of different strategies S_1, S_2 and S_3 against events N_1, N_2, N_3 and N_4.

	N_1	N_2	N_3	N_4
S_1	4000	–100	6000	18000
S_2	20000	5000	400	0
S_5	20000	15000	–2000	1000

Obtain the decision taken under the following approaches

(*i*) Pessimistic (*ii*) Optimistic

(*iii*) Regret criterion (iv) Equal probability

4. Attempt any two parts from following **(10 × 2 =20)**

(*a*) A baking company sells cakes by kilogram weight. It makes a profit of Rs. 5 per kg on each kilogram sold on the day it is baked. It disposes of all cakes not sold on the date it is baked at a loss of Rs. 1.20 per kg. If the demand is known to be rectangular between 2000 and 3000 kg, determine the optimal daily amount baked.

(*b*) A newspaper boy buys paper for Rs. 2.60 and sells them for Rs. 3.60 each. He cannot return unsold newspapers. Daily demand has the following distribution.

No. of customers	23	24	25	26	27	28	29	30	31	32
Probability	0.01	0.03	0.06	0.1	0.2	0.25	0.15	0.10	0.05	0.05

If each day's demand is independent of previous day's demand, how many papers should he order each day?

(*c*) Discuss in detail about necessity for maintaining inventories. What are different costs considered in inventory control? What are the causes of poor inventory control?

5. **Attempt any two parts.** **(10 × 2 =20)**

(*a*) Describe any two methods for generating random numbers. Also discuss different simulation languages.

(*b*) Telephone users arrive at a booth following a poisson distribution with an average time of 5 minutes between one arrival and the next. The time taken for a telephone call is on an average 3 minutes and it follows an exponential distribution. What is the probability that the booth is busy. How many more booths should be established to reduce the waiting time to less than or equal to half of the present waiting time.

(*c*) At what average rate must a clerk at a supermarket work in order to ensure a probability of 0.90 that the customer will not have to wait longer than 12 minutes? It is assumed that there is only one counter to which customers arrive in a Poisson fashion at an average rate of 15 per hour. The length of service by the clerk has an exponential distribution.

❑❑❑

Model Question Paper-3

OPERATIONS RESEARCH (TME-601)

Time: 3 Hours | *Total Marks:* 100

Note : (*i*) Attempt all questions.
(*ii*) All questions carry equal marks.

1. Attempt any two parts: **(10 × 2 = 20)**

(*a*) Solve the following LPP by simplex

Maximize $Z = 4x_1 + 10x_2$

Subject to
$$2x_1 + x_2 \le 10$$
$$2x_1 + 5x_2 \le 20$$
$$2x_1 + 3x_2 \le 18$$
$$x_1, x_2 \ge 0$$

(*b*) Use simplex method to maximize Z

$$Z = x_1 + 2x_2 + 3x_3$$

Subject to
$$x_1 + 2x_2 + 3x_3 \le 10$$
$$x_1 + x_2 \le 5$$
$$x_1, x_2, x_3 \ge 0$$

(*c*) Solve the following linear programming problem graphically:

Maximize $Z = 4x_1 + 6x_2$

Subject to
$$x_1 + x_2 = 5$$
$$x_1 \ge 2$$
$$x_2 \le 4$$
$$x_1, x_2 \ge 0$$

2. Attempt any two parts **(10 × 2 = 20)**

(*a*) Determine an initial basic feasible solution to the following transportation problem using north-west corner method.

From/to	W_1	W_2	W_3	W_4	*Supply*
F_1	14	25	45	5	6
F_2	65	25	35	55	8
F_3	35	3	65	16	16
Requirement	4	7	6	13	

(*b*) Solve the following assignment problem

	a	*b*	*c*	*d*	*e*	*f*
A	9	22	58	11	19	27
B	46	78	72	50	63	48
C	41	28	91	37	45	33
D	74	42	27	49	39	32
E	36	11	57	22	25	18
F	3	56	53	31	17	28

(*c*) Explain the concept of Dynamic programing. How does it differ from linear programming? Describe its applications also.

3. **Attempt any two parts** **(10 × 2 = 20)**

(*a*) Solve the following 2 × 5 game by graphical method:

	1	2	3	4	5
1	–5	5	0	–1	8
2	8	–4	–1	6	–5

(*b*) Two companies A and B are competing for the same product. Their different strategies are given in the following pay off matrix.

		Company A A_1	A_2	A_3
Company B	B_1	2	–2	3
	B_2	–3	5	–1

Use linear programming to determine the best strategies for both the companies.

(*c*) A steel manufacturing company is concerned with the possibility of a strike. It will cost an extra Rs. 10000 to acquire an adequate stockpile. If there is a strike and the company has not stockpiled, then management estimates an additional expense of Rs. 50,000 in lost sales, late order charges and so forth. Should the company stockpile or not?

4. **Attempt any two parts.** **(10 × 2= 20)**

(*a*) A particular item has a demand of 9000 units per year. The cost of one procurement is Rs. 100 and the holding cost per unit is Rs. 2.40 per year. The replacement is instantaneous and no shortage is allowed. Determine

(*i*) the economic lot size

(*ii*) the no. of orders per year

(*iii*) the time between orders

(*iv*) the total cost per year if the cost of one unit is Rs. 1.00.

(*b*) A manufacturing company purchases 9000 parts of a machine for its annual requirements, ordering one month usage at a time. Each part costs Rs. 20. The ordering cost per order is Rs. 15 and the carrying charges are 15% of average inventory per year. Suggest a more economical purchase policy for the company. How much would it save the company per year.

(*c*) Let the probability density of demand of a certain item during a week be

$$f(x) = \begin{cases} 0.1 & 0 \leq x \leq 10 \\ 0 & \text{otherwise} \end{cases}$$

This demand is assumed to occur with a uniform pattern over the week. Let the unit carrying cost of the item in inventory be Rs. 2.00 per week and unit shortage cost be Rs. 8.00 per week. How will you determine the optimal order level of the inventory?

5. Attempt any two parts. **(10 × 2 = 20)**

(*a*) A TV repairman finds that the time spent on his jobs has an exponential distribution with mean 30 minutes. If he repairs TV sets in the order in which they come in and if the arrival of sets is approximately Poisson with average of 10 per 8 hour day, what is the repairman's expected idle time each day? How many jobs are ahead of the average set just brought in?

(*b*) A super market has two girls ringing up sales at the counters. If the service time for each customer is exponential with mean 4 minutes, and if people arrive in a Poisson fashion at the rate of 10 an hour, then calculate the

(*i*) probability of having to wait for service

(*ii*) expected percentage of idle time for each girl.

(*iii*) If a customer has to wait, what is the expected length of his waiting time.

(*c*) What is simulation? Write its limitations and advantages. Describe Monte Carlo technique also.

❑❑❑

Questions from GATE (ME and PI)

1. A linear programming problem has an unbounded solution when the replacement ratio values are: *(PI–2005)*
   (*a*) infinite or negative (*b*) infinite or positive
   (*c*) finite or negative (*d*) finite or positive.

2. For a stochastic decision making problem, the best alternative based on Expected Monetary Value (EMV) and Expected Opportunity Loss (EOL) is obtained when *(PI–2005)*
   (*a*) EMV is lowest possible or EOL is highest possible
   (*b*) EMV is lowest possible or EOL is lowest possible
   (*c*) EMV is highest possible or EOL is lowest possible
   (*d*) EMV is highest possible or EOL is highest possible.

3. A manager must assign three engineers to three different tasks. Estimates of the time (in hours) to complete each task for every engineer are given below: *(PI-2005)*

Engineer	Time to complete		
	Task 1	Task 2	Task 3
P	10	12	16
Q	12	24	20
R	14	20	24

   The total time (in hrs) for the best assignment and the engineer who completes his work the last are
   (*a*) 36, P (*b*) 46, Q
   (*c*) 46, R (*d*) 48, R

4. Consider a transportation problem with two supply points and three demand points. A feasible shipment schedule is given in the table below. The unit transportation costs are shown on the top left hand corner of the cells.

4 40	9	6 30
8 20	5 50	3 10

   The minimum cost of transportation is *(PI 2006)*
   (*a*) 450 (*b*) 590
   (*c*) 640 (*d*) 780

5. Consider a single item with annual demands of 1000 units. The ordering cost is Rs. 200 per order and the carrying cost is Rs. 4/unit/year. The unit price of the item is Rs. 10. The vendor is willing to give a 1% discount on the price if the order quantity exceeds 800 units. The total annual cost of inventory

including the cost of the item is *(PI 2006)*

(*a*) Rs. 1,03,000 (*b*) Rs. 1,03,100
(*c*) Rs. 1,04,000 (*d*) Rs. 1,04,100.

6. Cars arrive at a service facility following a Poisson distribution (λ = 10/hour) and service times are exponential) μ = 15/hour). The facility can accommodate a maximum of nine cars (including the one under service) and customers do not join the queue if the number in the system exceeds nine. The percentage of customers leaving without service is *(PI-2006)*

(*a*) 6 (*b*) 10
(*c*) 11.11 (*d*) 33.33.

7. Lead time demand of an item follows a uniform distribution between 100 and 200 units. A service level of 95% is required for this item. The EOQ for the item is 1000 units and the purchase price of the item is Rs. 100. The investment in cycle inventory and safety inventory for this item should not exceed Rs 30000. The order quantity under these conditions is *(PI 2006)*

(*a*) 500 (*b*) 510
(*c*) 555 (*d*) 1000.

8. Consider the linear programming problem (P)
Maximize $x_1 + x_2$
Subject to $3x_1 + 2x_2 \leq 8$.
$x_1 \geq 0$, x_2 unrestricted in sign. Problem P has *(PI 2006)*

(*a*) an unbounded solution (*b*) alternate optima
(*c*) an infeasible solution (*d*) a unique optimum.

9. Consider the dual (D) of problem P. The dual D has (PI–2006)

(*a*) a unique optimum (*b*) an unbounded solution
(*c*) alternate optima (***d***) an infeasible solution.

10. Let C_1 be the objective function coefficient of variable x_1, the values of C_1 for which x_1 will be in the optimum bases are *(PI 2006)*

(*a*) $C_1 \geq 1/2$ (*b*) $C_1 \geq 1$
(*c*) $C_1 \geq 3/2$ (*d*) $C_1 \geq 8/3$.

11. An item has an annual demand of 1200 units. It is decided to procure 300 units every quarter. The purchase price is Rs. 100/unit, the carrying cost is Rs. 10/unit/year and the ordering cost is Rs. 200/order. The total inventory cost (in Rs) for the current ordering policy is *(PI–2006)*

(*a*) 400 (*b*) 1500
(*c*) 1550 (*d*) 1900.

12. In problem 11, if the policy were utilized for ordering the above item, the annual savings in cost (in Rs) would be *(PI–2006)*

(*a*) 150 (*b*) 250
(*c*) 350 (*d*) 450.

13. In an Assignment problem for assigning *n* jobs to n machines, the number of decision variables and constraints are respectively: *(PI 2004)*

(*a*) n^2 and $2n$ (*b*) $2n$ and $2n$
(*c*) n^2 and n (*d*) $2n$ and n.

14. The number of Basic solutions in a linear programming problem with n decision variables and m inequality constraints (of ≤ type) with non-negative RHS values is *(PI-2004)*
(*a*) ${}^{n}C_{m}$ (*b*) ${}^{m}C_{n}$
(*c*) ${}^{(m+n)}C_{m}$ (*d*) ${}^{(m+n)}C_{n}$

15. The average weekly demand for an item is 10 unit/week. The fixed cost of ordering is Rs. 98 per order. The inventory holding cost is Rs. 0.58 per unit per week. Suppose the permitted values of reorder-interval are only: 1 or 2 or 4 or 8 weeks. Among these values of reorder interval (in weeks) which one yields the least total relevant cost per week? *(PI 2004)*
(*a*) 1 (b) 2
(*c*) 4 (*d*) 8.

16. Determine the maximum value of the objective function in the problem *(PI 2004)*

Maximize $x_1 + x_2 + x_3$
Subject to $x_1 + 2\,x_2 + 3\,x_3 = 18$
$x_1,\ x_2,\ x_3 \geq 0$

(*a*) 24 (*b*) 36
(*c*) 54 (*d*) 60.

17. There are two identical machines, each with a memoryless failure rate, λ = 0.2 failure/hour. These two machines are repaired by a common repairman, with a memoryless repair (service completion) rate, μ = 0.4 repairs/hour. Assume that the repairman follows FCFS order for service. The system is operating in steady state. What is the repairman's utilization (i.e., fraction of time he is busy)? *(PI–2004)*
(*a*) 0.50 (*b*) 0.60
(*c*) 0.67 (*d*) 0.75.

18. In problem 17, what is the utilization of each one of the machines?
(*a*) 0.5 (*b*) 0.6
(*c*) 0.7 (*d*) 0.8.

19. A newspaper stall owner at the beginning of every week buys a certain number of copies, S of a certain weekly for the new week, and disposes off the left over copies (if any) of the last week's issue. Suppose the demand for the weekly during the week has the probability mass function

$$(0.02)\,(0.98)^n;\ n \geq 0 \qquad \textit{(PI–2004)}$$

If S = 50, then what is the expected value of the demand in a week that is not met
(*a*) 17.84 (*b*) 21.39
(*c*) 24.92 (*d*) 48.38.

20. In problem 19, suppose that on every copy sold during the week, the owner makes a profit of Rs. 1.20, and on every copy not sold during the week he loses Rs 5.00. Then the optimal value of S is *(PI 2004)*

(*a*) 8 (*b*) 10
(*c*) 12 (*d*) 14.

21. Consider the linear programming problem:
Maximize $Z = 3x_1 + 4x_2$
subject to $x_1 + x_2 \le 12$
$2x_1 + 3x_2 \le 30$
$x_1 + 4x_2 \le 36,\ x_1, x_2 \ge 0$
Slack variables x_3, x_4 and x_5 are associated with the 1st, 2nd and 3rd in equalities respectively. The optimal solution is $x_1 = 6$, $x_2 = 6$ and $z = 42$. The variables in the optimal basis are *(PI 2004)*
(*a*) x_1, x_2, x_3 (*b*) x_1, x_2, x_4
(*c*) x_1, x_2, x_5 (*d*) x_3, x_4, x_5.

22. In problem 21, the marginal value of the third resource at the optimum is *(PI 2004)*
(*a*) 0 (*b*) 12
(*c*) 36 (*d*) 42.

23. If the right hand side (RHS) value in the third constraint changes to 30 in problem 21, then *(PI 2004)*
(*a*) the problem has an alternate optimum solution.
(*b*) the optimal solution is degenerate
(*c*) the dual has unbounded solution
(*d*) the dual is infeasible.

24. You are given a transportation problem with two supply points ($i = 1,2$) and three demand points. (J = 1 to 3). The unit costs of transportation are indicated in top left hand corner of each position in the matrix. A feasible solution ($x_{11} = 20$, $x_{12} = 10$, $x_{21} = 10$, $x_{23} = 30$) is also given in the table.

[12] 20	[9] 10	[11]
[6] 10	[Q]	[2] 30

The set of all the values of Q for which the given solution is the only optimum solution is *(PI 2004)*
(*a*) Q > 0 (*b*) Q ≥ 1
(*c*) Q > 2 (*d*) Q > 3.

25. In problem 24, for $Q = 2$, which of the following is not true? *(PI 2004)*
(*a*) The problem has a unique optimal.
(*b*) The given feasible solution is optimal.
(*c*) The problem has a degenerate optimal solution.
(*d*) The problem has non basic solutions which are optimal.

26. The number of customers arriving at a railway reservation counter is Poisson distributed with an arrival rate of eight customers per hour. The reservation clerk at this counter takes six minutes per customer on an average with an

exponentially distributed service time. The average number of the customers in the queue will be *(ME 2006)*

(*a*) 3 (*b*) 3.2

(*c*) 4 (*d*) 1.2.

27. Consider the following data for an item

Annual demand: 2500 units per year

Ordering cost: Rs 100 per order

Inventory holding rate : 25% of unit price

Price quoted by a supplier

Order quantity (Units)	Unit Price (Rs.)
< 500	10
≥ 500	9

The optimum order quantity (in units) is

(*a*) 447 (*b*) 471

(*c*) 500 (*d*) ≥ 600. *(ME 2006)*

28. A firm is required to procure three items (P, Q and R), The prices quoted for these items (in Rs.) by suppliers S_1, S_2 and S_3 are given in table. The management policy requires that each item has to be supplied by only one supplier and one supplier supplies only one item.

Item	Suppliers		
	S_1	S_2	S_3
P	110	120	130
Q	115	140	130
R	125	145	165

The minimum total cost (in Rs.) of procurement to the firm is *(ME 2006)*

(*a*) 350 (*b*) 360

(*c*) 385 (*d*) 395.

29. A stockist wishes to optimize the number of perishable items he needs to stock in any month in his store. The demand distribution for this perishable item is

Demand (in units)	2	3	4	5
Probability	0.10	0.35	0.35	0.20

The stockist pays Rs. 70 for each item and he sells each at Rs. 90. If the stock is left unsold in any month, he can sell the item at Rs 50 each. There is no penalty for unfulfilled demand. To maximize the expected profit, the optimal stock level is *(ME 2006)*

(*a*) 5 units (*b*) 4 units

(*c*) 3 units (*d*) 2 units.

30. Consider a single service queuing model with Poisson arrivals (λ = 4/hour) and exponential service (μ = 4/hour). The number in the system is restricted

to a maximum of 10. The probability that a person who comes in and leaves without joining the queue is *(ME 2005)*

(*a*) 1/11 (*b*) 1/10
(*c*) 1/9 (*d*) 1/2.

31. The distribution of lead time demand for an item is as follows:

Lead time demand	Probability
80	0.20
100	0.25
120	0.30
140	0.25

The reorder level is 1.25 times the expected value of the lead time demand. The service level is *(ME 2005)*

(*a*) 25% (*b*) 50%
(*c*) 75% (*d*) 100% .

32. A company has two factories S_1, S_2 and two ware houses D_1, D_2. The supplies from S_1 and S_2 are 50 and 40 units respectively. Warehouse D_1 requires a minimum of 20 units and a maximum of 40 units. Warehouse D_2 requires a minimum of 20 units and, over and above, it can take as much as can be supplied. A balanced transportation problem is to be formulated for the above situation. The number of supply points, the number of demand points, and the total supply (or total demand) in the balanced transportation problem respectively are *(ME 2005)*

(*a*) 2, 4, 90 (*b*) 2, 4, 110
(*c*) 3, 4, 90 (*d*) 3, 4, 110.

33. There are two products *P* and *Q* with the following characteristics

Product	Demand (units)	Order cost (Rs/order)	Holding cost (Rs/unit/year)
P	100	50	4
Q	400	50	1

The economic order quantity (EOQ) of products P and Q will be in the ratio *(ME 2004)*

(*a*) 1 : 1 (*b*) 1 : 2
(*c*) 1 : 4 (*d*) 1 : 8.

34. A maintenance service facility has Poisson arrival rates, negative exponential service time and operates on a 'first come first served' queue discipline. Breakdowns occur on an average of 3 per day with a range of zero to eight. The maintenance crew can service an average of 6 machines per day with a range of zero to seven. The mean waiting time for an item to be serviced would be *(ME 2004)*

(*a*) 1/6 day (*b*) 1/3 day
(*c*) 1 day (*d*) 3 day.

35. A company has an annual demand of 1000 units, ordering cost of Rs 100/order and carrying cost of Rs. 100/unit/year. If the stock-out costs are estimated to be nearly Rs 400 each time the company runs out-of-stock, the safety stock justified by the carrying cost will be *(ME 2004)*

(*a*) 4 (*b*) 20
(*c*) 40 (*d*) 100.

36. A company produces two types of toys: P and Q. Production time of Q is twice that of P and the company has a maximum of 2000 time units per day. The supply of raw material is just sufficient to produce 1500 toys (of any type) per day. Toy type Q requires an electric switch which is available @ 600 pieces per day only. The company makes a profit of Rs 3 and Rs 5 on type P and Q respectively. For maximization of profits, the daily production quantities of P and Q toys should respectively be *(ME 2004)*

(*a*) 100, 500 (*b*) 500, 1000
(*c*) 800, 600 (*d*) 1000, 1000.

❒❒❒

Previous Years Question Papers

B.Tech.
Eighth Semester Examination, 2003-2004
Operations Research (ME-803)

Time : 3 Hours *Total Marks : 100*

Note: Answer the following questions.

1. Maximize $f = 240x_1 + 104x_2 + 60x_3 + 19x_4$ subject to 20

$20x_1 + 9x_2 + 6x_3 + x_4 \leq 20$

$10x_1 + 4x_2 + 2x_3 + x_4 \leq 10$

$x_i \geq 0 ; \quad i = 1 \text{ to } 4$

Using simplex method, find the solution of the above problem.

OR

(*a*) Define an optimization problem. What constitutes an objective function ? Give the conditions, an optimization problem must satisfy so that it is called a Linear Programming Problem (LPP). 5

(*b*) What do you understand from sensitivity analysis in a LPP ? What is the advantage of carrying out sensitivity analysis? Give examples. 5

(*c*) Using graphical method or otherwise, 10

Minimize $f = 3x_1 + 2x_2$, subject to

$8x_1 + x_2 \geq 8$

$2x_1 + x_2 \geq 6$

$x_1 + 3x_2 \geq 6$

$x_1 + 6x_2 \geq 8$

$x_1 \geq 0 ; x_2 \geq 0.$

2. Attempt Two parts : (10×2=20)

(*a*) A product is produced by four factories *A*, *B*, *C* and *D*. The unit production costs in them are Rs. 2, 3, 1 and 5 respectively. Their production capacities are : Factory *A* — 50 units, *B* — 70 units, *C* — 30 units and *D* — 50 units. These factories supply the product to four stores, demand of which are 25, 35, 105 and 20 units respectively. Unit transportation cost in rupees from each factory to each store is given in the table below.

Determine the extent of deliveries from each of the factories to each of the stores so that the total production and transportation cost is the minimum.

	1	2	3	4
A	2	4	6	11
B	10	8	7	5
C	13	3	9	12
D	4	6	8	3

(*b*) A car rental company has one car at each of its five depots, D_1, D_2, D_3, D_4 and D_5. A customer in each of the five towns *A*, *B*, *C*, *D* and *E* requires a car. The distances between the depots and the towns are indicated below :

Towns/Depots	D_1	D_2	D_3	D_4	D_5
A	160	130	176	190	200
B	135	120	130	160	175
C	140	110	155	170	185
D	50	50	80	80	110
E	55	35	70	80	105

Find the optimal assignment of cars to the customers and find the total distance travelled by the five cars.

(*c*) What do you understand from multi-stage decision process. Give example. State the "principle of optimality" and explain it giving examples. What do you understand from the "Concept of sub-optimization"?

3. Attempt any Two parts : **(10×2=20)**

(*a*) A person is holding 10000 shares (Rs. 10 each) of M/s. ABC Corporation which are currently priced at Rs. 14 each. He is considering two alternate strategies :

(*i*) Holding the shares for one year and then selling them.

(*ii*) Selling the shares now and buying the shares of M/s XYZ Corporation currently priced at Rs. 80 with a view to sell them at the end of the year. He feels that the price of the XYZ stock at the end of the year will be Rs. 85, Rs. 80, or Rs. 70 with probabilities 0.5, 0.3 and 0.2 respectively.

The ABC Corporation's share price and dividends (paid at the end of year) are expected to depend on general performance of industry as given below :

General Performance of industry	Probability	Share price at the end of the year	Dividend (Rs. per share)
Good	0.1	up by 20%	Rs. 2
Fair	0.5	same	Rs. 1
Poor	0.4	down by 10%	Nil

Draw the decision-tree for this investment problem and determine the optimal policy.

(*b*) Define the terms, "Game"; "Two-person zero-sum game"; "Minimax criterion"; "Saddle point" and "Sub-games". Explain under what situation linear programming technique is useful in solving Game problems. Give examples.

(*c*) There are two business locations in the city. About 70% of the population of the city live near locality 1 and the remaining near locality 2. Two banks—one large and one small are planning to open their branches in the city. It is estimated that if both the banks are located in the same locality, the larger one will get 60% of the business. On the other hand, if the two banks are located in two different localities, the larger one will get 80% of the business of the locality where it is located and 40% of the business of the other locality. What should be the optimal strategy for the two banks?

4. Attempt any Two parts : **(10×2=20)**

(*a*) Find the optimal order quantity for a product for which the price-breaks are given below :

Quantity	Purchase Cost
$0 \le Q_1 < 100$	Rs. 20 per unit
$100 \le Q_2 < 200$	Rs. 18 per unit
$200 \le Q_3$	Rs. 16 per unit

The monthly demand for the product is 400 units. The storage cost is 20% of unit cost of the product and the cost of ordering is Rs. 25.

(*b*) The probability distribution of monthly sales of a certain item is given below:

Monthly Sale	Probability
0	0.02
1	0.05
2	0.03
3	0.27
4	0.20
5	0.10
6	0.06

The cost of carrying inventory is Rs. 10 per unit per month. The current policy is to maintain a stock of four items in the beginning of each month. Assuming that the cost of shortage is proportional to both time and quantity short, obtain the imputed cost of shortage of one item for one time unit.

(*c*) Name various inventory models available with continuous and discrete demands. Discuss, in detail, any one such model. Give applications.

(*d*) A trader stocks woollen sweaters at the beginning of the winter and cannot reorder. The item costs him Rs. 50 each and he sells at Rs. 100 per unit. For each item, that cannot be met on demand, he loses a goodwill of Rs. 30. Each unsold piece will have a salvage value of Rs. 20. Holding cost during the period is 10% of item cost. The probability distribution of demand is as follows :

Unit stocked (in hundreds)	4	6	8	10	12
Demand Probability $p(D \le Q)$	0.30	0.20	0.30	0.15	0.05

Compute the optimal number of items to be stocked. If a stock level is to be maintained at 10 units, find the value of under-stocking cost.

5. Attempt any Two parts : (10×2=20)

(*a*) What is a "Queuing Model" ? What are its objectives ? Giving basic structure, explain (M/M/I) model.

(*b*) What do you understand from "Simulation"? What are its elements ? Discuss "Monte Carlo Simulation Technique" giving its advantages and applications.

(*c*) At a car service centre, cars arrive according to Poisson process with mean 3 per hour. The time for servicing each car is exponentially distributed with mean 15 minutes per car. The facility can handle one car at a time. Compute: (*i*) average queue length; (*ii*) average number of cars in the queuing system; (*iii*) waiting time in the system, and (*iv*) percentage utilization of service.

❑❑❑

B.Tech.

Eighth Semester Examination, 2004-2005

Operations Research (ME-803)

Time : 3 Hours *Total Marks : 100*

Note: (*i*) Attempt All questions.

(*ii*) All questions carry equal marks.

(*iii*) In case of numerical problems assume data wherever not provided.

1. Attempt any two parts of the following: **(10 × 2 = 20)**

(*a*) Explain the general structure of a linear programming model.

(*b*) Solve the following LP problem using graphical method:

Maximize $Z = 6x_1 + x_2$

Subject to $5x_1 + 10x_2 \leq 60$

$4x_1 + 4x_2 \leq 40$

x_1 and $x_2 \geq 0$

(*c*) Consider two different types of food stuffs say F_1 and F_2. Assume these food stuffs contain vitamin V_1 and V_2. Minimum daily requirement of V_1 is 40 units and vitamin V_2 is 50 units. Food F_1 contains 2 units of Vitamin V_1 and 3 units of vitamin V_2. Further, F_2 contains 4 units of V_1 and 2 units of V_2. The cost of 1 unit of F_1 is Rs. 3 and of F_2 is Rs. 2.5. Find minimum cost of diet that would supply the body atleast minimum requirements of each vitamins. Formulate L.P.P. and solve it graphically.

2. Attempt any two parts of the following: **(10 × 2 = 20)**

(*a*) Consider the following transportation problem involving three sources and three destinations. The cell entries represent the cost of transportation per unit in Rupees.

		Destination			
		A	B	C	Supply ↓
	1	6	8	4	14
Source	2	4	9	3	12
	3	1	2	6	5
Demand →		6	10	15	

Obtain the optimum cost of transportation.

(*b*) The table shown below is a profit matrix, entries of which indicate the profit generated by sales person working in various zones.

		Sales persons		
		S1	S2	S3
Sales	1	10	17	14
Zones	2	16	8	11
	3	14	15	16

Find assigned zones of sales persons, so that profit is maximum.

(*c*) Define dynamic programming problem. List and explain the terminologies of dynamic programming problem, also applications of this technique.

3. Attempt any two parts of the following : (10×2=20)

(*a*) A news paper boy has the following probabilities of selling a magazine:

No. of copies	Probability
10	0.10
11	0.15
12	0.20
13	0.25
14	0.30
	1.00

Cost of a copy is Rs. 30 and sales price is Rs. 50. he cannot return unsold copies. How many copies should he order?

(*b*) Solve the game whose pay-off matrix is:

		B I	B II	B III
A	I	– 3	– 2	6
A	II	2	0	2
A	III	5	– 2	– 4

(*c*) Define the following:

(*i*) Pure strategy (*ii*) Saddle point

(*iii*) Min-Max principle (*iv*) Rule of Dominance

4. Attempt any two parts of the following: (10×2=20)

(*a*) Enlist various inventory models available. Explain any one of them in detail.

(*b*) A manufacture of electric motors is required to purchase 2400 castings per year. These are subject to quantity discount of Rs. 0.75 from a quoted price of Rs. 10, if quantity purchased is 3000 at any given time. If the ordering cost is Rs. 100/- and inventory carrying cost is 24% per year, find:

(*i*) What should be the ordering quantity?

(*ii*) Number of orders placed in a year.

(*c*) A merchant in a market wishes to stock some chrismas trees for sales during the chrismas season. He has the following probability distribution of demand.

Demand	1	2	3	4	5	6
Probability	0.05	0.15	0.20	0.40	0.10	0.10

He pays Rs. 20 for each tree and he sells each tree for Rs.50/-. If he has any tree left-over at the end of the season, he can sell them for Rs. 5/- each. Find the optimum number of trees he should stock.

5. Attempt any two parts of the following: (10×2=20)

(*a*) What do you understand by simulation. Explain different types of simulation models, in brief, with application.

(*b*) Write sort note on Queuing theory.

(*c*) A person repairing TV set finds that the time spent on the set has been exponential distribution with mean 20 minutes. If the TVs are repaired in the order in which they come and their arrival is approximately Poisson with an average rate of 15 for 8 hour day. What is the repairman's expected idle time each day.

❑❑❑

B.Tech.

Eighth Semester Examination, 2005-2006

Operations Research (ME-803)

Time: 3 Hours *Total Marks: 100*

Note: (*i*) Attempt **ALL** questions
(*ii*) All questions carry equal marks.
(*iii*) In case of numerical problems assume data wherever not provided.
(*iv*) Be precise in your answer.

1. Attempt any four parts of the following: **(5 × 4 = 20)**

(*a*) What do you understand by 'Graphical Method'? Discuss its limitations.

(*b*) What are the basic characteristics of a linear programming model? Discuss its applications.

(*c*) Discuss the duality theory in context to the linear programming. Explain the relationship between a primal and dual.

(*d*) A company has two grades of inspectors, I and II, who are to be assigned for a quality control inspection. It is required that atleast 2000 pieces be inspected per 8 hrs day. Grade I inspector can check pieces at the rate of 40 per hour with an accuracy of 97 %. The wage rate of Grade I inspector is Rs. 4.50 per hour. Grade II inspector can check pieces at the rate of 40 per hour with an accuracy of 90 % and the wage rate Rs. 2.50 per hour. Each time an error is made by an inspector, the cost to the company is Rs. 1. The company has 10 grade I and 5 grade II inspectors. Formulate the problem to minimize the total cost.

(*e*) Solve graphically, the following L.P.P.

Minimize $Z = 3x_1 + 5x_2$

with constraints

$$-3x_1 + 4x_2 \leq 12$$
$$2x_1 - x_2 \geq -2$$
$$2x_1 + 3x_2 \geq 12$$
$$x_1 \leq 4$$
$$x_2 \leq 2$$
$$x_1, x_2 \geq 0$$

(*f*) Consider the problem

Maximize $Z = 4x_1 + 3x_2 + 6x_3$

Subject to

$$3x_1 + x_2 + 3x_3 \leq 30$$
$$2x_1 + 2x_2 + 3x_3 \leq 40$$
$$x_1, x_2, x_3 \geq 0$$

Construct the primal-dual table and the dual problem.

2. Attempt any four parts of the following: **(5 × 4 = 20)**

(*a*) Show that an assignment problem can be treated as a particular case of transportation problem.

(*b*) Prove that an assignment problem does not change in structure if each row or column is reduced by a constant. How this property can be used to solve an assignment problem?

(*c*) Explain Vogel's approximation method of solving transportation problem.

(*d*) Determine an initial feasible solution for the following transportation problem:

	To					Available
From	3	4	6	8	9	20
	2	10	1	5	8	30
	7	11	20	40	3	15
	2	1	9	14	16	13
Demand	40	6	8	18	6	

(*e*) Use row minima method and determine initial feasible solution to the following problem:

	To				Available
From	5	2	4	3	22
	4	8	1	6	15
	4	6	7	5	8
Demand	7	12	17	9	

(*f*) Find initial basic feasible solution using column minima method:

	To			Available
From	10	13	6	10
	16	7	13	12
	8	22	2	8
Demand	6	11	13	

3. Attempt any two parts of the following: (10 × 2 = 20)

(*a*) Discuss the difference between decision making under risk and under uncertainty in statistical decision theory. Explain with suitable examples.

(*b*) A newspaper boy has the following probabilities of selling a magazine.

No. of copies sold	Probabilities
10	0.10
11	0.15
12	0.20
13	0.25
14	0.30

Cost of one copy is Rs. 30 and sale price is Rs. 50. He cannot return unsold copies. How many copies should he order?

(*c*) Solve the following game by using dominance property

I	3	5	4	9	6
II	5	6	3	7	8
III	8	7	9	8	7
IV	4	2	8	5	3

4. Attempt any two parts of the following: (10 × 2 = 20)

(*a*) Define inventory. What are the advantages and disadvantages of maintaining inventory? Discuss the cost associated with inventory.

(*b*) Explain the following terms: (i) Lead time (ii) re-order point (iii) EOQ (iv) Buffer stock.

(*c*) Drive the expression for economic order quantity if there is instantaneous replenishment and no shortage is allowed. Clearly state the assumptions made. Discuss the impact on the lot size if shortages are allowed.

5. Attempt any two parts of the following: **(10 × 2 = 20)**

(a) What is simulation? Explain the theory of Monte-Carlo simulation. Write the names and applications of various simulation languages.

(b) What do you understand by a queue? State the important distributions of arrival interval and service times. Explain the following terms:

(*i*) Queue discipline

(*ii*) Input and holding times.

(c) Arrivals at a telephone booth are considered to be following Poisson distribution with average time of 10 minutes between one arrival and the next. Length of phone call is assumed to be distributed exponentially with mean 3 minutes: Find,

(*i*) What is the probability that a person arriving at the booth will have to wait?

(*ii*) What is average length of the queue that is formed time to time?

(*iii*) A new booth can be installed if a person has to wait at least 3 minutes for the phone. By how much must the flow of arrival be increased in order to justify a second booth?

❐❐❐

B. Tech.
Sixth Semester Examination, 2007-2008
Operations Research (TME-601)

Time : 3 Hours *Total Marks : 100*

Note: Attempt **all** questions which carry **equal** marks.

1. Attempt any **two** parts of the following : **10 × 2 = 20**

(*a*) Describe various models and then applications of operations research.

(*b*) A manufacturing company can sell a maximum of 10, 000 units of a products, which are quality approved. The company have 3 processing plants, each with a processing capacity of 5000 units, cost of processing and percentage of rejections varies for various plants.

Plant	*Cost of Processing/unit*	*% age rejections*
A	*100*	*5%*
B	*125*	*3%*
C	*150*	*2%*

The only cost other then processing cost is Raw Material Cost, which is Rs. 200 per unit. All quality approved goods are sold for Rs. 600 each and rejections are sold per Rs. 50 each. The company have policy of not having more than overall 3% rejections.

Formulate LPP to decide optimum production by various plants.

(*c*) What is a 'dual'? How does, it help in linear programming? Write the dual of following LPP

Minimize $Z = x_1 + \frac{1}{2} x_2$

Subject to
$$6x_1 + 2x_2 \geq 24$$
$$3x_1 + 2x_2 \geq 18$$
$$x_1 + 3x_2 \geq 12$$
$$x_1, x_2 \geq 0$$

2. Attempt any two parts of the following : 10 × 2 = 20

(*a*) Determine the optimal solution of the transportation matrix and check the optimality. The matrix is cost matrix.

Requirement	Availability	30	30	30
	Market Godown	a	b	c
28	X	7	9	3
50	Y	2	5	7

(*b*) Write the Hungarian algorithm and prove that Assignment Model is a special case of transportation model.

(*c*) Following profit matrix indicates profit in stages *A*, *B*, and *C*, when they make 1, 2, 3 or 4 costings.

Determine best productions schedule using DP. Minimum 1 costing is must for every stage

	A	B	C
1	4	9	8
2	12	10	15
3	20	22	20
4	20	30	30

3. Attempt any **two** parts of the following : **10 × 2 = 20**

(*a*) The annual demand for a seasonal product follows the distribution as follows:

Demand	3000	3500	4000	4500	5000
Probability	0.1	0.2	0.3	0.3	0.1

The manufactures of this item can produce it by one of three methods :

(*i*) Using existing equipment at a cost of Rs. 8 per unit.

(*ii*) Buy special equipment for Rs. 22, 000, whose salvage value of the end of year would by 2000 the variable cost per unit would be reduced to Rs. 2

(*iii*) Buy a superspecial equipment for Rs. 90, 000 which would be useful for 4 years and would be depreciated on straight line. The variable cost would further be reduced to Rs. 1.20/unit, which equipment should be used?

(*b*) For the following pay off matrix, find the value of the game and strategy using linear Programming :

	Player B		
Player A	1	2	3
1	3	– 1	4
2	6	7	–2

(*c*) Describe the principles of decision making in various conditions. Differentiate between Laplace Criterian, Hurcwitz criteria, Minimax and maximin using a suitable example.

4. Attempt any **two** parts of the following : **10 × 2 = 20**

(*a*) Derive expression for EOQ for Inventory models with planed shortages.

(*b*) A manufacturing company needs 2500 units of a particular component every year. The company buys it at the rate of Rs. 30/unit. The order processing cost is estimated at Rs. 15 and cost of carrying a part in stock comes out to be Rs. 4 per years.

The company can manufacture this part. In this case, it saves 20% of price of product. It estimates a set up cost of Rs. 250 per production run. The annual production rate would be 9800 units. Inventory holding cost remains unchanged:

(*i*) Determine EOQ and frequency of orders

(*ii*) Determine production lot-size and average production run

(*iii*) Suggest company on make or buy

(*c*) What are assumptions in EOQ model? What are consequences if each one of them are relaxed at a time? How can they be taken care of in such consequences.

5. Attempt any **two** parts of the following : **10 × 2 = 20**

(*a*) A single counter reservation office serves passengers with an average service time of 5 minutes (exponentially distributed) Passengers comes for reservation in poisson at an average rate of 8 per hour.

(*i*) What is average queue length

(*ii*) What is probability that there ae 3 passengers in the system

(*iii*) Probability that a passenger shall wait for more than 10 minutes before getting served

(*iv*) What is average queue length if average no. of passengers increase to 12 per Hr? Why?

(*b*) A car garage have an average inter arrival time of 2 hrs in Poisson. The garage does service a car in 3 hrs, exponentially distributed. If cost of retention of cars and operating a new channel of service is same, determine optimum no. of Service channels, the garage should employ.

(*c*) There are 3 machines to perform an operation each on various jobs. If time consumed, for various variciles of job on each shop is given below, determine % age ideal time for machine *B* and *C*. The operations are to be done in sequence on machine *A*, *B* and *C* :

		Time in minutes		
Job	Probability	M/C A	M/C B	M/C C
P	0.2	10	15	20
Q	0.3	20	15	10
R	0.4	15	15	15
S	0.1	12	16	15

Assume the workshop to work round the clock. Use Monte Carlo simulation (2 sets) and use random number as follows :

Set 1	7	3	5	2	8	9	9	0	1	3
Set 2	7	4	2	1	9	1	3	2	8	4

OR

(*c*) Describe and explain Monte Carlo simulation technique.

❑❑❑

B.Tech.
Sixth Semester Examination, 2008-2009
Operations Research (TME-601)

Time : 3 Hours | *Total Marks : 100*

Note: (1) Attempt **all** questions.
(2) Attempt **two** parts from each question.
(3) All questions carry **equal** marks.
(4) Assume suitably missing data if any.

1. (*a*) What is 'Operations Research'? Describe OR Methodology with an example.
(*b*) Air-Company forecasts need of air-hostess for its operations for next 4 months:

Month	*April*	*May*	*June*	*July*
Air-hostess	80	60	110	90

It needs one month training for an air-hostess to be prepared to get ready for service, during which the trainee gets Rs. 6000 as stipend. The salary of airhostess is Rs. 25,000 per month. All recruitments are made on first of every month, whereas all terminations and resignations are on last data of the month. If company has to pay Rs. 20, 000 to air hostess if her services are terminated; create an LPP for deciding recruitment and termination plan of the company. Today it is 28 Feb. and there are 90 air-hostesses on roll. It is observed that 10% air-hostesses leave the job on their own every month.

(*c*) Convert the following LPP into its dual and solve using Simplex method:

Maximize $Z = 5x_1 + 10x_2 + 8x_3$

subject to $3x_1 + 5x_2 + 2x_3 \leq 60$

$4x_1 + 4x_2 + 4x_3 \leq 72$

$2x_1 + 4x_2 + 5x_3 \leq 100$

$x_1, x_2, x_3 \geq 0.$

2. (*a*) Optimize the following Profit Matrix for transportation using VAM. Test the optimality using stepping stone algorithm :

	Warehouse			
Markets	A	B	C	Demand
Agra	32	20	17	600
Lucknow	30	16	13	600
Allahabad	40	12	18	600
Availability	700	700	700	

(*b*) Solve the following assignment matrix using Hangarian Algorithm (Maximization). Also write the following as an LPP :

Person	Job P	Q	R	S
A	20	16	30	12
B	24	20	32	10
C	28	18	34	4
D	32	20	36	6
E	36	16	18	8

(*c*) There are 10 truck loads to be transported. There are 3 companies providing services. No company can provide more than 5 trucks. A company should by given at least twice load as compared to B. Optimise the cost using DP : Cost in Rs.

	A	B	C
1	10, 000	8, 000	15, 000
2	18, 000	14, 000	25, 000
3	25, 000	20, 000	25, 000
4	31, 000	28, 000	25, 000
5	40, 000	30, 000	30, 000

3. (*a*) Holiday Homes Ltd. is planning to build a 700-room complex. It has been suggested that because existing hotel averages 70% occupancy, the new complex should have only 500 rooms. It has been estimated that cost per room per annum is Rs. 21, 000.

The following data are forecasted :

	No. of days	Daily Demand	Avg. Price/ Occupied room/day
Peak season	200	800	Rs. 100
In between	80	600	Rs. 80
Off season	85	500	60

Prepare :

(*i*) Pay-off table for a complex with 500, 600, 700, 800 rooms.

(*ii*) Advise Management on the no. of rooms to be included.

(*b*) Describe the algorithm for solution of $m \times n$ game, as an LPP.

(*c*) Discuss 'Saddle Point' and 'Dominance Rule'. Also describe applications and limitations of 'theory of games' with suitable example.

4. (*a*) ABC manufactures consumes 1000 kL of furnace oil per annum. The ordering cost per order is Rs 200 and carrying cost is 21% per annum. The furnace oil costs Rs. 20 per ltr.

(*i*) Determine EOQ **2 + 2 + 6**

(*ii*) Frequency of ordering

(*iii*) If suppliers offer a discount of 2% if every order is for more than 2500 kL and a further discount of 1% if ordered quantity is more than 5000 ltrs, should the company avail the discount?

(*b*) What is 'selective inventory control'? What are various approaches of selective inventory control? Describe their application with live examples.

(*c*) Derive an expression for Economic run quantity, of an industry, whose intermittent product is produced at rate '*P*' and consumed at rate '*D*' Enumerate all the assumptions used.

5. (*a*) The workshop operates on 8 hours basis. **2 + 2 + 3 + 3**

Determine :

(*i*) Probability that number of motors in workshop is 5.

(*ii*) If workshop has capacity to hold 5 motors, what should be average service time to ensure 98% probability that an incoming motor does not wait outside the workshop?

(*iii*) If service time can not be reduced in (*ii*), how much capacity should be provided to ensure 98% probability in (*ii*)?

(*iv*) What is probability, that a motor waits 2 hours before getting served?

(*b*) A railway reservation office has 2 counters. The arrival of passengers is in Poisson and 8 passengers arrive per hr. It takes 10 minutes to serve a passenger (exponentially distributed). Operation of counter costs Rs. 100/- per hr. and time of passenger is estimated @ Rs. 400/- hr. Determine optimum no. of service counters.

(*c*) What is process of 'Monte-Carlo Simulation'? What are its applications? Why it is preferred over mathematical models at places?

❑❑❑

M.B.A.

Second Semester Examination, 2008-2009

Operations Research (MBA-213)

Time : 3 Hours *Total Marks : 100*

Note: Attempt **all** questions which carry **equal** marks.

1. Attempt any **two** parts of the following : **2 × 10 = 20**

(*a*) What is meant by Statistical Decisions Analysis? How is it different from other methods used in decision-making? Describe any two methods, which are useful in decision-making under uncertainty.

(*b*) Describe the use, scope and applications of operation research in managerial decision-making.

(*c*) A Company is currently working with a process, which after paying for materials, labour etc. brings a profit of Rs. 12, 000. The Company has the following alternatives.

(*i*) The Company can conduct research R_1 which expected to cost Rs. 10, 000 and having 90% probability of success. If successful, the gross income will be Rs. 26, 000.

(*ii*) The Company can conduct research R_2, expected to cost Rs. 6, 000 and having a probability of 60% success. If successful, the gross income will be Rs. 24, 000.

(*iii*) The Company can pay Rs. 5, 000 as royalty of a new process which will bring a gross income Rs. 20, 000.

(*iv*) The Company may continue the current process.

Because of limited resource, only one of the two types of research can be carried out at a time. Draw the decision tree and find the optimal strategy for the Company.

2. (*a*) The ABC company wishes to plan its advertising strategy. There are two media under consideration, call them magazine I and II respectively. Magazine I has a reach of 2, 000 potential customers and magazine II has a reach of 2, 500 potential customers. The cost per page of advertising is Rs. 400 and Rs. 600 in magazines I and II respectively. The firm has a monthly budget of Rs. 6, 000. There is an important requirement that the total reach for the income group under Rs. 20, 000 per annum should not exceed 4, 000 potential customers. The reach in magazines I and II for this income group is 400 and 200 potential customers. How many pages should be brought in the two magazines to maximize the total reach.

OR

(*b*) Determine an initial basic feasible solution to the following transportation problem using Vogel's Approximation method and North-West Corner rule:

Destination

Origin	D_1	D_2	D_3	D_4	Availability
O_1	5	2	6	4	20
O_2	3	3	2	1	40
O_3	4	6	5	8	20
O_4	4	3	6	4	20
Requirement	20	40	30	10	

3. Attempt any **four** parts of the following : **4 × 5 = 20**

(a) What is dominance property in Game Theory?

(b) Find the optimum strategies for the two Players from the following payoff matrix.

	Player B	
Player A	4	2
	3	6

(c) What do you mean by Maximin and Minimax Strategy of Game?

(d) Explain how to process n jobs through 3 machines.

(e) A ready made garment manufacturer has to process the following items through two stages of production viz., cutting and sewing. The time taken for each of these items at the different stages is given below in appropriate units.

Item	1	2	3	4	5	6
Cutting Time	30	120	50	20	90	110
Sewing Time	80	100	90	60	30	10

Find and order in which these items are to be processed through these stages so as to minimize the total processing time.

(f) Give Johnson's procedure for determining an optimal sequence for processing n items on two machines. Give justification of the rule used in the procedure.

4. Attempt any **four** parts of the following : **4 × 5**

(a) Specify the characteristics of M/M/I queue model.

(b) Briefly state the limitations of the queuing theory.

(c) A tailor master with one man shop takes exactly 1 day to stitch a suit. Customer's arrival follow poisson distribution and the mean rate of arrival is 1 in every two days. How long on an average a customer is expected to wait in such a situation?

(d) Explain how the theory of replacement is used in the replacement of items whose maintenance cost varies with time.

(e) A machine costs Rs. 500. Operation and maintenance costs are zero for the first year and increase by Rs. 100 every year. If money is worth 5% every year, determine the best age at which the machine should be replaced.

The resale value of the machine is negligibly small. What is the weighted average cost of owning and operating the machine?

(f) The maintenance cost and resale value per year of machine whose purchase price is Rs. 7, 000 is given below.

Year	1	2	3	4
Maintenance cost in Rs.	900	1200	1600	2100
Resale value in Rs.	4000	2000	1200	600

When should the machine be replaced ?

5. Attempt any **four** parts of the following : **4 × 5**

(a) What is simulation? Describe the advantages and disadvantages of the simulation Technique.

(*b*) Explain Monte Carlo method of simulation Technique with suitable illustration.

(*c*) Describe the rules for drawing the network diagram with suitable illustration.

(*d*) Specify the limitation of Program Evaluation Review Technique.

(*e*) Write short notes on Independent float and free float.

(*f*) A certain project is composed of nine activities whose time estimates are as given below.

Activity	Duration
1-2	1
1-3	3
1-4	2
2-5	1
3-5	3
4-6	2
5-6	4
6-7	6
5-7	3

Draw the project network and find out the critical path.

❑❑❑

B.Tech.

TMA-013

SIXTH SEMESTER THEORY EXAMINATION, 2009-2010

PRINCIPLES OF OPERATION RESEARCH

Time: **3** *Hours* *Total Marks:* **100**

Note: *Attempt* ***All*** *the questions.*

Q.1. Attempt any two parts of the following: **(2×10=20)**

(*a*) Write simplex algorithm stepwise. Solve the following linear programming problem:

Maximize $x_1 + 5x_2 + 7x_3$

Subject to
$$3x_1 + x_2 + 2x_3 \leq 10$$
$$2x_1 + 5x_2 + x_3 \leq 18$$
$$x_1 + 2x_2 + 7x_3 \leq 28$$
and
$$x_1, x_2, x_3 \geq 0$$

(*b*) Solve using two-phase method

Maximize $5x_1 + 4x_2 + 3x_3$

Subject to
$$2x_1 + x_2 + 6x_3 = 200$$
$$6x_1 + 5x_2 + 10x_3 = 1000$$
$$8x_1 + 3x_2 + 6x_3 = 800$$
and
$$x_1, x_2, x_3 \geq 0$$

(*c*) Define the dual of a linear programming problem (LPP). Prove that the dual of the dual of a LPP is the LPP itself.

Q. 2. Attempt any two parts of the following: **(2×10=20)**

(*a*) Describe an algorithm to solve an integer LPP. Solve the following LPP:

Maximize $4x_1 + 2x_2 + 7x_3$

Subject to
$$x_1 + 3x_2 + 5x_3 \leq 15$$
$$2x_1 + x_2 + 3x_3 \leq 10$$
and x_1, x_2, x_3 non-negative integers.

(*b*) Solve the following transportation problem with cost matrix as follows:

	1	2	3	4	5	
A	7	5	7	7	3	360
B	9	11	6	11	5	250
C	11	10	6	6	8	350
D	7	15	9	2	12	40
	100	200	400	175	125	1600

(c) Formulate the following capital Budgeting problem as a 0-1 integer linear programming problem given in the following data:

There are four projects under consideration. Assume that the projects run into three years. Total available funds are Rs 75000, (to be used at the rate of

Rs. 25000/-each year). The expected profit and cost break-up is as follows:

Project	Expected Profit	Cost in Rs. Year 1	Year 2	Year 3
1	90000	8000	10000	120000
2	6000	2000	5000	8000
3	1,80000	15000	10000	5000
4	1,00000	10000	5000	5000

Q. 3. Attempt any two parts of the following: **(2×10=20)**

(*a*) The following table gives the activities of a construction project and duration:

Activity	1-2	1-3	2-3	2-4	3-4	4-5
Duration (Days)	20	25	10	12	6	10

(*i*) Draw the network for the project

(*ii*) Find the critical path

(*b*) A small project is composed seven activities whose time estimates are listed in the table as follows:

Activity *i*	*j*	Estimated duration (Week) Optimistic	Most likely	Pessimistic
1	2	1	1	7
1	3	1	4	8
1	4	2	2	1
2	5	1	1	14
3	5	2	5	7
4	6	2	5	8
5	6	3	6	15

(*i*) Draw the project network.

(ii) Find the expected duration and variance of each activity.

(*iii*) Calculate the early and late occurrence times for each event.

(*c*) For a project consisting of several activities, the duration and the required resources for carrying out each of the activities and their availabilities are given below:

Activity	Resource Required Equipment	Operators	Duration (Days)
1 - 2	X	30	4
1 - 3	Y	20	3
1 - 4	Z	20	6
2 - 4	X	30	4
2 - 5	Z	20	8
3 - 4	Y	20	4
3 - 5	Y	20	4
4 - 5	X	30	6

(*i*) Draw the network, identify critical path and compute the total float for each of the activities.

(*ii*) Find the project completion time under given resource-constraints
Resource availability: No. of operators 50
Equipment X = 1   Equipment Y = 1   Equipment Z = 1.

Q. 4. Attempt any two parts of the following: **(2×10=20)**

(a) Define ECQ problem with no shortages and several production runs of unequal length.

An oil engine manufacturer purchases lubricants at the rate of Rs. 42 per piece from a vendor. The requirement of these lubricants is 1800 per year. What should be the order quantity per order, if the cost per placement of an order is Rs. 16 and inventory carrying charge per rupee per year is only 20 paise.

(*b*) A firm is considering replacement of a machine, whose cost price is Rs. 12200 and the scrap value Rs. 200. The running (maintenance and operating) costs in Rs. are found from experience to be as follows:

Year	1	2	3	4	5	6	7	8
Running Cost	200	500	800	1200	1800	2500	3200	4000

When should the machine be replaced?

(*c*) The demand of an item is uniform at a rate of 20 units per month. The fixed cost is Rs. 10 each time a production run is made. The production cost is Rs. 1 per item and inventory carrying cost is Rs. 0.25 per item per month. If the shortage cost is Rs. 1.25 per item per month, determine how often to make a production run and of what lot size it should be?

Q. 5. Attempt any two parts of the following: **(2×10=20)**

(a) Divide a quantity b into n parts, so as to maximize their product.

Let $f_n(b)$ denote the maximum value, show that:

$$f_1(b) = b \text{ and}$$
$$f_n(b) = \text{Max } \{z, f_{n-1}(b-z)\}$$

Hence find f_n (b) and the division that maximizes it.

(*b*) Using dynamic programming approach solve:

$$\text{Max } z = x_1 + x_2 + x_3$$
$$x_1 + x_2 + x_3 \leq 5$$
$$x_1, x_2, x_3 \geq 0$$

(*c*) A company uses 10000 units per year of an item. The purchase price is Rs. 1 per item. Ordering cost = Rs. 25 per order. Carrying cost per year is 12% of the inventory value.

(*i*) Find the EOQ.

(*ii*) Find the number of orders/year.

(*iii*) If the lead item is 4 week and assuming 50 working weeks per year, find the re-order point.

❑❑❑

B.Tech

TME-601

SIXTH SEMESTER THEORY EXAMINATION, 2009-2010

OPERATIONS RESEARCH

Time: **3** *Hours* *Total Marks:* **100**

Note: *(i) Attempt **All** the questions.*

(ii) Assume suitably missing data if any.

Q.1. Attempt any two of the following:

(*a*) Baba Furniture Company employs four carpenters for 10 days to assemble table and chairs. It takes 2 hours to assemble a table and 30 minutes to assemble a chair. Customers usually buy between four and six chairs with each table. The profits are Rs. 135 per table and Rs. 50 per chair. The company operates on 8-hours shift a day. Determine the 10 days optimal production mix graphically. **10**

(*b*) (*i*) Write the dual of the following primal problem: **4**

$$\begin{aligned} \text{Maximize} \quad & z = 5x_1 + 6x_2 \\ \text{Subject to} \quad & x_1 + 2x_2 = 5 \\ & -x_1 + 5x_2 \geq 3 \\ & 4x_1 + 7x_2 \leq 8 \\ & x_1 \text{ unrestricted} \\ & x_2 \geq 2 \end{aligned}$$

(*ii*) Discuss post optimal or sensitivity analysis. **6**

(*c*) Solve the following LPP by using artificial variables or Dual Simplex Method. **10**

$$\begin{aligned} \text{Minimize} \quad & z = 3x_1 + 2x_2 \\ \text{Subject to} \quad & 3x_1 + x_2 \geq 3 \\ & 4x_1 + 3x_2 \geq 6 \\ & x_1 + x_2 \leq 3 \\ & x_1, x_2 \geq 0 \end{aligned}$$

Q. 2. Attempt any one of the following:

(*a*) (*i*) In the transportation problem in Table, the total demand exceeds the total supply. Suppose that the penalty costs per unit of unsatisfied demand are \$5, \$3 and \$2 for destination 1, 2 and 3 respectively. Determine the optimum solution. **14**

5	1	7	10
6	4	6	80
3	2	5	15
70	20	50	

(*ii*) Determine the optimal assignment for the following problem. The data indicate that worker 1 cannot work on job 3 and worker 3 cannot work on job 4. **6**

	Job 1	Job 2	Job 3	Job 4
Worker 1	50	50	–	20
Worker 2	70	40	20	30
Worker 3	90	30	50	–
Worker 4	70	20	60	70

(b) A four period inventory model operates with following data. Find the optimal policy and total cost using dynamic programming. The initial inventory $x_1 = 15$ units. The unit production cost Rs 20, and holding cost per period is Rs. 10 for all periods. **20**

Period *ix*	**Demand Di (Units)**	**Set up cost Ki Rs.**
1	76	980
2	26	1140
3	90	1850
4	67	700

Q. 3. Attempt any one of the following:

(a) (i) Discuss four different criteria of decision making under uncertainty. **12**

(ii) Formulate the following game as a LPP. **8**

	B_1	B_2	B_3
A_1	3	–1	–3
A_2	–2	4	–1
A_3	–5	–6	2

(b) (i) Solve the following game. The payoff is for player A **12**

	B_1	B_2	B_3	B_4
A_1	2	2	3	–1
A_2	4	3	2	6
A_3	3	2	2	5

(ii) Discuss decision making with utilitizes.

Q. 4. Attempt any one of the following:

(a) (i) A shop is about to order some heaters for a forecast spell of cold weather. The shop pays Rs. 1000 for each heater, and during the cold spell they sell for Rs. 2000 each. Demand for the heater declines after the cold spell, and any unsold unit sold at Rs. 500. Previous experience suggests the likely demand for heaters is as follows:

Demand :	10	20	30	40	50
Probability :	0.20	0.30	0.30	0.10	0.10

How many heaters should the shop buy?

(ii) Compare Reorder Point (ROP) and Periodic review systems. **8**

(b) (i) Explain how simulation can be applied in the case of inventory control where the demand is probabilistic and lead time is random. **10**

(ii) How different type of inventory classification helps in deciding service level? Explain taking an example.

Q. 5. Attempt any one of the following:

(a) (i) Discuss process oriented simulation language. **10**

(ii) A television repairman finds that the time spent on his jobs has an exponential distribution with a mean of 30 minutes. If he repairs sets in the order in which they came in and if the arrival of sets follows a Poisson distribution approximately with an average rate of 10 per - 8 hour day, what is the repairman's expected idle time each day? How many jobs are ahead of the average set just brought in ? **10**

(b) (i) What are some of the operating characteristics of a queuing system? How can they be used in the evaluation, or design system? **8**

(ii) Discuss the Monte Carlo method of solving a problem illustrating it by outlining a procedure to solve a specified problem of your choice by the same? **12**

❐❐❐

INDEX

INDEX

❑❑❑

Win Prizes

ATTENTION READERS

Win Prizes

We request for your frank assessment regarding various aspects of the book as given below :

OPERATIONS RESEARCH

By: Prof. Vivek Kumar

Please Fill up the Sheet in Neat and Clean Writing

(i) What topics of your syllabus, important for your examination point of view, are not covered in this book ?

....

....

....

....

(ii) What are the chapters wherein the treatment of the subject matter is not systematic or organised or updated ?

....

....

....

....

(iii) Have you come across misprints/mistakes/factual inaccuracies in the book? Please specify the chapters and the page numbers.

....

....

....

....

(iv) What other books on the same subject have you found/heard better than the present book? Please specify in terms of price and quality.

....

....

....

....

(v) Further suggestions and comments for the improvement of this book.

....

....

....

....

(PS: If need be please attach a separate sheet to write your views)

OTHER DETAILS

(i) Who recommended you this book ? (Please tick (✓) in the box near the option relevant to you.

☐ Teacher ☐ Friends ☐ Bookseller

....
....
....
....

(ii) Name of the recommending teacher, his designation and address :

....
....
....
....

(iii) Name and address of the bookseller you purchased the book from :

....
....
....
....

(iv) Name and address of your institution (Please mention the University or Board as the case may be)

....
....
....
....

(v) Your name and complete postal address :

....
....
....
....

(vi) Course you are enrolled in :

....
....
....
....

The best assessment will be awarded each month: The award will be in the form of our publications as decided by the Editorial Board.

Please mail the filled up coupon at your earliest to :

S.K. KATARIA & SONS®
4885/109, Prakash Mahal, Dr. Subhash Bhargav Lane,
Daryaganj, New Delhi-110002
☎: +91-11-23243489, 43551243
E-mail : katariabooks@yahoo.com
katariabook@gmail.com